Arvid Naess

Applied Extreme Value Statistics

With a Special Focus on the ACER Method

Springer

Arvid Naess
Faculty of Information Technology,
Mathematics and Electrical Engineering
Norwegian University of Science and
Technology
Trondheim, Norway

ISBN 978-3-031-60768-4 ISBN 978-3-031-60769-1 (eBook)
https://doi.org/10.1007/978-3-031-60769-1

This Springer imprint is published by the registered company Springer Nature Switzerland AG
The registered company address is: Gewerbestrasse 11, 6330 Cham, Switzerland

To Dorothea and Vemund

value distribution by necessity must approach the relevant asymptotic form in the limit.

Even if my own work in developing methods for use in applied extreme value analysis more or less avoided direct use of the asymptotic extreme value distributions, I have always clearly understood their importance as an unavoidable foundation. The publication in 1983 of the important book *Extremes and Related Properties of Random Sequences and Processes* by Leadbetter, Lindgren, and Rootzén happened when my own interest in extreme value analysis more or less started. I, therefore, read this book very carefully, and it gave me a very good grip on the asymptotic extreme value theory. Of course, I also read parts of the seminal book by E. J. Gumbel, published in 1958, which also has a focus on asymptotic results. However, by the mid-1980s, that book, which was written in a pre-computer era, appeared as more or less obsolete when compared to the book by Leadbetter et al.

I have written this book, not because I want to convince people to abandon the classical asymptotic approaches, which, unfortunately, too often in practice are reduced to blindfolded curve fitting exercises to asymptotic parametric distributions with no real analysis to back it up. No, I have written the book because I would also like to show that it is now possible to make a more rationally based extreme value analysis of observed data. I want to show that the ACER method very often provides a unique practical diagnostic tool for a rational extreme value analysis. If, as a result, asymptotic distributions turn out to be more or less acceptable, then their use would at least have a reasonable justification.

It is also important to emphasize that the book is not a comprehensive treatment of methods for applied extreme value analyses, but to a large extent a collection of methods that I have personally worked with on and off over a period of three decades, and which I have found to be relevant and useful. I have made an effort to write the book as much as possible like an introduction to extreme value statistics with emphasis on applications. Therefore, the book also contains introductory chapters to the classical asymptotic theories and the threshold exceedance models, as well as many illustrative examples. The mathematical level is elementary, and detailed mathematical proofs have been avoided in favor of heuristic arguments to increase readability. Hopefully, this makes the book useful and appealing to a large audience of people representing a wide range of diverse applications.

Since the topic of this book is applied extreme value statistics, an inevitable component to go along with it, is access to computer programs for carrying out the analysis of available data. For the methods based on the asymptotic results described in Chaps. 2 and 3, there are several excellent programs easily available. Specific recommendations are not given here. Whichever program is chosen, good results can be obtained within the framework of asymptotic distributions. On the other hand, the ACER method has not yet attained a comparable level of software development. References to computer programs for univariate and bivariate analyses by the ACER method have therefore been given in this book. These programs can be freely downloaded.

To Dorothea and Vemund

Preface

This book grew out of many years of involvement with the practical applications of extreme value analysis to measured or simulated data. This is a fascinating area of research because of the fundamental dichotomy inherent in this problem area. On the one hand, you have beautiful mathematical results for asymptotic extreme value distributions. On the other hand, you have real-life data, which are hardly asymptotic. So, the unavoidable question becomes: To what extent can you use the asymptotic distributions to analyze real-life data? Personally, I have always felt uncomfortable with the use of the parametric classes of asymptotic extreme value distributions in applications. This was largely due to the fact that the justification for applying them generally seemed dubious, and amazingly enough, the problem of justification is rarely discussed at all in papers using asymptotic distributions on real-life data. The problems of justification and other issues related to the fundamental dichotomy are discussed in Chap. 1.

Of course, I was not the only one who disliked asymptotics for use on real-life data. A consequence of this situation was that alternative procedures for extreme value analysis were developed in several engineering disciplines. Some of these alternative procedures were based on ideas similar to those developed in Chap. 4. This chapter contains what was largely my world view on applied extreme value statistics for quite some time, and to some extent, it still is. However, the development of the ACER method, which is a central theme in this book, cf. Chap. 5, allows for a much wider perspective. Its use in practice basically involves two separate steps. The first step is based exclusively on the data and ends up with a nonparametric representation of the extreme value distribution inherent in the data. This is the crucial element of the ACER method. The second step consists of an optimization procedure to fit a parametric function to the nonparametric distribution. This step is necessary in order to be able to predict extremes larger than those contained in the data, which is typically demanded by applications. To develop a rational method of optimized fitting, the asymptotic extreme value theory by necessity becomes an essential ingredient which guides the construction of the parametric functions used in the optimization procedure. It is therefore necessary to identify the correct asymptotic extreme value distribution, since the fitted extreme value distribution by necessity must approach the relevant asymptotic form in the limit.

Even if my own work in developing methods for use in applied extreme value analysis more or less avoided direct use of the asymptotic extreme value distributions, I have always clearly understood their importance as an unavoidable foundation. The publication in 1983 of the important book *Extremes and Related Properties of Random Sequences and Processes* by Leadbetter, Lindgren, and Rootzén happened when my own interest in extreme value analysis more or less started. I, therefore, read this book very carefully, and it gave me a very good grip on the asymptotic extreme value theory. Of course, I also read parts of the seminal book by E. J. Gumbel, published in 1958, which also has a focus on asymptotic results. However, by the mid-1980s, that book, which was written in a pre-computer era, appeared as more or less obsolete when compared to the book by Leadbetter et al.

I have written this book, not because I want to convince people to abandon the classical asymptotic approaches, which, unfortunately, too often in practice are reduced to blindfolded curve fitting exercises to asymptotic parametric distributions with no real analysis to back it up. No, I have written the book because I would also like to show that it is now possible to make a more rationally based extreme value analysis of observed data. I want to show that the ACER method very often provides a unique practical diagnostic tool for a rational extreme value analysis. If, as a result, asymptotic distributions turn out to be more or less acceptable, then their use would at least have a reasonable justification.

It is also important to emphasize that the book is not a comprehensive treatment of methods for applied extreme value analyses, but to a large extent a collection of methods that I have personally worked with on and off over a period of three decades, and which I have found to be relevant and useful. I have made an effort to write the book as much as possible like an introduction to extreme value statistics with emphasis on applications. Therefore, the book also contains introductory chapters to the classical asymptotic theories and the threshold exceedance models, as well as many illustrative examples. The mathematical level is elementary, and detailed mathematical proofs have been avoided in favor of heuristic arguments to increase readability. Hopefully, this makes the book useful and appealing to a large audience of people representing a wide range of diverse applications.

Since the topic of this book is applied extreme value statistics, an inevitable component to go along with it, is access to computer programs for carrying out the analysis of available data. For the methods based on the asymptotic results described in Chaps. 2 and 3, there are several excellent programs easily available. Specific recommendations are not given here. Whichever program is chosen, good results can be obtained within the framework of asymptotic distributions. On the other hand, the ACER method has not yet attained a comparable level of software development. References to computer programs for univariate and bivariate analyses by the ACER method have therefore been given in this book. These programs can be freely downloaded.

Writing on the technical level necessary for this book requires a lot of attention to details. It is in practice impossible to avoid errors and mistakes, poorly formulated explanations, or misprints in initial versions of such a book. Fortunately, I have some very good friends and colleagues who have helped me identify and correct many such shortcomings, and for this, I am forever thankful. Any mistakes, which may still remain, are entirely my own responsibility. The first group of people that I would like to mention for their important contributions to improving the book are Professors Bernt J. Leira, Bo H. Lindqvist, and Sverre Haver and Dr. Karl W. Breitung. Previous collaborators and PhD students that have been important in helping me in various ways are Professors Torgeir Moan, Oleg Gaidai, Sjur Westgaard, Marc Maes, Nilanjan Saha, Wei Chai, and Arild Brandrud Næss; Drs. Oleksandr Batsevich, Oleh Karpa, Ali Cetin, Hans K. Karlsen, and Kai Erik Dahlen; and Morten Skjong. I am also very grateful to my many good students that I have had the pleasure of working with over the years, who have also inevitably been part of my own never-ending education as a researcher.

Trondheim, Norway
March, 2023

Arvid Naess

Contents

Chapter 1
Challenges of Applied Extreme Value Statistics

1.1 Introduction

This book provides an introduction to the calculation of extreme value statistics for measured or simulated data. "Extreme" here means "the largest", interpreted in a way that follows from the context. As opposed to books on asymptotic extreme value statistics, the focus is also on methods specifically developed to work for real-life data. A consequence of this, is that the book contains much less theoretical issues about the asymptotic properties of extreme value statistics than is usual. However, the most important elements from the asymptotic extreme value statistics will be discussed, since they are still widely used in practical applications.

Although two of the asymptotic methods described in this book have been used extensively over several decades for prediction of the extreme value statistics of many natural phenomena, the prerequisites for their application are often not satisfied, and in some cases, not even approximately. Under such circumstances, there would appear to be a problem. It is this situation that will be highlighted in this chapter.

1.2 A Brief Summary of Status, Problems and Challenges

Statistical distributions of the extreme values of large samples of data were derived almost one hundred years ago by Fisher and Tippett (1928), cf. also Fréchet (1927); Gnedenko (1943); de Haan (1970). The main prerequisite for the existence of the derived results were that the data could be considered as outcomes of independent and identically distributed random variables. As it turned out, in non-degenerate cases there are only three possible types of limiting extreme value distributions with increasing amounts of data. It means that these results are asymptotic, as the technical term goes. On the positive side, the fact that we know explicitly what the

A. Naess, *Applied Extreme Value Statistics*,
https://doi.org/10.1007/978-3-031-60769-1_1

possible distributions look like, even if only in the limit of large samples increasing indefinitely, is very satisfactory. And there are criteria that can tell us which type of distribution applies if the underlying distribution of the data is known (Leadbetter et al. 1983). However, on the negative side, it is not possible to know to what extent one of the three types of limiting distributions actually applies to a real-life case with only a limited amount of data, even though there may be reasons to expect that the true extreme value distribution should not deviate too much from one of the limiting forms. Unfortunately, there are no useful convergence results that are precise enough to really help us decide quantitatively on this issue. Still, the common practice has been to assume an appropriate limiting form as the extreme value distribution to use. This can easily be understood from the simple fact that the limiting distributions are known explicitly, while the exact extreme value distributions inherent in the data, are largely unknown. The procedure to identify the appropriate limiting distribution is to optimize the fit of the extreme values derived from the observed data to the asymptotic forms. Typically, the extreme values from the data are taken as the maxima of specified blocks of data, e.g. annual maxima.

The three asymptotic types of extreme value distributions are essentially characterized by the value of one parameter, γ say, called the shape parameter. As will be seen later, the most important case for us in this book is when $\gamma = 0$. This is called a Type I, or Gumbel, distribution. For positive values of γ, Type II, or Fréchet, distributions are obtained, while for negative values of γ, the distributions are of Type III, or Weibull (for maxima). As it turns out, all three distribution types may be expressed in terms of one parametric form called the generalized extreme value (GEV) distribution. A standard recommendation is then to use the GEV parametric form for the sake of optimized fitting of the obtained extreme value sample. There is, however, one serious flaw with this procedure. The extreme value sample being extracted from limited amounts of data, are hardly a sample from an asymptotic distribution. Hence, one cannot expect that the estimated parameters will point to the correct asymptotic distribution. This is an issue of importance for extrapolation to out-of-sample long return period levels. For example, a practical task may be to say something about a 100 year return period level on the basis of 25 years of measured data. Then the correct asymptotic distribution is of paramount importance because the different types of extreme value distribution may lead to quite different extrapolation results. An additional issue is, of course, that with limited amounts of data follows considerable uncertainty on the estimated quantities. It may, in fact, happen that the estimated value of γ is slightly negative, pointing to a Type III distribution, but with the confidence interval accounted for, also $\gamma = 0$, or even $\gamma > 0$ are possible candidates for the value of γ. Hence, all three types of extreme value distributions seems to be possible alternatives in such a case. Since these asymptotic distributions have very different behaviour when extrapolated to high quantile values, the previous comments on the importance of this aspect, would often necessitate a more careful analysis of the situation to decide which asymptotic distribution to apply.

The peaks-over-threshold (POT) method for extreme value analysis will be discussed to some extent in this book. This method is also based on asymptotics.

The data extracted for its use, are the exceedances above high thresholds. Asymptotically, these data are assumed to follow a generalized Pareto (GP) distribution, which is then the equivalent of the GEV for the block maxima method. The POT method also has three classes of distributions, again characterized by the γ parameter. For example, the singular case $\gamma = 0$ corresponds to the exponential distribution. It is a rather popular method, mainly because it uses more of the data for inference. Unfortunately, it has certain deficiencies, which will be highlighted in this book.

There is an important and interesting observation to be made at this initial stage of our exposition of extreme value statistics. As already been stated, for all negative values of the shape parameter γ, the Type III class of extreme value distributions apply, while for all positive values of γ, it is the Type II class of extreme value distributions that is obtained. This would seem to indicate that there are two huge classes of extreme value distributions that would tend to make the singular case $\gamma = 0$ a rather special and maybe uninteresting case. The fact of the matter is quite the opposite. For almost all environmental processes that will be dealt with in this book, it is the Gumbel distribution that has prevailed as the correct asymptotic extreme value distribution. There has over the years been some suggestions to the other types as well, but these have almost all been finally rejected in the face of overwhelming evidence for Type I distributions. Of course, it is impossible to fully answer the fundamental question: To what extent do our statistical models apply to real-life data? But so far, it seems that these statistical methods work rather well on such data, but being overconfident in these methods is perhaps an unwarranted position to take.

One important reason that the singular asymptotic Gumbel case is so important in practice, is that from the perspective of a sub-asymptotic world, the picture of the size of the extreme value distribution classes looks very different. When only limited amounts of data are available, the asymptotic limiting distributions, strictly speaking, do not apply, except in very special cases. Hence, we are in a sub-asymptotic situation. As will be seen in large parts of this book, there is a huge class of extreme value distributions that apply to a range of different problems, which all end up at the Gumbel distribution asymptotically. So, the apparent singularity of the Gumbel case is an artefact of the asymptotic limiting process, and does not reflect the situation in the sub-asymptotic regime.

In an effort to resolve the inconsistency between real-life data and asymptotic distributions, a new method has been developed that is based on the concept of the average conditional exceedance rate (ACER). The method proceeds by establishing a cascade of empirical, non-parametric distribution functions that converge to the extreme value distribution inherent in the data. The advantages of the method is that no assumptions about independent and stationary data have to be made. For example, seasonal variations of the data do not require special modelling. The method also has a unique diagnostic feature in how it displays the effect of dependence between the data on the extreme value distribution. This may be of significance for the choice of which data can be included in the analysis. The ACER method will be discussed in detail in this book.

Whatever method of extreme value statistics is chosen for the analysis of the available data, the goal is almost always to predict extreme values with return periods larger, and often much larger, than the period of data collection. This inevitably requires extrapolation techniques to be used. The seemingly stochastic mechanism generating the sampled data is often sufficiently well understood to support the assumption of the validity of extrapolation. Unfortunately, this may not always be the situation. Ideally, in such cases, the predicted extreme values obtained by extrapolation should then be accompanied by a cautionary note. However, this is rarely done, simply because more credible alternatives for the prediction process are not available.

The extrapolation procedure is, in general, based on obtaining estimates of the parameters that determine the extreme value distribution type adopted for the data at hand. If an asymptotic approach is used, the GEV distribution is often preferred for parameter estimation in the case of the block maxima method, or the GP distribution for the POT method. Since these are parametrized forms covering all three types of asymptotic extreme value distributions, it is often recommended to use these forms, allowing the data to determine which type of extreme value distribution to use. As already mentioned, such a procedure may not always be a good idea. Also for the ACER method, a parametrized family of functions is proposed for the purpose of extrapolation, which is tailored to reflect the sub-asymptotic character of the data.

The parameter estimates calculated for the examples in this book, are based on either the method of moments or the maximum likelihood method in the case of the GEV or the GP distributions. For the ACER method, the optimized fitting is obtained by using a Levenberg-Marquardt method on an objective function expressed as a weighted mean square deviation measure between the empirical and the proposed parametric ACER functions on the log level. Uncertainty quantification is also a very important aspect of any statistical inference. In this book, the use of bootstrapping will serve to illustrate this issue, since it has some attractive properties.

Chapter 2
Classical Extreme Value Theory

2.1 Introduction

Classical extreme value statistics is concerned with the distributional properties of the maximum of a number of independent and identically distributed (iid) random variables when the number of variables becomes large. A partial result was obtained by Fréchet (1927), while Fisher and Tippett (1928) discovered that there are three types of possible limiting or asymptotic distributions, which are now contained in the Extremal Types Theorem, which is discussed in the next section. These three asymptotic distributions are typically referred to as the Gumbel, Fréchet, and Weibull distributions. It is also common practice to refer to them as Type I, Type II, and Type III, in the same order. Important contributions to this theory were later made by Gnedenko (1943), Gumbel (1958), and de Haan (1970).

2.2 The Asymptotic Limits of Extreme Value Distributions

The classical extreme value theory starts by looking at a sequence of independent and identically distributed (iid) random variables $X_1, X_2, \ldots$ with common distribution function $F_X(x)$. The extreme value of a finite number $X_1, \ldots, X_n$ is then $M_n = \max\{X_1, \ldots, X_n\}$. The distribution of M_n can be easily derived as

$$\begin{aligned} F_{M_n}(x) &= \text{Prob}(M_n \le x) = \text{Prob}(X_1 \le x, \ldots, X_n \le x) \\ &= \text{Prob}(X_1 \le x) \cdot \ldots \cdot \text{Prob}(X_n \le x) = \big(F_X(x)\big)^n. \end{aligned} \tag{2.1}$$

This relation is not very helpful in practice, because in most cases the distribution function $F_X(x)$ is not known exactly. Therefore, it would have to be estimated from recorded data. However, small discrepancies in the estimates of $F_X(x)$ can lead

A. Naess, *Applied Extreme Value Statistics*,
https://doi.org/10.1007/978-3-031-60769-1_2

to substantial discrepancies, in a relative sense, in the values of $\left(F_X(x)\right)^n$ for large values of n. In classical extreme value theory, one proceeds by studying the behavior of $\left(F_X(x)\right)^n$ as $n \to \infty$, but with a twist. Obviously, for any x such that $F_X(x) < 1$, $\left(F_X(x)\right)^n \to 0$ as $n \to \infty$. This necessitates a rescaling. Specifically, instead of studying M_n, one introduces a renormalized version of M_n:

$$M_n^* = \frac{M_n - b_n}{a_n} \tag{2.2}$$

for suitable sequences of constants $a_n > 0$ and b_n that are chosen to stabilize the location and scale of M_n^* as $n \to \infty$. It is then proven that there are, in fact, only three types of limiting distributions for this renormalized M_n^*. This is the famous Extremal Types Theorem (Leadbetter et al. 1983), which can be expressed as follows.

If there exist sequences of constants $a_n > 0$ and b_n such that

$$\text{Prob}\left(\frac{M_n - b_n}{a_n} \leq x\right) \to G(x)\,, \quad n \to \infty, \tag{2.3}$$

where $G(x)$ is a nondegenerate distribution function, then $G(x)$ belongs to one of the following three families:

$$\text{I} \qquad G(x) = \exp\left\{-\exp\left[-\left(\frac{x-b}{a}\right)\right]\right\}\,, \quad -\infty < x < \infty\,; \tag{2.4}$$

$$\text{II} \qquad G(x) = \begin{cases} 0 & , \quad x \leq b\,, \\ \exp\left\{-\left(\frac{x-b}{a}\right)^{-c}\right\} & , \quad x > b\,; \end{cases} \tag{2.5}$$

$$\text{III} \qquad G(x) = \begin{cases} \exp\left\{-\left(\frac{b-x}{a}\right)^{c}\right\} & , \quad x < b\,, \\ 1 & , \quad x \geq b\,; \end{cases} \tag{2.6}$$

for parameters $a > 0$, b and for families II and III, $c > 0$.

These three types of extreme value distributions are also commonly referred to as Gumbel, Fréchet, and Weibull, respectively. Note that the Weibull distribution given here is not the same as the commonly known Weibull distribution, which corresponds to the type III extreme value distribution for minima. Also, carefully note that even if the Weibull distribution is the only type of extreme value distribution with a finite upper limit on its values, this does not mean that extremes of limited data must follow this distribution. For such data, it may very well happen that the rescaling constant $a_n \to 0$ as n increases. Hence, even the Gumbel distribution may be the appropriate asymptotic limit for the extreme values of bounded data.

It may be verified that it is, in fact, possible to express all three types of extreme value distributions in a common form, which is known as the generalized extreme value (GEV) distribution. This is achieved as follows:

$$G(x) = G(x; \mu, \sigma, \gamma) = \exp\left\{ -\left[1 + \gamma\left(\frac{x-\mu}{\sigma}\right)\right]^{-1/\gamma} \right\}, \tag{2.7}$$

defined on the set $\{x \ : \ 1 + \gamma\big((x-\mu)/\sigma\big) > 0\}$, where the parameters satisfy $-\infty < \mu < \infty$, $\sigma > 0$, $-\infty < \gamma < \infty$. This distribution has three parameters: a location parameter μ, a scale parameter σ, and a shape parameter γ. The type II distributions correspond to $\gamma > 0$, while type III corresponds to $\gamma < 0$. The case $\gamma = 0$ must be interpreted as a limiting case when $\gamma \to 0$, which leads to the Gumbel distribution:

$$G(x) = \exp\left\{ -\exp\left[-\left(\frac{x-\mu}{\sigma}\right)\right]\right\}, \quad -\infty < x < \infty. \tag{2.8}$$

The statistical moments of the GEV distributions can now be calculated based on the explicit formulas of Eqs. (2.7) and (2.8). Denoting the random variable determined by a GEV distribution by M, its first two moments are

$$\mathbb{E}(M) = \mu + (e_1 - 1)\frac{\sigma}{\gamma}, \quad \gamma \neq 0, \gamma < 1 \tag{2.9}$$

and

$$\mathrm{Var}(M) = \left(e_2 - e_1^2\right)\frac{\sigma^2}{\gamma^2}, \quad \gamma \neq 0, \gamma < 1/2, \tag{2.10}$$

where $e_k = \Gamma(1 - k\gamma)$, $k = 1, 2$, and $\Gamma(\cdot)$ is the gamma function. For $\gamma \geq 1$, $\mathbb{E}(M) = \infty$, while $\mathbb{E}(M) = \mu + \lambda_E\,\sigma$ when $\gamma = 0$, that is, for the Gumbel case. Here, $\lambda_E = 0.5772...$ denotes Euler's constant. For $\gamma \geq 1/2$, $\mathrm{Var}(M) = \infty$, while $\mathrm{Var}(M) = \sigma^2\pi^2/6$, when $\gamma = 0$.

For statistical inference on experimental data, the unified form expressed by Eq. (2.7) has the advantage that the data themselves determine which type of distribution is appropriate, thereby avoiding a prior subjective judgment about any specific tail behavior. The uncertainty in the estimated value of γ is also a reflection of the uncertainty about the correct distribution for the data. Unfortunately, in practice, it may very well happen that the uncertainty in γ may cover all three types of extreme value distribution, which would necessitate a more careful analysis of the data. Also note that the data used for estimation purposes are never truly asymptotic, thereby introducing additional uncertainty when trying to identify the correct asymptotic distribution. Since the results of extrapolation to determine long return period design values may depend very much on the asymptotic extreme value distribution used, identifying the correct one is clearly important in such cases.

2.3 The Block Maxima Method

In practical application of the GEV distributions to a long time series of observed data, it is assumed that the maximum observation of a reasonably large chunk of the time series follows a GEV distribution. This is recognized by observing that from (2.3) we would assume that for large n,

$$\text{Prob}\Big(\frac{M_n - b_n}{a_n} \le x\Big) \approx G(x)\,. \tag{2.11}$$

But this may be rewritten as ($y = a_n x + b_n$)

$$\text{Prob}\big(M_n \le y\big) \approx G\Big(\frac{y - b_n}{a_n}\Big) = G^*(y)\,, \tag{2.12}$$

where G^* is also a member of the GEV family of distributions. Hence, if the main theorem applies, that is, by (2.3), $M_n^* = (M_n - b_n)/a_n$ approximately follows a GEV distribution, then M_n itself will approximately follow a GEV distribution, but with different parameters. Anyway, in practice, it is the parameters of G^* that would be of most interest.

This leads to the following approach, which is often referred to as the block maxima method. Assume that a sequence of independent observations $x_1, x_2, \ldots$ from a stationary time series is long enough to allow segmenting it into blocks of data of length n, for some large value of n, generating a series of m block maxima, $M_{n,1}, \ldots, M_{n,m}$, say, to which a GEV distribution is tentatively fitted. A typical application of the block maxima method would be to yearly extreme value observations of an environmental parameter, e.g., wind speed. In such a case, it is also often referred to as an annual maxima method. There is a practical argument behind extracting the maximum over the period of 1 year, because by choosing shorter periods, the assumption that the sampled maxima are outcomes of a common distribution would easily be violated due to seasonal variations. Still, of course, the underlying assumption that the block maxima are extracted from a set of iid random variables is clearly violated. Fortunately, by experience, this does not seem to pose a serious obstacle to the practical use of the block maxima method.

A quantity of specific interest in applications is the return period level x_p, where $G(x_p) = 1 - p$. For the annual maxima method, x_p has a return period of $1/p$ years. That is, x_p would be exceeded on the average every $1/p$ years. Inverting (2.7), it is found that for $\gamma \neq 0$,

$$x_p = \mu - (\sigma/\gamma)\big[1 - \big(-\log(1-p)\big)^{-\gamma}\big]\,, \tag{2.13}$$

while for $\gamma = 0$,

$$x_p = \mu - \sigma \log\big(-\log(1-p)\big)\,. \tag{2.14}$$

Coles (2001) discusses how to estimate confidence intervals on x_p using profile likelihood methods, which seem to provide reasonable accuracy. In this book the focus is on the bootstrap method, cf. Sect. 2.8.

2.4 Outline Proof of the Extremal Types Theorem

The proof of the Extremal Types Theorem is not a very complicated proof, but it is rather lengthy and technical (Leadbetter et al. 1983). Since it is not central to the focus of this book, only a sketch will be given here to illustrate the main ingredients. The concept of max-stability is needed. It is defined as follows:

A distribution G is called max-stable if, for every $m = 2, 3, \ldots$, there are constants $\alpha_m > 0$ and β_m such that

$$G^m(\alpha_m x + \beta_m) = G(x)\,. \tag{2.15}$$

G^m is the distribution function of $M_m = \max\{Z_1, \ldots, Z_m\}$, where the Z_i are iid random variables with distribution function G. Therefore, max-stability is a property satisfied by distributions that are invariant under the operation of taking sample maxima, except for a change of scale and location. The following result brings forward the connection between max-stability and extreme value distributions (Leadbetter et al. 1983),

A distribution is max-stable if, and only if, it is a GEV distribution.

To check that a GEV distribution is max-stable is a straightforward exercise in algebra. The converse is much harder. Anyway, this result can now be used to prove the Extremal Types Theorem. Consider first $M_{nk} = \max\{X_1, \ldots, X_{nk}\}$ of a sample of nk iid random variables X_i, for some large value of n. This large sample can be divided into k subsamples of n variables in each. Hence, there will be k iid random variables like $M_n = \max\{X_1, \ldots, X_n\}$. n is chosen large enough to claim that

$$\text{Prob}\Big(\frac{M_n - b_n}{a_n} \le x\Big) \approx G(x)\,, \tag{2.16}$$

for suitable constants a_n and b_n and for the limiting distribution G. Hence, for any integer $k \ge 2$, since $nk > n$,

$$\text{Prob}\Big(\frac{M_{nk} - b_{nk}}{a_{nk}} \le x\Big) \approx G(x)\,, \tag{2.17}$$

Eq. (2.16) leads to $\text{Prob}\big(M_n \le z\big) \approx G\big((z - b_n)/a_n\big)$, while Eq. (2.17) gives $\text{Prob}\big(M_{nk} \le z\big) \approx G\big((z - b_{nk})/a_{nk}\big)$. However, M_{nk} is obviously the maximum of k variables having the same distribution as M_n. But then,

$$\text{Prob}\big(M_{nk} \le z\big) = \Big[\text{Prob}\big(M_n \le z\big)\Big]^k\,. \tag{2.18}$$

From this, it is deduced that (in the limit)

$$G\Big(\frac{z-b_{nk}}{a_{nk}}\Big) = G^k\Big(\frac{z-b_n}{a_n}\Big). \tag{2.19}$$

From this it follows that G and G^k are identical apart from location and scale parameters. Hence, G is max-stable, and by the result above, it is a member of the GEV family of distributions.

2.5 Domains of Attraction for the Extreme Value Distributions

In practice, the exact statistical distribution of the data being analyzed is rarely known. However, in many cases there may be rather strong evidence as to what type of distribution to expect. For instance, average wind speeds over periods of 10 minutes in northern Europe have been found to follow a Weibull type distribution. Then it would be useful to know what kind of extreme value distribution to expect for such data. The answer to such questions is the subject of the theory of domains of attraction for extreme value distributions. It is beyond the scope of our treatment of this topic here to go into much detail, but some useful results seem worthwhile presenting. A more thorough discussion is given by Gnedenko (1943) and Leadbetter et al. (1983).

A time series $X_1, X_2, \ldots$ of iid random variables with distribution function F, and with a density function f, is considered. x_F is defined to be the right endpoint of F by $x_F = \sup\{x;\ F(x) < 1\}$ $(x_F \leq \infty)$. Then the following sufficient conditions due to von Mises apply:

Suppose that F is absolutely continuous with density f. Then sufficient conditions for F belonging to each of the three possible domains of attraction are as follows:

Type I: f has a negative derivative f' for all x in some interval (x_0, x_F), $(x_F \leq \infty)$, and

$$\lim_{x \nearrow x_F} \frac{f'(x)(1-F(x))}{f^2(x)} = -1.$$

Type II: $f(x) > 0$ for $x \geq x_0$ finite, and for some constant $\alpha > 0$,

$$\lim_{x \to \infty} \frac{x f(x)}{1-F(x)} = \alpha.$$

Type III: $f(x) > 0$ for all x in some finite interval (x_0, x_F), $f(x) = 0$ for $x > x_F$, and for some constant $\alpha > 0$,

$$\lim_{x \nearrow x_F} \frac{(x_F - x) f(x)}{1 - F(x)} = \alpha.$$

Using these results, it is straightforward to verify that the following list of distributions belongs to the domain of attraction of the Type I (Gumbel) case, just to mention a few well-known cases: normal, lognormal, exponential, Weibull, gamma, and, of course, the Gumbel distribution itself.

Distributions belonging to the domain of attraction of Type II are, e.g., the Pareto, the generalized Pareto for positive shape parameter, and the Type II extreme value distribution itself. For Type III may be mentioned, e.g., the uniform distributions, distributions truncated on the upper side (provided a smooth density function), and the Type III extreme value distribution itself.

2.6 Parameter Estimation for the GEV Distributions

The practical application of the block maxima method involves the need to decide on how to divide the observed data into blocks. Obviously, there will be two conflicting issues that have to be dealt with. The desire to have large blocks so that the distribution of the block maxima will approximate a GEV distribution may easily lead to a sample of few block maxima. Statistical inference on small samples may entail large uncertainties. On the other hand, increasing the sample of block maxima by choosing smaller blocks may violate the asymptotic approximation by assuming a GEV distribution for the block maxima. These issues may be further complicated by the issues of independence and stationarity, which were discussed in Sect. 2.3. While establishing general rules for the choice of block size relative to the amount of data available is hardly feasible, for some practical cases the accumulated experience has led to what may be called a consensus. For example, in wind engineering, the choice of 1 year as a block size has become very close to a standard procedure. An important consideration for this choice is that the data may then reasonably be assumed to belong to the same population since seasonal effects have effectively been removed.

When the sample of block maxima has been determined, the next step would be to estimate the parameters of the GEV model, or one of the three types, if that can be ascertained a priori. In this book the focus is on two rather popular estimation methods, which is the method of moments (primarily for the Gumbel model) and the maximum likelihood method. The probability weighted moment method has also been used to some extent. For this method, please cf. Hosking et al. (1985).

To simplify notation, the block maxima are denoted by $Z_1, \ldots, Z_k$, assuming k blocks. These random variables are assumed to be iid with a common GEV distribution, the parameters of which are to be estimated from the outcomes of the block maxima, that is, the observed data.

2.6.1 Estimation by the Method of Moments

The exposition of the method of moments for parameter estimation is limited to the Gumbel model. It is for this case that it seems to be most popular, maybe due to its simplicity in this case. The general Gumbel model has two parameters. Since the first two statistical moments $m_1 = \mathbb{E}(Z)$ and $m_2 = \mathbb{E}(Z^2)$ of a Gumbel distributed variable Z can be expressed in terms of these two parameters, for estimation the following two empirical moments are calculated:

$$\hat{m}_j = (1/k)\sum_{i=1}^{k} z_i^j \,, j = 1, 2, \tag{2.20}$$

where $z_1, \ldots, z_k$ are the observed data.

Assuming that Z has the general Gumbel distribution $G(z) = \exp\{-\exp[-(z - \mu)/\sigma]\}$, then $m_1 = \mathbb{E}(Z) = \mu + 0.5772\sigma$ and $m_2 = \mathbb{E}(Z^2) = m_1^2 + \pi^2\sigma^2/6$, cf. Sect. 2.2. Denote by μ_k and σ_k the estimated values of the parameters based on the k observations of block maxima. It is then obtained that

$$\mu_k = \hat{m}_1 - 0.5772\sigma_k \tag{2.21}$$

and

$$\sigma_k = (\sqrt{6}/\pi)\sqrt{\hat{m}_2 - \hat{m}_1^2}. \tag{2.22}$$

2.6.2 Maximum Likelihood Estimation

The maximum likelihood (ML) method is very popular and has widespread use in almost every branch of statistics. It turns out that the application of the ML methods for estimation on GEV models requires some caution. Fortunately, it seems that for the applications relevant for this book, the restrictions that need to be observed are rarely an issue. Specifically, for values of the shape parameter $\gamma > -0.5$, the ML estimators behave regularly. The only thing to note is that there are some small sample issues related to the use of ML estimators also for GEV models, cf. Coles and Dixon (1999).

Based on the assumption that $Z_1, \ldots, Z_k$ are iid random variables having a common GEV distribution, then the log-likelihood function for the GEV parameters when $\gamma \neq 0$ has the following expression:

$$\ell(\mu, \sigma, \gamma) = -k \log \sigma - (1 + 1/\gamma) \sum_{i=1}^{k} \log \left[1 + \gamma \left(\frac{z_i - \mu}{\sigma} \right) \right]$$
$$- \sum_{i=1}^{k} \left[1 + \gamma \left(\frac{z_i - \mu}{\sigma} \right) \right]^{-1/\gamma}, \tag{2.23}$$

provided that

$$1 + \gamma \left(\frac{z_i - \mu}{\sigma} \right) > 0, \quad \text{for} \quad i = 1, \ldots, k. \tag{2.24}$$

If the last condition is violated, the likelihood becomes zero and the log-likelihood therefore $-\infty$. The case $\gamma = 0$ needs to be considered separately, using the Gumbel model. In this case the log-likelihood becomes

$$\ell(\mu, \sigma) = -k \log \sigma - \sum_{i=1}^{k} \left(\frac{z_i - \mu}{\sigma} \right) - \sum_{i=1}^{k} \exp \left\{ - \left(\frac{z_i - \mu}{\sigma} \right) \right\}. \tag{2.25}$$

To obtain the numerical maximum likelihood estimates from the observed data by using (2.23) and (2.25), standard numerical optimization programs may be utilized. If (2.23) is used, care must be exercised to avoid numerical problems in cases where the optimization algorithms tend to parameter estimates in the close vicinity of $\gamma = 0$. Then it is strongly advisable to use (2.25).

Confidence intervals on the estimated parameter values can be calculated exploiting that the approximate distribution of the estimators $(\hat{\mu}, \hat{\sigma}, \hat{\gamma})$ is multivariate normal with mean value (μ, σ, γ). This is discussed by Coles (2001).

2.7 Model Validation

As is well known from basic courses in statistics, the use of probability (or PP) plots and quantile (or QQ) plots may reveal very useful information about the extent of agreement between an assumed or estimated probability distribution and the empirical distribution of the data. These are also highly useful tools for a visual check of fitted GEV models in particular cases. For a thorough discussion of the use of these plots, cf. Beirlant et al. (2004).

A probability or a PP plot is a direct comparison of the fitted distribution model to the empirical distribution. Assume that the sample of block maxima has been ordered by increasing value: $z_{(1)} \leq z_{(2)} \leq \ldots \leq z_{(k)}$. The empirical distribution function, $\tilde{G}$ say, evaluated at $z_{(i)}$ is given by

$$\tilde{G}(z_{(i)}) = i/(k+1). \tag{2.26}$$

The proposed GEV model distribution is obtained by substituting the parameter estimates into (2.7)

$$\hat{G}(z_{(i)}) = \exp\left\{-\left[1+\hat{\gamma}\left(\frac{z_{(i)}-\hat{\mu}}{\hat{\sigma}}\right)\right]^{-1/\hat{\gamma}}\right\}, \tag{2.27}$$

provided $\hat{\gamma} \neq 0$. If $\hat{\gamma} = 0$, the plot is constructed using the Gumbel distribution. If the GEV model is a good approximation, then

$$\hat{G}(z_{(i)}) \approx \tilde{G}(z_{(i)}) \tag{2.28}$$

for each index i, so that the PP plot consisting of the points

$$\left(\hat{G}(z_{(i)}), \tilde{G}(z_{(i)})\right) \quad i = 1, \ldots, k \tag{2.29}$$

should follow approximately the unit diagonal.

For the case of extreme value distributions, a quantile or QQ plot is usually considered to be more informative than a PP plot because it shows more clearly the agreement at high values of the observed data, which is of primary concern when fitting extreme value models. Assuming again that $\hat{\gamma} \neq 0$, the QQ plot is traced out by the point graph

$$\left(\hat{G}^{-1}(i/(k+1)), z_{(i)}\right), \quad i = 1, \ldots, k, \tag{2.30}$$

where

$$\hat{G}^{-1}(i/(k+1)) = \hat{\mu} - \frac{\hat{\sigma}}{\hat{\gamma}}\left[1 - \left\{-\log\left(i/(k+1)\right)\right\}^{-\hat{\gamma}}\right]. \tag{2.31}$$

This graph should also approximately follow a straight line. These procedures are discussed at greater length in Chap. 9.

2.8 Estimating Confidence Intervals by Bootstrapping

The bootstrapping method is a statistical technique of fairly recent origin that can be used for estimating confidence intervals on quantities derived from a statistical distribution on the basis of a limited sample generated by that same distribution (Efron and Tibshirani 1993; Davison and Hinkley 1997). It is based on resampling from a distribution determined by the available sample of data. Despite the fact that the name of the method alludes to lifting oneself up by the bootstraps (Baron von Munchausen), the method appears to be reasonably effective for the specific purpose of estimating confidence bands. For convenience, a brief discussion of some basic features of the bootstrapping method is provided here.

Assume that $\mathbf{z} = (z_1, z_2, \ldots, z_n)$ is a sample or a vector consisting of n independent observations of a random variable Z and that this is the only empirical information available about Z. Confidence intervals for a statistical quantity require the estimation of quantiles from the distribution of a relevant estimator. There are in principle two available options for obtaining bootstrap estimates of such quantiles. One is the nonparametric approach, where a purely empirical distribution function is established for Z on the basis of the observed data by allocating a probability of $1/n$ to each of the observed data points. The other is the parametric bootstrap, which is obtained by assuming that Z has a specified distribution function $F_Z(z; \theta) = \text{Prob}(Z \leq z)$, where θ denotes a vector of unknown parameters, which determine the distribution. These parameters are then estimated from the observed data $\mathbf{z}$, giving $\hat{\theta}$, and $F_Z(z; \hat{\theta})$ is adopted as the distribution of Z.

In this section on the block maxima method using GEV models, only the parametric bootstrap is used. The goal is to estimate some statistical quantity V, e.g., a high quantile like $100(1-\alpha)\%$ $(0 < \alpha << 1)$, given by the unknown distribution. Let $\hat{V}$ denote the estimate of V obtained from the fitted model distribution $F_Z(z; \hat{\theta})$, which is a GEV distribution. The parametric bootstrapping technique for estimating confidence intervals on V is based on resampling from the GEV model obtained.

This is done as follows: Let Z^* denote the random variable with distribution function $F_Z(z; \hat{\theta})$. ℓ bootstrap samples $\mathbf{z}_j^*$, $j = 1, \ldots, \ell$, with n independent observations of Z^* in each sample are now generated. Each sample $\mathbf{z}_j^*$ is used to fit a new GEV model from which an estimate V_j^* of V is obtained.

Simple estimates for confidence intervals on V are derived by calculating the sample standard deviation s_V^*:

$$s_V^* = \sqrt{\frac{1}{\ell - 1} \sum_{j=1}^{\ell} (V_j^* - \bar{V}^*)^2}, \tag{2.32}$$

where $\bar{V}^* = (1/\ell)\sum_{j=1}^{\ell} V_j^*$. An approximate confidence interval at level $1-q$ is then obtained as

$$(\hat{V} - w_{q/2}\, s_V^*\,,\ \hat{V} + w_{q/2}\, s_V^*\,), \tag{2.33}$$

where $w_{q/2}$ denotes the $100(1-q/2)\%$ standard normal fractile. To get stable results for this method, usually 20–30 bootstrap samples are sufficient. However, to avoid making the assumption that the bootstrap estimates V_j^* are generated by a normal distribution, which is the basis for Eq. (2.33), the true distribution may be approximated by generating a large number of bootstrap samples, usually several thousands are needed, especially for small values of q. If ℓ samples were generated, the V_j^* are rearranged in increasing order $V_{(1)}^* \leq V_{(2)}^* \leq \ldots \leq V_{(\ell)}^*$. A $100(1-q)\%$ confidence interval for V is then

$$(\,V_{(L)}^*\,,\ V_{(M)}^*\,), \tag{2.34}$$

where $(L) = [q\ell/2]$ and $(M) = [(1-q/2)\ell]$ ($[a]$ means the integer part of a). Such estimates may be further improved as described by Davison and Hinkley (1997). However, such details will not be discussed here.

2.9 The Asymptotic Extreme Value Distributions for Dependent Sequences

The assumed sequence of iid random variables underlying the classical approach to extreme value distributions is obviously not a very practical model for many physical phenomena where dependence effects are obvious. Fortunately, it turns out that the extremal types theorem still applies, provided some conditions are satisfied. These conditions relate to the long range dependence structure of the sequence of random variables. The typical model adopted is one of a stationary time series $X_1, X_2, \ldots$. Stationarity means that the joint probability law of a group of random variables from the sequence is invariant with respect to time shifts. That is, e.g., X_1, X_2 has the same joint distribution as X_{51}, X_{52}.

The condition that has to be satisfied by the long range dependence of the stationary time series to allow for an extremal types theorem, can be formulated as follows:

A stationary time series $X_1, X_2, \ldots$ is said to satisfy the $D(u_n)$ condition if, for all $i_1 < \ldots < i_p < j_1 < \ldots < j_q$ with $j_1 - i_p > l$,

$$\Big|\mathrm{Prob}\Big(X_{i_1} \leq u_n, \ldots, X_{i_p} \leq u_n, X_{j_1} \leq u_n, \ldots, X_{j_q} \leq u_n\Big) - \mathrm{Prob}\Big(X_{i_1} \leq u_n, \ldots, X_{i_p} \leq u_n\Big)\mathrm{Prob}\Big(X_{j_1} \leq u_n, \ldots, X_{j_q} \leq u_n\Big)\Big| \leq \alpha(n, l), \tag{2.35}$$

where $\alpha(n, l_n) \to 0$ for some sequence l_n such that $l_n/n \to 0$ as $n \to \infty$.

A scrutiny of this condition conveys the understanding that it is required that block maxima tend to become independent random variables if the blocks are sufficiently far apart. Provided that this condition is satisfied, the following theorem applies (Leadbetter et al. 1983):

Let $X_1, X_2, \ldots$ be a stationary time series, and define $M_n = \max\{X_1, \ldots, X_n\}$. If there exist sequences of constants $a_n > 0$ and b_n such that

$$\text{Prob}\Big(\frac{M_n - b_n}{a_n} \leq x\Big) \to G(x)\,, \quad n \to \infty, \tag{2.36}$$

where $G(x)$ is a nondegenerate distribution function, and the $D(u_n)$ condition is satisfied with $u_n = a_n z + b_n$ for every real z, and $G(z) > 0$, then $G(x)$ belongs to the class of generalized extreme value distributions.

The practical significance of this theorem is, in fact, substantial since very few time series met in practice would consist of independent data. Hence, without it, the application of GEV distributions to practical problems would be very hard to justify. Its application, of course, presupposes that the $D(u_n)$ condition is satisfied, which may seem like a highly nontrivial criterion to check for a given stationary time series. Fortunately, it turns out that for most applications it may be routinely assumed to be satisfied. For example, a stationary Gaussian time series $X_1, X_2, \ldots$ with an autocovariance function $\rho_n = \mathbb{E}[(X_i - \mu)(X_{i+n} - \mu)]$, where $\mu = \mathbb{E}X_i$, will satisfy $D(u_n)$ if $\rho_n \log n \to 0$ when $n \to \infty$. Time series of measured data in engineering applications where ρ_n decays slower than $1/\log n$ is actually very hard to imagine. However, it should be clearly understood that the convergence to the appropriate asymptotic limit may depend very much on the dependence structure of the time series. For instance, if there is strong dependence between consecutive data points of the time series, the convergence to the asymptotic limit may be very slow.

From the original time series $X_1, X_2, \ldots$, a time series of independent variables $\tilde{X}_1, \tilde{X}_2, \ldots$ may be constructed. Let $\tilde{M}_n = \max\{\tilde{X}_1, \ldots, \tilde{X}_n\}$. Then the following theorem has been proved by Leadbetter (1983),

If there exist sequences of constants $a_n > 0$ and b_n and a nondegenerate distribution function $\tilde{G}(x)$ such that

$$\text{Prob}\Big(\frac{\tilde{M}_n - b_n}{a_n} \leq x\Big) \to \tilde{G}(x)\,, \quad n \to \infty, \tag{2.37}$$

if the $D(u_n)$ condition is satisfied with $u_n = a_n z + b_n$ for every real z, and $G(z) > 0$, and if $\text{Prob}\big((M_n - b_n)/a_n \leq x\big)$ converges for some x, then

$$\text{Prob}\Big(\frac{M_n - b_n}{a_n} \leq x\Big) \to G(x) = \big(\tilde{G}(x)\big)^{\theta}\,, \quad n \to \infty, \tag{2.38}$$

for some constant $\theta \in [0, 1]$.

The constant θ is called the *extremal index*, and unless it is equal to one, the limiting distributions for the independent and the original stationary sequences are not the same. If $\theta > 0$, then $G(x)$ is an extreme value distribution, but with different parameters than $\tilde{G}(x)$. If (μ, σ, γ) are the parameters of $G(x)$ and $(\tilde{\mu}, \tilde{\sigma}, \tilde{\gamma})$ are the parameters of $\tilde{G}(x)$, then their relationship is

$$\gamma = \tilde{\gamma}, \quad \mu = \tilde{\mu} - \frac{\tilde{\sigma}}{\gamma}(1 - \theta^{\gamma}), \quad \sigma = \tilde{\sigma}\,\theta^{\gamma}, \tag{2.39}$$

or, if $\gamma = 0$, taking limits, it is obtained that $\mu = \tilde{\mu} + \sigma \log\theta$ and $\sigma = \tilde{\sigma}$. Note that the shape parameter γ remains the same.

The results cited previously are derived under the assumption of a sequence of random variables. For many applications in this book, a sequence of random variables $X_1, X_2, \ldots$ is initially not considered, but rather a stochastic process in continuous time $X(t)$. To fit into the framework discussed in this section, one could envisage sampling the considered process at discrete time points and obtaining a sequence $X_j = X(t_j)$, $j = 1, 2, \ldots$. In practice, this is often done by extracting local peak values from an observed realization of the process. As has been seen, if the obtained time series is stationary, then, under suitable conditions, the extremal types theorem still holds true. Unfortunately, no satisfactory general theory of extremes is available for the case of continuous time stochastic processes, but some results have been proven (Leadbetter et al. 1983). Of course, in order for a realization of a stochastic process to be stored in computer memory, it has to be sampled. In that sense, what is available for further analysis is, in fact, a time series of outcomes of random variables. Hence, the sampling frequency relative to the characteristics of the stochastic process determines to what extent one may consider the stored time series to be a good replica of the realization of the stochastic process, which may then be used for further processing, like extraction of peak values.

Chapter 3
The Peaks-Over-Threshold Method

3.1 Introduction

A common approach to practical extreme value analysis is to use the GEV form of the asymptotic extreme value distributions to fit to the observed extreme values. The typical data that are used in this process are the extreme values observed over specified periods of time, e.g., over one year periods. Such a procedure, extracting only extremes over blocks of data, would immediately appear to be wasteful, since potentially very useful data might be discarded. The Peaks-Over-Threshold method, or simply the POT method, represents an approach to extreme value analysis that tries to avoid this waste of data by considering all data that exceed a prescribed high threshold. It will be shown that also for the POT method there are three limiting forms of the exceedance probability distributions corresponding to The Extremal Types Theorem. The Generalized Pareto (GP) distribution will take the place of the GEV distribution.

3.2 The Peaks-Over-Threshold Method

The basic assumption of the POT method is that the observed time series $x_1, x_2, \dots$ are the outcomes of a sequence of independent and identically distributed (iid) random variables $X_1, X_2, \dots$ with common distribution function $F_X(x)$, which belongs to the domain of attraction of one of the extreme value distributions. Instead of extracting the extreme observations over blocks of data, the focus of the POT method is on all data that exceed a given high threshold. If the chosen threshold u is high enough, it would be natural to consider also the data exceeding u as extremes.

A. Naess, *Applied Extreme Value Statistics*,
https://doi.org/10.1007/978-3-031-60769-1_3

Denoting an arbitrary term of the X_i sequence by X, the exceedance probability essential to the POT method is the following conditional probability:

$$\text{Prob}(X > u + y \mid X > u) = \frac{1 - F(u + y)}{1 - F(u)}, \quad y > 0, \tag{3.1}$$

where y denotes the size of the exceedance above the threshold u.

If the parent distribution F were fully known, the distribution of the threshold exceedances in Eq. (3.1) would also be known. However, in practical applications this is rarely the case. Typically, only estimated approximations to the parent distribution based on the observed data would be available. A consequence of this is that the estimates of $F(x)$ for $x \le u$ in general become very uncertain for large to extreme values of u. This makes the direct application of Eq. (3.1) useless for practical purposes. Hence, limiting forms that would parallel the GEV distributions for extremes are sought.

The main result needed is expressed by the following theorem:

Let $X_1, X_2, \ldots$ be a sequence of iid random variables with common distribution function F, and let $M_n = \max\{X_1, \ldots, X_n\}$. Assume that the conditions of the Extremal Types Theorem are satisfied, so that for large n,

$$\text{Prob}(M_n \le z) \approx G(z),$$

where

$$G(x) = \exp\left\{-\left(1 + \gamma\left(\frac{x - \mu}{\sigma}\right)\right)^{-1/\gamma}\right\}, \tag{3.2}$$

defined on the set $\{x : 1 + \gamma\big((x - \mu)/\sigma\big) > 0\}$, where the parameters satisfy $-\infty < \mu < \infty, \sigma > 0, -\infty < \gamma < \infty$. Then, for large enough u,

$$H(y) = \text{Prob}(X \le u + y \mid X > u) \approx 1 - \left(1 + \gamma\frac{y}{\tilde{\sigma}}\right)^{-1/\gamma}, \tag{3.3}$$

defined on $\{y : y > 0 \text{ and } (1 + \gamma y/\tilde{\sigma}) > 0\}$, where $\tilde{\sigma} = \sigma + \gamma(u - \mu)$.

Note that the special case $\gamma = 0$ has to be interpreted as a limit, in analogy with the GEV case $\gamma = 0$, viz.

$$H(y) = 1 - \exp\left(-\frac{y}{\tilde{\sigma}}\right), \quad y > 0, \tag{3.4}$$

which corresponds to an exponential distribution with parameter $1/\tilde{\sigma}$.

The family of distribution functions defined by Eq. (3.3) is called the Generalized Pareto (GP) distributions. The theorem above implies that if the block extremes have the limiting distribution G, then threshold excesses have a corresponding limiting distribution within the GP family of distributions. And it follows from this theorem that the parameters of the GP distribution of threshold excesses are uniquely determined by those of the associated GEV distribution of block extremes. In particular, the shape parameter γ of the GP distribution is identical to that of the corresponding GEV distribution. It is important to note, however, that this statement is valid only asymptotically. The practical significance of this is that, in general, the γ-parameter estimated on the basis of a finite set of data rarely equals that of the correct asymptotic extreme value distribution. Also relevant in this context is the observation made by Fisher and Tippett (1928) that the best fit to a finite set of extreme value data from a stationary Gaussian process is provided by a Type III distribution and not by the asymptotically correct Type I distribution.

A full proof of the theorem above will not be given here, but a sketch of the proof may serve to illustrate the main ideas. A more precise argument is provided by Leadbetter et al. (1983).

For large n and for suitable values of the argument z,

$$\mathrm{Prob}(M_n \leq z) = F^n(z) \approx \exp\left\{-\left(1+\gamma\left(\frac{z-\mu}{\sigma}\right)\right)^{-1/\gamma}\right\}.$$

Hence,

$$n \log F(z) \approx -\left(1+\gamma\left(\frac{z-\mu}{\sigma}\right)\right)^{-1/\gamma}, \tag{3.5}$$

By a Taylor expansion, it is found that $\log(1-\varepsilon) \approx -\varepsilon$ for small positive ε. Consequently, for large values of z, it follows that $\log F(z) \approx -(1-F(z))$. Substituting into Eq. (3.5), followed by a rearrangement, leads to

$$1 - F(u) \approx \frac{1}{n}\left(1+\gamma\left(\frac{u-\mu}{\sigma}\right)\right)^{-1/\gamma},$$

for large values of u. Similarly, for $y > 0$,

$$1 - F(u+y) \approx \frac{1}{n}\left(1+\gamma\left(\frac{u+y-\mu}{\sigma}\right)\right)^{-1/\gamma},$$

Hence, it follows that

$$\begin{aligned}\mathrm{Prob}\Big(X > u+y \mid X > u\Big) &\approx \left(\frac{1+\gamma(u+y-\mu)/\sigma}{1+\gamma(u-\mu)/\sigma}\right)^{-1/\gamma} \\ &= \left(1+\gamma\frac{y}{\tilde{\sigma}}\right)^{-1/\gamma}, \end{aligned} \tag{3.6}$$

where $\tilde{\sigma} = \sigma + \gamma(u-\mu)$, as required by Eq. (3.3).

3.3 Threshold Selection

Selection of an appropriate threshold to use when applying the POT method on measured data is an essential ingredient of this method. Its importance is reflected by the fact that the predicted extreme values will in many cases show a significant dependence on this choice. Unfortunately, in practice, there are no fully reliable methods to guide the selection of an appropriate threshold. In spite of this, a couple of methods have seen extensive use. These two methods will be discussed in this section.

The practical application of the POT method would typically proceed as follows. The observed data $x_1, x_2, \ldots, x_n$ are assumed to be outcomes of independent and identically distributed random variables $X_1, X_2, \ldots, X_n$, which, of course, would require some degree of justification. The extreme events to be used are the observed exceedances above a high threshold u, that is, the data $x_i : x_i > u$. Denote these exceedances as $x_{(1)}, x_{(2)}, \ldots, x_{(k)}$, and define the threshold excesses by $y_j = x_{(j)} - u$ for $j = 1, \ldots, k$. According to our main result in the previous section, the excess data y_j may be regarded as independent realizations of a random variable whose distribution function is approximately of the GP type. By fitting the excess data to the GP distribution, an approximate distribution of the excess variable is obtained.

In the choice of threshold, there are clearly two conflicting issues that have to be dealt with. Choosing a high threshold is desirable from the point of view of not violating too much the asymptotic basis for the theory. However, if too few data are retained, there will be high uncertainty in every estimate. On the other hand, if the threshold is too low, the apparent uncertainty of estimates may be reduced, but the assumption that the excess values follow a GP distribution may be seriously in error. Hence, in practice, there is a need to balance the two issues. There seems to be two procedures that are used for this purpose. One is carried out prior to model estimation by investigating how a diagnostic follows an expected pattern. The other is an assessment of the stability of parameter estimates based on the fitting of models for a range of different thresholds.

The first of the two methods uses the expected value of a GP variable Y with distribution function

$$H(y) = 1 - \left(1 + \gamma y/\sigma\right)^{-1/\gamma}. \tag{3.7}$$

The expected value is then

$$\mathbb{E}[Y] = \frac{\sigma}{1-\gamma}, \tag{3.8}$$

provided $\gamma < 1$; otherwise the mean value is infinite. Assume that the GP distribution is a valid model for the excesses of a fixed threshold u_0. Let X denote an arbitrary term among $X_1, X_2, \ldots, X_n$. Then, by (3.8),

$$\mathbb{E}[X - u_0 \mid X > u_0] = \frac{\sigma_{u_0}}{1-\gamma}, \tag{3.9}$$

provided $\gamma < 1$, where the convention of using σ_u to denote the scale parameter corresponding to excesses of the threshold u has been adopted. If the GP model holds for the threshold u_0, by necessity it also holds for any $u > u_0$. Hence, for $u > u_0$, it is obtained that

$$\mathbb{E}[X - u \mid X > u] = \frac{\sigma_u}{1-\gamma} = \frac{\sigma_{u_0} + \gamma(u - u_0)}{1-\gamma}, \tag{3.10}$$

where the last equation follows from the main result of the previous section. This tells us that for $u > u_0$, $\mathbb{E}[X - u \mid X > u]$ is a linear function of u. This can then be verified by calculating the mean value of the excesses for a range of thresholds. Thus, a plot of the following empirical point estimates,

$$\left\{\left(u, \frac{1}{n_u}\sum_{i=1}^{n_u}(x_{(i)} - u)\right) : u < x_{\max}\right\}, \tag{3.11}$$

where $x_{(1)}, x_{(2)}, \ldots, x_{(n_u)}$ consist of the n_u observations that exceed u, and $x_{\max}$ denotes the largest observation, should be an approximately straight line for a range of u-values where the GP model is applicable. This empirical graph is commonly called the *mean residual life plot*. Confidence intervals can be added to the graph by using the approximate normality of the sample means. We shall have occasion to illustrate this diagnostic for the applicability of the POT method for real-life data, and it will be clear that it is not always of much help.

3.4 Return Periods

The return period R of a given wind speed, in years, is defined as the inverse of the probability that the specified wind speed will be exceeded in any 1 year. If λ denotes the mean crossing rate of the threshold u per year (i.e., the average number of data points above the threshold u per year), the return period R of the value of X corresponding to the level $x_R = u + y_R$ is given by the relation

$$R = \frac{1}{\lambda\,\mathrm{Prob}(X > x_R)} = \frac{1}{\lambda\,\mathrm{Prob}(Y > y_R)}. \tag{3.12}$$

Hence, it follows that

$$\text{Prob}(Y \le y_R) = 1 - 1/(\lambda R). \tag{3.13}$$

Invoking Eq. (3.7) for $\gamma \neq 0$ leads to the result

$$x_R = u + \sigma[(\lambda R)^\gamma - 1]/\gamma. \tag{3.14}$$

Similarly, for $\gamma = 0$, it is found that

$$x_R = u + \sigma \ln(\lambda R), \tag{3.15}$$

where u is the threshold used in the estimation of γ and σ. A discussion of how to estimate confidence intervals on x_R using profile likelihood methods is provided by Coles (2001). In this book the focus is on the bootstrap method, cf. Sect. 3.7.

3.5 Parameter Estimation for the GP Distributions

There is a range of possible estimation methods available for the parameters of the GP distribution. In our own work, we have mostly used three methods that serve the purpose reasonably well when the shape parameter does not deviate too much from zero ($\gamma < 0.5$): The Dekkers–Einmahl–de Haan estimators, the moment estimators, and the maximum likelihood estimators. These three methods will be discussed below. Alternative methods include the Hill estimators and the probability weighted moment (PWM) estimators, cf. Beirlant et al. (2004). PWM estimators have been popular in flood frequency analysis (Hosking et al. 1985). Further development is provided by the L-moment estimators (Hosking and Wallis 1997). It has been shown that L-moments have some issues related to their lack of sensitivity to the tail behavior of the underlying statistical distribution. This is clearly of importance for properly representing extreme value distributions, cf. Winterstein and MacKenzie (2013).

3.5.1 Dekkers–Einmahl–de Haan Estimators

Let n denote the total number of data points, while the number of observations above the threshold value u is denoted by k. The threshold u then represents the $(k+1)$th highest data point(s). An estimate for λ is $\hat{\lambda} = k/n_{yrs}$, where n_{yrs} denotes the length of the record in years. The highest, second highest, ..., kth highest, $(k+1)$th highest variates are denoted by $X_n^*, X_{n-1}^*, \ldots, X_{n-k+1}^*, X_{n-k}^* = u$, respectively.

The parameter estimators proposed by Dekkers et al. (1989) and de Haan (1994) are based on the following two quantities:

$$H_{k,n} = \frac{1}{k}\sum_{i=0}^{k-1}\{\ln(X^*_{n-i}) - \ln(X^*_{n-k})\} \tag{3.16}$$

and

$$H^{(2)}_{k,n} = \frac{1}{k}\sum_{i=0}^{k-1}\{\ln(X^*_{n-i}) - \ln(X^*_{n-k})\}^2 . \tag{3.17}$$

Estimators for σ and γ are then given by the relations

$$\hat{\sigma} = \rho\, X^*_{n-k}\, H_{k,n} = \rho\, u\, H_{k,n} \tag{3.18}$$

and

$$\hat{\gamma} = H_{k,n} + 1 - \frac{1}{2}\left\{1 - \frac{(H_{k,n})^2}{H^{(2)}_{k,n}}\right\}^{-1} , \tag{3.19}$$

where $\rho = 1$ if $\hat{\gamma} \geq 0$, while $\rho = 1 - \hat{\gamma}$ if $\hat{\gamma} < 0$. These two estimators will be referred to as the Dekkers–Einmahl–de Haan estimators.

Subject to general conditions on the underlying probability law, Dekkers et al. (1989) and de Haan (1994) showed that $\hat{\gamma} \to \gamma$ and $\hat{\sigma} \to \sigma$ as $n \to \infty$ (in probability).

Closely related to the Dekkers–Einmahl–de Haan estimators are the Hill estimators. Their application to the problem of estimating extreme wind speeds was investigated by Naess and Clausen (1999). Their conclusion was that the Hill estimators lead to results that are quite similar to those provided by the Dekkers–Einmahl–de Haan estimators. Because the Hill estimators require considerably higher numerical efforts than the Dekkers–Einmahl–de Haan estimators and rarely provide significantly better results, the Hill estimators were excluded from the present discussion. The interested reader is referred to Naess and Clausen (1999), Beirlant et al. (2004) for details.

3.5.2 Moment Estimators

In terms of the mean value $\mathbb{E}(Y)$ and the standard deviation $s(Y)$ of the exceedance variate Y, it can be shown that (Hosking and Wallis 1987)

$$\sigma = \frac{1}{2}\mathbb{E}(Y)\{1 + [\mathbb{E}(Y)/s(Y)]^2\} \tag{3.20}$$

and

$$\gamma = \frac{1}{2}\{1 - [\mathbb{E}(Y)/s(Y)]^2\}. \tag{3.21}$$

Hence, empirical estimates of the first two moments of Y provide estimates of σ and γ. The resulting estimators are referred to as the moment estimators for σ and γ.

3.5.3 *Maximum Likelihood Estimators*

The maximum likelihood estimators (MLEs) are often preferred due to their asymptotic efficiency. Let $y_1, \ldots, y_k$ denote the observed sample of exceedances above the threshold u from an observed sample of peak values $x_1, \ldots, x_n$. The log-likelihood function $\ell(\sigma, \gamma | y_1, \ldots, y_k)$ for the sample $y_1, \ldots, y_k$ is given by

$$\ell(\sigma, \gamma | y_1, \ldots, y_k) = -k \ln \sigma - \left(\frac{1}{\sigma} + 1\right) \sum_{i=1}^{k} \ln\left(1 + \gamma \frac{y_i}{\sigma}\right), \tag{3.22}$$

provided $(1 + \gamma y_i/\sigma) > 0$ for $i = 1, \ldots, k$. If $\gamma = 0$, the log-likelihood assumes the form

$$\ell(\sigma, 0 | y_1, \ldots, y_k) = -k \ln \sigma - \frac{1}{\sigma} \sum_{i=1}^{k} y_i . \tag{3.23}$$

The MLE $\hat{\sigma}$ and $\hat{\gamma}$ are obtained by maximizing $\ell(\sigma, \gamma | y_1, \ldots, y_k)$ with respect to σ and γ. These values are found by numerical methods, except for the special case $\gamma = 0$, for which a simple, closed-form solution exists for $\hat{\sigma}$. It is given as $\hat{\sigma} = \sum_{i=1}^{k} y_i / k$.

3.6 Model Validation

Probability (PP) and quantile (QQ) plots may be useful tools to get a grip on the suitability of the GP model for the excesses beyond a chosen threshold u, cf. Beirlant et al. (2004). Denote the threshold excesses by $y_1, y_2, \ldots, y_k$ and the estimated GP model by $\hat{H}$. Ordering the threshold excesses by increasing magnitude: $y_{(1)} \leq y_{(2)} \leq \ldots \leq y_{(k)}$, the PP plot consists of the point graph

$$\big(i/(k+1), \hat{H}(y_{(i)})\big), \; i = 1, \ldots, k, \tag{3.24}$$

where

$$\hat{H}(y) = 1 - \left(1 + \hat{\gamma} y/\hat{\sigma}\right)^{-1/\hat{\gamma}}, \tag{3.25}$$

provided $\hat{\gamma} \neq 0$. If $\hat{\gamma} = 0$, the plot is constructed using the exponential distribution.

Assuming again that $\hat{\gamma} \neq 0$, the QQ plot is traced out by the point graph

$$\left(\hat{H}^{-1}(i/(k+1)), y_{(i)}\right),\ i = 1, \ldots, k, \tag{3.26}$$

where

$$\hat{H}^{-1}(y) = u + \frac{\hat{\sigma}}{\hat{\gamma}}\left(y^{-\hat{\gamma}} - 1\right). \tag{3.27}$$

If the GP model is a reasonable model for the distribution of the excesses of u, both the PP and the QQ plot should follow approximately a straight line. These procedures are discussed more thoroughly in Chap. 9.

3.7 Estimating Confidence Intervals by Bootstrapping

The principle of the bootstrapping method was briefly explained in the previous chapter, cf. Sect. 2.8. In this section on the POT method, only the nonparametric bootstrap is used. The goal of this method is to estimate some statistical quantity V given by the unknown distribution function on the basis of an observed sample $\mathbf{y} = (y_1, y_2, \ldots, y_n)$, which is a sample or vector consisting of n independent observations of a random variable Y. Let $\hat{V}$ denote the estimate of V based on the given sample. The nonparametric bootstrapping technique for estimating confidence intervals on V is based on resampling (with replacement) from the empirical distribution function (EDF) provided by the observed sample $\mathbf{y}$, cf. Sect. 2.8.

This is done as follows: The EDF gives rise to an empirical random variable Y^*. ℓ bootstrap samples $\mathbf{y}_j^*$, $j = 1, \ldots, \ell$, with n independent observations of Y^* in each sample are now generated. Each sample $\mathbf{y}_j^*$ gives rise to an estimate V_j^* of V.

Simple estimates for confidence intervals on V are derived by calculating the sample standard deviation s_V^*:

$$s_V^* = \sqrt{\frac{1}{\ell - 1} \sum_{j=1}^{\ell} (V_j^* - \bar{V}^*)^2}, \tag{3.28}$$

where $\bar{V}^* = (1/\ell) \sum_{j=1}^{\ell} V_j^*$. An approximate confidence interval at level $1 - q$ is then obtained as

$$(\hat{V} - w_{q/2}\, s_V^*\, ,\ \hat{V} + w_{q/2}\, s_V^*\,), \tag{3.29}$$

where $w_{q/2}$ denotes the $100(1 - q/2)\%$ standard normal fractile. To get stable results for this procedure, usually 20–30 bootstrap samples are sufficient. To avoid making the assumption that the bootstrap estimates V_j^* are normally distributed, the true distribution may be approximated by generating a large number of bootstrap samples, usually several thousands are needed, especially for small values of q. If ℓ samples were generated, the V_j^* are rearranged in increasing order $V_{(1)}^* \leq V_{(2)}^* \leq \ldots \leq V_{(\ell)}^*$. A $100(1 - q)\%$ confidence interval for V is then

$$(\, V_{(L)}^*\, ,\ V_{(M)}^*\,)\, , \tag{3.30}$$

where $(L) = [q\ell/2]$ and $(M) = [(1 - q/2)\ell]$ ($[a]$ means the integer part of a). Davison and Hinkley (1997) describe possible improvements of such estimates. However, they are not discussed here as they are considered to be of less practical interest.

Chapter 4
A Point Process Approach to Extreme Value Statistics

4.1 Introduction

In Chap. 1, some challenges to the problem of estimating extreme value distributions from limited amounts of data were discussed. In the current chapter, this problem will be approached by exploiting the concept of the mean upcrossing rate. It will be shown that this opens the door to very robust and reasonably accurate approximations to the extreme value distributions of stochastic processes, provided some reasonable conditions are satisfied. In the majority of books on extreme value statistics, which generally focus on asymptotic results for sequences of data, this approach is usually discussed under a heading that typically goes like the title of the current chapter. By stopping short of the asymptotic limits, this approach offers a uniquely applicable methodology for approximate extreme value analysis of a host of engineering problems.

In all sections of this chapter, except the last, it is assumed that the stochastic process model is stationary. This is typically referred to as a short-term condition, which is highlighted because the environmental processes causing loads and motions of structures of interest to us, are changing their characteristics with time. For example, in offshore engineering, the practical time window for sea states to be considered stationary, is typically chosen to be three hours. Hence, in order to properly handle the estimation of extreme values over the design life of a structure, it is necessary to derive methods that allow us to obtain extreme value distributions for the long-term condition. This problem is discussed in Sect. 4.10.

4.2 Average Rate of Level Crossings

As an example, let us assume that the forces at a given location in a structure due to wind loads can be modeled as a stationary stochastic process $X(t)$ with

A. Naess, *Applied Extreme Value Statistics*,
https://doi.org/10.1007/978-3-031-60769-1_4

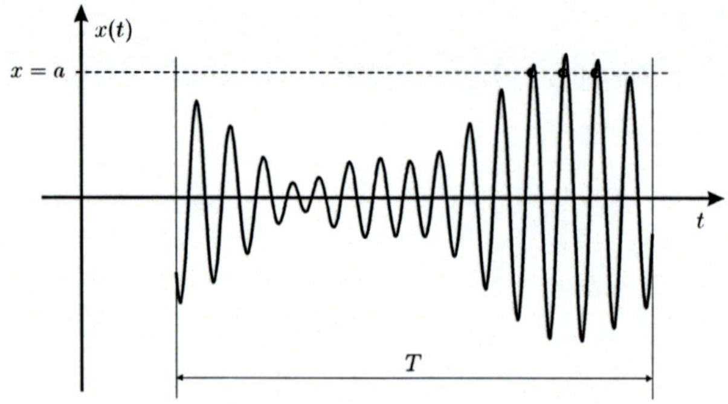

Fig. 4.1 A realization of a narrow banded process with three upcrossings of the level a during time T

smooth realizations, cf. Chap. 9. In fact, in practice, it may be assumed that the variance spectrum, see Chap. 9, of the process has compact support, implying that the realizations are infinitely smooth. It is now desirable to calculate how often an arbitrary realization of $X(t)$ can be expected to exceed a given force level a. An equivalent formulation of the same problem is the following: what is the average number of a-upcrossings per unit time by $X(t)$? An a-upcrossing means that the level a is exceeded with positive slope. Figure 4.1 shows part of a realization of a narrow banded process, where there are three upcrossings of the indicated level a. Each upcrossing is marked with a small circle in Fig. 4.1.

Let $N^+(a, \Delta t)$ denote the random number of times that $X(t)$ upcrosses the level a during the time interval $(t, t + \Delta t)$. It has been assumed that $X(t)$ has smooth realizations, which means that they are differentiable and that the differentiated realizations are continuous. This means that $x(t+\delta) \approx x(t) + \dot{x}(t)\,\delta$ for $0 \leq \delta \leq \Delta t$ (Δt small) for any realization $x(t)$ of $X(t)$. In other words, any realization $x(t)$ can be approximated by a straight line in the interval $(t, t + \Delta t)$. This implies that $x(t)$ crosses the level a at most once in this interval, see Fig. 4.2. The conditions for one upcrossing of the level a in the interval $(t, t + \Delta t)$ then become

$$x(t) \leq a \tag{4.1}$$

and

$$x(t + \Delta t) \approx x(t) + \dot{x}(t)\,\Delta t > a. \tag{4.2}$$

To satisfy Eqs. (4.1) and (4.2), it is clearly necessary that $\dot{x}(t) > 0$. The conditions for one upcrossing can therefore be written in the following way:

$$a - \dot{x}(t)\,\Delta t < x(t) \leq a \tag{4.3}$$

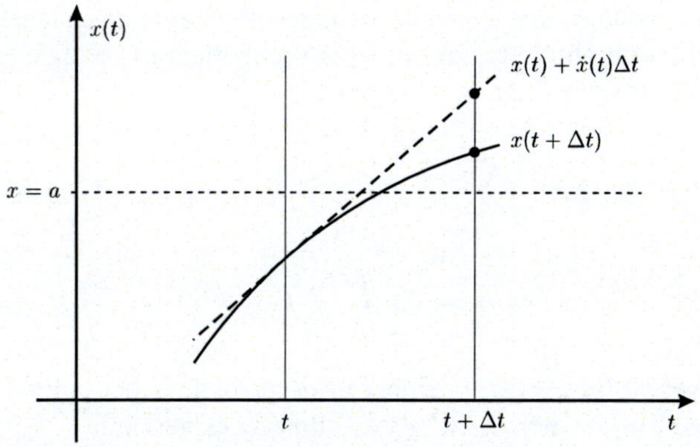

Fig. 4.2 Local approximation by a straight line at an upcrossing

and

$$\dot{x}(t) > 0\,. \tag{4.4}$$

It is seen that to calculate the probability for an upcrossing, the joint density $f_{X(t)\dot{X}(t)}(x, \dot{x})$ of $X(t)$ and $\dot{X}(t)$ is needed. The way stationarity has been defined; $f_{X(t)\dot{X}(t)}(x, \dot{x})$ is automatically independent of t. Hence, we simply write $f_{X\dot{X}}(x, \dot{x})$ to indicate this independence. From Eqs. (4.3) and (4.4), it is obtained that, for sufficiently small Δt,

$$\text{Prob}\{N^+(a, \Delta t) = 1\} = \int_0^\infty \int_{a-\dot{x}\Delta t}^{a} f_{X\dot{X}}(x, \dot{x})\, dx\, d\dot{x}\,. \tag{4.5}$$

Hence, for small Δt,

$$\int_{a-\dot{x}\Delta t}^{a} f_{X\dot{X}}(x, \dot{x})\, dx = \dot{x}\, \Delta t\; f_{X\dot{X}}(a, \dot{x}), \tag{4.6}$$

provided that $f_{X\dot{X}}(x, \dot{x})$ is continuous. This implies that

$$\text{Prob}\{N^+(a, \Delta t) = 1\} = \Delta t \int_0^\infty \dot{x}\; f_{X\dot{X}}(a, \dot{x})\, d\dot{x}\,. \tag{4.7}$$

Since it is assumed that the realizations can be approximated locally by a straight line, then $p_n = \text{Prob}\{N^+(a, \Delta t) = n\}$ is negligible for $n = 2, 3, \ldots$ compared to p_1 for sufficiently small Δt. It follows that

$$\mathbb{E}[N^+(a, \Delta t)] = \sum_{n=0}^{\infty} n\, p_n \approx 0 \cdot p_0 + 1 \cdot p_1 + 2 \cdot 0 + 3 \cdot 0 + \ldots$$
$$= p_1 = \Delta t \int_0^{\infty} \dot{x}\, f_{X\dot{X}}(a, \dot{x})\, d\dot{x}. \tag{4.8}$$

The expected (or average) number of a-upcrossings per unit of time, which is denoted by $\nu_X^+(a)$, is then given by the following expression:

$$\nu_X^+(a) = \lim_{\Delta t \to 0} \frac{1}{\Delta t} \mathbb{E}[N^+(a, \Delta t)] = \int_0^{\infty} \dot{x}\, f_{X\dot{X}}(a, \dot{x})\, d\dot{x}. \tag{4.9}$$

$\nu_X^+(a)$ is referred to by several names. In this book we use mostly average (or mean) (a-)upcrossing rate and average (or mean) (a-)upcrossing frequency. It is seen that $\nu_X^+(a)$ depends only on the level a. Because $\nu_X^+(a)$ is independent of t, $\mathbb{E}[N^+(a, T)] = \nu_X^+(a)\, T$ for any value of T, cf. Eq. (4.8), which was derived under the assumption that Δt is small.

Equation (4.9) is a useful formula. It is often referred to as the Rice formula after its creator S. O. Rice (1954). To get a feeling for the physical content of the Rice formula, one may note that the right hand side of Eq. (4.9) expresses a sort of expectation value of the slope at upcrossing of the level a coupled with the probability of being at that level. The greater the average slope at a given level, the more often an arbitrary realization will upcross that level. In other words, large average positive slope of the time histories implies shorter cycles and thereby more frequent level crossings. At the same time one must expect that the number of upcrossings of a given level is coupled to the probability of reaching that level. The Rice formula therefore appears to have a fairly plausible form when it is subjected to closer scrutiny.

In the same way that an a-upcrossing was defined, an a-downcrossing can be defined in a similar way. An a-downcrossing implies that the level a is passed with negative slope. The expected number of a-downcrossings per unit of time of a stationary process $X(t)$ is denoted by $\nu_X^-(a)$. To derive the formula for $\nu_X^-(a)$, the following observation is made: it is obvious that an a-downcrossing for the process $X(t)$ is equivalent with a $(-a)$-upcrossing for the process $Y(t) = -X(t)$. Hence $\nu_X^-(a) = \nu_Y^+(-a)$. From the relation $Y(t) = -X(t)$ follows that $f_{Y\dot{Y}}(y, \dot{y}) = f_{X\dot{X}}(-y, -\dot{y})$. This, together with Eq. (4.9), gives

$$\nu_X^-(a) = \nu_Y^+(-a) = \int_0^{\infty} \dot{y}\, f_{Y\dot{Y}}(-a, \dot{y})\, d\dot{y} = \int_0^{\infty} \dot{y}\, f_{X\dot{X}}(a, -\dot{y})\, d\dot{y}$$

$$= \int_0^{-\infty} \dot{x}\, f_{X\dot{X}}(a, \dot{x})\, d\dot{x} = -\int_{-\infty}^{0} \dot{x}\, f_{X\dot{X}}(a, \dot{x})\, d\dot{x}. \tag{4.10}$$

A common way of rewriting Eq. (4.10) is

$$\nu_X^-(a) = \int_{-\infty}^{0} |\dot{x}|\, f_{X\dot{X}}(a, \dot{x})\, d\dot{x}. \tag{4.11}$$

An a-crossing is either an a-upcrossing or an a-downcrossing. The expected number of a-crossings per unit of time, denoted by $\nu_X(a)$, must satisfy the equation $\nu_X(a) = \nu_X^+(a) + \nu_X^-(a)$. From Eqs. (4.9) and (4.11), it is then obtained that

$$\nu_X(a) = \int_{-\infty}^{\infty} |\dot{x}|\, f_{X\dot{X}}(a, \dot{x})\, d\dot{x}. \tag{4.12}$$

For a stationary process, any a-upcrossing must by necessity be followed by an a-downcrossing, and conversely. If that was not the case, the realizations would have a mean drift in the positive or negative direction; that is, the mean value would not be constant. For a stationary process, the following relations apply:

$$\nu_X^-(a) = \nu_X^+(a) = \frac{1}{2}\,\nu_X(a). \tag{4.13}$$

4.3 Distribution of Peaks of a Narrow Banded Process

Assume that $X(t)$ is a stationary process with zero mean value, which is also narrow banded. What characterizes a realization of a narrow banded process is that the amplitude and length (period) of subsequent cycles vary slowly, as illustrated in Fig. 4.1. This implies that almost invariably, there is only one maximum or peak value between an upcrossing and a subsequent downcrossing of any level a, see Fig. 4.3. Because a zero mean value was assumed, the mean number of peaks per unit of time will therefore be approximately equal to the mean rate of zero-upcrossings, that is, $\nu_X^+(0)$.

Let X_p denote the size or height of an arbitrary peak of $X(t)$. X_p becomes a random variable. The probability distribution of X_p for a narrow banded process $X(t)$ with zero mean value is now defined as

$$\text{Prob}\{X_p > a\} = \frac{\nu_X^+(a)}{\nu_X^+(0)}, \quad (a \geq 0). \tag{4.14}$$

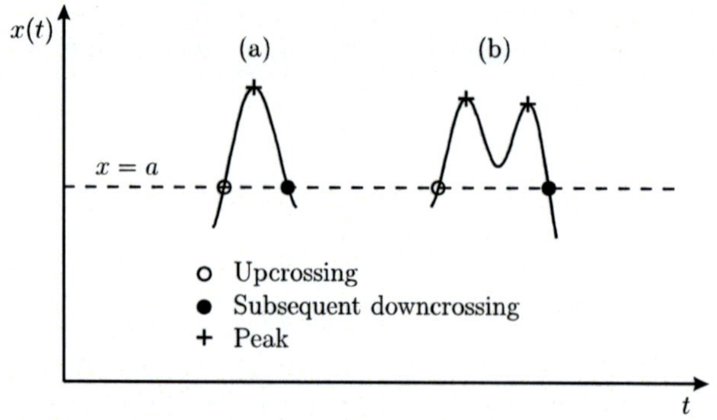

Fig. 4.3 (**a**) Peak between an upcrossing and a subsequent downcrossing of the level a. (**b**) Two peaks between an upcrossing and a subsequent downcrossing

The distribution $F_{X_p}(a)$ is then given as

$$F_{X_p}(a) = 1 - \frac{\nu_X^+(a)}{\nu_X^+(0)}, \quad a \geq 0, \tag{4.15}$$

while $F_{X_p}(a) = 0$ for $a < 0$.

If $m_X \neq 0$, the following definition applies: $F_{X_p}(a) = 1 - \nu_X^+(a)/\nu_X^+(m_X)$ for $a \geq m_X$ and $F_{X_p}(a) = 0$ for $a < m_X$. Because $\nu_X^+(a)$ is assumed to equal the mean number of peaks per unit of time above the level a, then clearly $\nu_X^+(a)$ will decrease with increasing a. For all processes of interest to us, it may be assumed that $\nu_X^+(a) \to 0$ when $a \to \infty$. This implies that $F_{X_p}(a)$ gets the properties that a distribution must have; that is, $F_{X_p}(a)$ is a nondecreasing function for increasing a, $F_{X_p}(a) \to 0$ when $a \to -\infty$ and $F_{X_p}(a) \to 1$ when $a \to \infty$. It is tacitly assumed that $\nu_X^+(a)$ gets its largest value when $a = m_X$, which is usually the case. As observed, it is true for a Gaussian process. Note that it must also apply to processes that are characterized by having only one maximum between an upcrossing and a subsequent downcrossing of the mean value level, that is, for infinitely narrow banded processes. The assumption made is therefore quite reasonable.

If $\nu_X^+(a)$ can be differentiated with respect to a, the density for peaks is obtained, assuming that $m_X = 0$,

$$f_{X_p}(a) = -\frac{1}{\nu_X^+(0)} \frac{d\nu_X^+(a)}{da}, \quad a \geq 0, \tag{4.16}$$

while $f_{X_p}(a) = 0$ for $a < 0$. Equations (4.15) and (4.16) are sometimes called the "peak formulas." It is emphasized that they only apply to narrow banded processes.

For the sake of completeness, it should be mentioned that the peak distributions can also be defined in the general case. However, this would require that the exact mean number of peaks per unit of time were used in definitions. The expressions then obtained would be of limited practical use because they are difficult, if not impossible, to calculate. The simplifications that are sometimes introduced to make the expressions amenable to calculations are often, in fact, ill defined.

4.4 Average Upcrossing Rate and Distribution of Peaks of a Gaussian Process

A random variable X is normally distributed if the density of X is given as,

$$f_X(x) = \frac{1}{\sqrt{2\pi}\sigma_X} \exp\left\{ -\frac{1}{2}\left(\frac{x - m_X}{\sigma_X}\right)^2 \right\} \tag{4.17}$$

for $\sigma_X > 0$. If $\sigma_X = 0$, then $X = m_X$. In this case, X may be considered as a degenerate normal variable.

Two random variables X and Y are called *jointly normally distributed* if the joint density of X and Y is given by the equation,

$$f_{XY}(x, y) = \frac{1}{2\pi\, \sigma_X\, \sigma_Y \sqrt{1 - \rho_{XY}^2}} \exp\left\{ -\frac{1}{2\,(1 - \rho_{XY}^2)} \left[\left(\frac{x - m_X}{\sigma_X}\right)^2 \right.\right.$$
$$\left.\left. - 2\rho_{XY} \left(\frac{x - m_X}{\sigma_X}\right)\left(\frac{y - m_Y}{\sigma_Y}\right) + \left(\frac{y - m_Y}{\sigma_Y}\right)^2 \right] \right\}, \tag{4.18}$$

where $\rho_{XY} = \mathbb{E}[(X - m_X)(Y - m_Y)]/\sigma_X\sigma_Y$ is the correlation coefficient for X and Y. Invariably, $|\ \rho_{XY}\ | \leq 1$, but to be precise, in Eq. (4.18) it is assumed that $|\rho_{XY}| < 1$ and that $\sigma_X > 0, \sigma_Y > 0$.

If X and Y are uncorrelated, $\rho_{XY} = 0$ by definition. For this case, Eq. (4.18) assumes the form,

$$\begin{aligned} f_{XY}(x, y) &= \frac{1}{2\pi\, \sigma_X\, \sigma_Y} \exp\left\{ -\frac{1}{2}\left[\left(\frac{x - m_X}{\sigma_X}\right)^2 + \left(\frac{y - m_Y}{\sigma_Y}\right)^2 \right] \right\} \\ &= \frac{1}{\sqrt{2\pi}\, \sigma_X} \exp\left\{ -\frac{1}{2}\left(\frac{x - m_X}{\sigma_X}\right)^2 \right\} \cdot \frac{1}{\sqrt{2\pi}\, \sigma_Y} \exp\left\{ -\frac{1}{2}\left(\frac{y - m_Y}{\sigma_Y}\right)^2 \right\} \\ &= f_X(x) \cdot f_Y(y). \end{aligned} \tag{4.19}$$

According to Eq. (4.19), X and Y are (statistically) independent variables. We have therefore shown that two uncorrelated normally distributed (real) variables are automatically independent. However, one should make a note of the fact that this does not apply to other types of random variables.

A stochastic process $X(t)$ is called *Gaussian* or *normally distributed* if the random variable $Z = \sum_{j=1}^{n} c_j\, X(t_j)$ is normally distributed for any (arbitrary) choice of n $(= 1, 2, \ldots)$, constants $c_1, \ldots, c_n$, and times $t_1, \ldots, t_n$. If $Y(t)$ is the response of a linear, time-invariant system where the input process $F(t)$ is Gaussian, $F(t)$ and $Y(t)$ are jointly normally distributed variables for any time t, and $Y(t)$ also becomes a Gaussian process.

If $X(t)$ is a stationary and differentiable Gaussian process, $X(t)$ and $\dot{X}(t)$ are jointly normally distributed for any t. It can be shown that $\rho_{X(t)\,\dot{X}(t)} = 0$, cf. Wong and Hajek (1985), Naess and Moan (2013); hence, $X(t)$ and $\dot{X}(t)$ are independent variables, and it follows that,

$$f_{X(t)\dot{X}(t)}(x, \dot{x}) = f_{X(t)}(x) \cdot f_{\dot{X}(t)}(\dot{x})$$
$$= \frac{1}{2\pi\, \sigma_X\, \sigma_{\dot{X}}} \exp\Big\{ -\frac{1}{2}\Big[\Big(\frac{x - m_X}{\sigma_X}\Big)^2 + \Big(\frac{\dot{x}}{\sigma_{\dot{X}}}\Big)^2\Big]\Big\}, \tag{4.20}$$

where the fact that $m_{\dot{X}} = 0$ for the derivative $\dot{X}(t)$ of any stationary and differentiable process $X(t)$ has been used. This can be seen as follows:

$$\mathbb{E}[\dot{X}(t)] = \lim_{N\to\infty} \frac{1}{N} \sum_{j=1}^{N} \dot{x}_j(t) = \frac{d}{dt} \lim_{N\to\infty} \frac{1}{N} \sum_{j=1}^{N} x_j(t) = \frac{d}{dt}\mathbb{E}[X(t)]. \tag{4.21}$$

Because $m_X = \mathbb{E}[X(t)] = \text{constant}$ for a stationary process, it follows immediately that $m_{\dot{X}} = 0$. Note also that $f_{X(t)\dot{X}(t)}(x, \dot{x})$ is independent of t.

Let us calculate the mean upcrossing rate $\nu_X^+(a)$. Substituting from Eq. (4.20) into Eq. (4.9) gives

$$\nu_X^+(a) = \int_0^\infty \frac{\dot{x}}{2\pi\, \sigma_X\, \sigma_{\dot{X}}} \exp\Big\{ -\frac{1}{2}\Big[\Big(\frac{a - m_X}{\sigma_X}\Big)^2 + \Big(\frac{\dot{x}}{\sigma_{\dot{X}}}\Big)^2\Big]\Big\}\, d\dot{x}$$
$$= \frac{1}{2\pi}\frac{\sigma_{\dot{X}}}{\sigma_X} \exp\Big\{ -\frac{1}{2}\Big(\frac{a - m_X}{\sigma_X}\Big)^2\Big\} \int_0^\infty \frac{\dot{x}}{\sigma_{\dot{X}}} \exp\Big\{ -\frac{1}{2}\Big(\frac{\dot{x}}{\sigma_{\dot{X}}}\Big)^2\Big\}\, d\Big(\frac{\dot{x}}{\sigma_{\dot{X}}}\Big)$$
$$= \frac{1}{2\pi}\frac{\sigma_{\dot{X}}}{\sigma_X} \exp\Big\{ -\frac{1}{2}\Big(\frac{a - m_X}{\sigma_X}\Big)^2\Big\}. \tag{4.22}$$

It is seen that $\nu_X^+(a)$ decreases rapidly (with σ_X as reference scale) at each side of the mean value, where it assumes its largest value

$$\nu_X^+(m_X) = \frac{1}{2\pi}\frac{\sigma_{\dot{X}}}{\sigma_X}. \tag{4.23}$$

In many situations one would prefer to define the origin so that $m_X = 0$. The expression on the rhs of Eq. (4.23) is therefore often referred to as the mean zero-upcrossing rate, under the tacit assumption that $m_X = 0$. Another corresponding parameter that is often met in the literature, is the mean zero-crossing period T_z, which is defined by,

$$T_z = (\nu_X^+(0))^{-1} = 2\pi \frac{\sigma_X}{\sigma_{\dot{X}}}. \tag{4.24}$$

For a stationary Gaussian process $X(t)$ with mean value zero, $\nu_X^+(a)$ is completely determined by the two standard deviations σ_X and $\sigma_{\dot{X}}$. If the variance spectrum $S_X(\omega)$ of $X(t)$ is known, σ_X and $\sigma_{\dot{X}}$ can be calculated by using the formulas,

$$\sigma_X^2 = \int_{-\infty}^{\infty} S_X(\omega)\, d\omega \tag{4.25}$$

and

$$\sigma_{\dot{X}}^2 = \int_{-\infty}^{\infty} \omega^2 S_X(\omega)\, d\omega. \tag{4.26}$$

If $X(t)$ is also assumed to be narrow banded, the density $f_{X_p}(a)$ of the peaks of $X(t)$ may be calculated. From Eqs. (4.16) and (4.22), it is found that ($m_X = 0$),

$$f_{X_p}(a) = \begin{cases} \frac{a}{\sigma_X^2} \exp\left\{ -\frac{a^2}{2\sigma_X^2} \right\} & a \geq 0 \\ 0, \quad a < 0. \end{cases} \tag{4.27}$$

A density of this type is called a ***Rayleigh density***, and X_p becomes a Rayleigh distributed variable. An example of $f_{X_p}(a)$ is shown in Fig. 4.4

Gaussian processes have great practical significance. This is primarily due to the following two reasons. In many cases, important physical phenomena that give rise to loads on structures can be modeled as Gaussian processes with a reasonable degree of accuracy. Moreover, weakly damped structures usually make the response more Gaussian than the load. In addition comes the fact that a Gaussian process is particularly amenable to analytical treatment.

From Eq. (4.22), it is seen that $\nu_X^+(a)$ is proportional to $f_X(a)$, which is a consequence of the fact that $X(t)$ and $\dot{X}(t)$ are independent random variables for any t. Of course, in general this is not the case. Returning to Eq. (4.9), it may be rewritten as,

$$\begin{aligned} \nu_X^+(a) &= \int_0^{\infty} \dot{x}\, f_{X\dot{X}}(a, \dot{x})\, d\dot{x} = \int_0^{\infty} \dot{x}\, f_{\dot{X}|X}(\dot{x} \mid a)\, f_X(a)\, d\dot{x} \\ &= \mathbb{E}[\dot{X}^+ \mid X = a]\, f_X(a). \end{aligned} \tag{4.28}$$

Here $\mathbb{E}[\dot{X}^+ \mid X = a]$ denotes the average positive slope of $X(t)$ at the level $X(t) = a$, which in general depends on the level a. However, as pointed out by Naess and

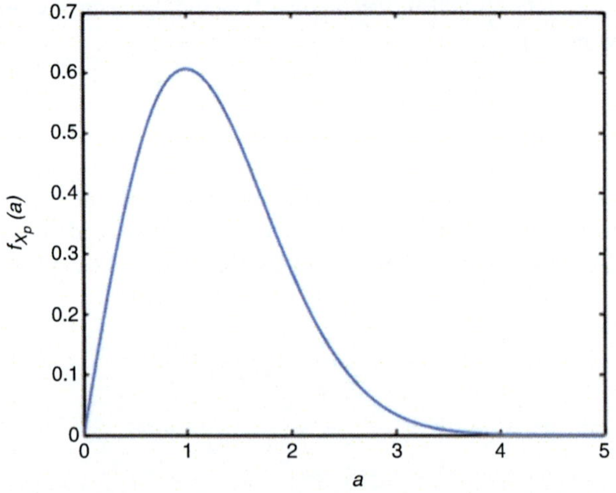

Fig. 4.4 The PDF $f_{X_p}(a)$ with $\sigma_X = 1.0$ of Rayleigh distributed peaks

Gaidai (2008), for a wide range of response processes, this dependence appears to be surprisingly weak.

4.5 Extreme Value Distributions by the Upcrossing Rate Method

The starting point is that a response quantity has been modeled as a stationary stochastic process. The goal now is to calculate the probability distribution of the largest value of the response process $X(t)$ during a specified time period. It is also a goal to determine the probability distribution of the time to the first exceedance of a given response level. These problems are very difficult to solve exactly, but by simplifying somewhat, one may often find reasonably accurate approximate solutions.

Let us denote the largest value that $X(t)$ assumes during the time T by $M(T)$. That is, $M(T) = \max\{X(t); 0 \leq t \leq T\}$. Also, let $\Theta(a)$ denote the time to the first exceedance of the level a. $M(T)$ and $\Theta(a)$ are random variables. If it is convenient to emphasize that $M(T)$ and $\Theta(a)$ refer to the process $X(t)$, the notation $M_X(T)$ and $\Theta_X(a)$ is used. Clearly,

$$\text{Prob}\{M(T) \leq a\} = \text{Prob}\{\Theta(a) > T\} \tag{4.29}$$

because both events $\{M(T) \leq a\}$ and $\{\Theta(a) > T\}$ express the same, namely, that there are no exceedances of the level a during time T. Let us call this event $\mathscr{E}$. Then $\mathscr{E} = \{X(t) \leq a \text{ for all } t \in (0, T)\}$, but this event can also be expressed as

$\mathscr{E} = \{X(0) \le a \text{ and } N^+(a,T) = 0\}$. This is so, because if $X(0) \le a$ and there are no subsequent upcrossings of a, there can be no exceedances. Hence, $\text{Prob}\{\mathscr{E}\} = \text{Prob}\{X(0) \le a \text{ and } N^+(a,T) = 0\} \to \text{Prob}\{N^+(a,T) = 0\}$ when $a \to \infty$ because of the law of marginal probability. When it is written that $a \to \infty$ here, it means that a assumes values that are large compared to the typical values for the process considered; it should not be strictly interpreted as meaning that a grows beyond all limits. In the chapter on asymptotic extreme value distributions, this will be different. Because the extreme values in most cases are much larger than the typical values, the approximation $\text{Prob}\{\mathscr{E}\} = \text{Prob}\{N^+(a,T) = 0\}$ is introduced.

To determine $\text{Prob}\{N^+(a,T) = 0\}$, the following simplifying assumption is introduced: upcrossings of high levels are statistically independent events. If the process $X(t)$ is not too narrow banded, that is, neighboring peaks tend to be of similar size, this is a reasonable approximation. This simplification implies that the random number of upcrossings in an arbitrary time interval of length T is Poisson distributed with parameter $\mathbb{E}[N^+(a,T)] = \nu_X^+(a)\,T$. In particular, this leads to the result

$$\text{Prob}\{N^+(a,T) = 0\} = \exp\{-\nu_X^+(a)\,T\}. \tag{4.30}$$

A derivation of Eq. (4.30) is given in the appendix at the end of this chapter.

From Eq. (4.30), it is then obtained that, for large values of a,

$$F_{M(T)}(a) = \text{Prob}\{M(T) \le a\} = \exp\{-\nu_X^+(a)\,T\}\,, \tag{4.31}$$

and

$$F_{\Theta(a)}(\theta) = 1 - \text{Prob}\{\Theta(a) > \theta\} = 1 - \exp\{-\nu_X^+(a)\,\theta\}\,. \tag{4.32}$$

Often, $\nu_X^+(a)\,T \ll 1$ for a relevant level a and time interval $(0, T)$, such that the probability of exceedance of a during the time T can be approximated as

$$\text{Prob}\{\text{Exceedance}\} = F_{\Theta(a)}(T) \approx \nu_X^+(a)\,T \tag{4.33}$$

because $e^x \approx 1 + x$ for $|x| \ll 1$.

The derivations above represent a successful attempt in deriving an approximate expression for the distribution of the extreme value $M(T)$ and for the time to first passage $\Theta(a)$ for large values of a, and it was seen that these distributions are determined by the mean level-upcrossing rate $\nu_X^+(a)$. The only significant simplification that has been adopted is the assumption that upcrossings of high levels are independent. Regarding the response of a lightly damped structure, the response maxima will have a tendency to occur in clumps. In particular, large peaks will tend to occur simultaneously as illustrated in Fig. 4.1. The assumption about independent upcrossings will then tend to be less valid. In such cases, Eqs. (4.31) and (4.32) may give significant deviations from the correct values, but always on the safe side in the sense that Eq. (4.31), for instance, leads to larger extreme value estimates than the correct ones. It will be seen in the next chapter that the ACER method provides a practical and elegant solution to this specific problem.

When the upcrossing rate is estimated from time series of limited length, there will be uncertainty due to sample variability. Hence, uncertainty quantification is an important issue in extreme value analysis, which is sensitive to small model changes. Since the upcrossing rate is, in fact, largely equivalent to one of the ACER functions, this problem is discussed in the next chapter.

In some situations, the relevant extreme values will be connected to the smallest or minimum values of a process. Such a case can, however, be easily recast to a study of maximum values by observing that $\min\{X(t); 0 \le t \le T\} = -\max\{-X(t); 0 \le t \le T\}$.

4.6 Extreme Values of Gaussian Processes

The particular case of a Gaussian process warrants special attention. Hence, let $X(t)$ be a stationary Gaussian process with a mean level upcrossing rate given by Eq. (4.22). For simplicity, it is assumed that $m_X = 0$. In any case, since changing the mean value is equivalent to a constant shift of all realizations, nothing is lost by this assumption.

The distribution of $M(T)$ (≥ 0) for large values of a is then, according to Eqs. (4.22) and (4.31), given by the expression

$$F_{M(T)}(a) = \exp\left\{ - \nu_X^+(0)\, T \exp\left(- \frac{a^2}{2\sigma_X^2}\right)\right\}, \tag{4.34}$$

where the mean zero-upcrossing rate enters, that is,

$$\nu_X^+(0) = \frac{1}{2\pi}\frac{\sigma_{\dot{X}}}{\sigma_X}. \tag{4.35}$$

The density of $M(T)$, $f_{M(T)}(a)$, can be calculated from Eq. (4.34) by $f_{M(T)}(a) = dF_{M(T)}(a)/da$, and it is given as follows (for large values of a):

$$\begin{aligned} f_{M(T)}(a) &= \frac{a}{\sigma_X^2}\, \nu_X^+(0)\, T \exp\left(- \frac{a^2}{2\sigma_X^2}\right) \exp\left\{ - \nu_X^+(0)\, T \exp\left(- \frac{a^2}{2\sigma_X^2}\right)\right\} \\ &= \frac{a}{\sigma_X^2}\, \nu_X^+(0)\, T \exp\left(- \frac{a^2}{2\sigma_X^2}\right) F_{M(T)}(a). \end{aligned} \tag{4.36}$$

Assuming that Eq. (4.36) is valid for all values of a, Fig. 4.5 shows the density of $M(T)$ for various values of $\nu_X^+(0)\, T$.

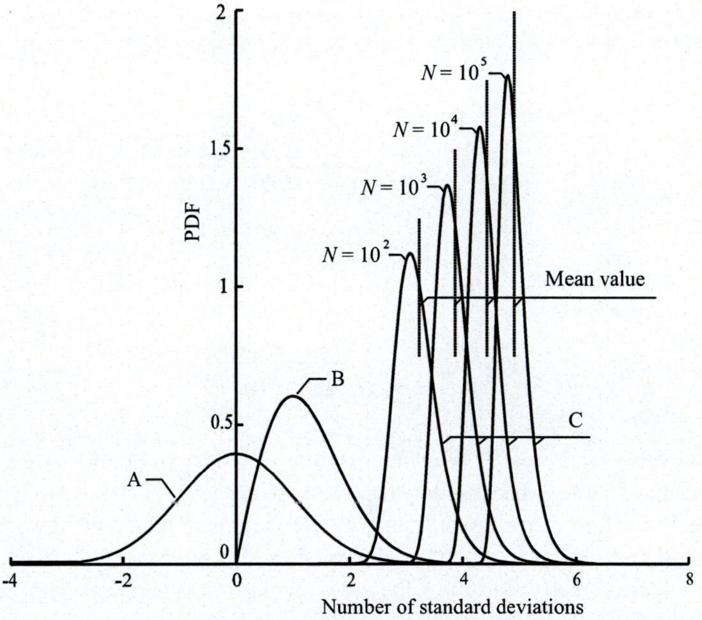

Fig. 4.5 Various densities for a stationary Gaussian process $X(t)$. A: Density of $X(t)$. B: Density of the peaks of $X(t)$ (narrow banded case). C: Density of $M(T)$ for various values of $N = \nu_X^+(0)\,T$

A quantity of particular interest in connection with design of structures is the level $a = \xi_p = \xi_p(T)$, which with probability p is not exceeded during the time T, that is,

$$F_{M(T)}(\xi_p) = p. \tag{4.37}$$

Because $p = \exp(\ln p)$, it follows from Eq. (4.34) that

$$\exp\left(-\frac{\xi_p^2}{2\sigma_X^2}\right) = -\frac{\ln p}{\nu_X^+(0)\,T}. \tag{4.38}$$

This leads to the formula

$$\xi_p(T) = \sigma_X \sqrt{2\ln\left(\frac{\nu_X^+(0)\,T}{\ln(1/p)}\right)}. \tag{4.39}$$

This formula can then be used to find the response level that has a probability of 1%, say, of being exceeded ($p = 0.99$) during time T.

Table 4.1 Table of ξ_p/σ_X-values

$\nu_X^+(0)\,T$	10	100	1000	10, 000
$p = 0.37$	2.1460	3.0349	3.7169	4.2919
$p = 0.50$	2.3105	3.1533	3.8143	4.3765
$p = 0.90$	3.0176	3.7028	4.2797	4.7876
$p = 0.95$	3.2474	3.8924	4.4448	4.9357
$p = 0.99$	3.7156	4.2908	4.7975	5.2556

The most probable extreme value, denoted by $\hat{\xi} = \hat{\xi}(T)$, is given to good approximation by the formula

$$\hat{\xi} = \sigma_X \sqrt{2 \ln\left(\nu_X^+(0)\,T\right)} \tag{4.40}$$

$\hat{\xi}$ is the value where the PDF of the extreme value $M(T)$ attains its maximum. It follows that $\hat{\xi} \approx \xi_{0.37}$ because $\ln(1/0.37) \approx 1.0$ $(e^{-1} \approx 0.37)$.

Specific values of the quantiles ξ_p/σ_X for various numbers of zero-upcrossings are listed in Table 4.1. Apart from providing a set of useful reference values, Table 4.1 also clearly illustrates the slow increase in typical extreme values with increasing time for a stationary Gaussian process.

Another quantity that is often used as a measure of extreme values is the expected largest value during a given time T, that is, $\mathbb{E}[M(T)]$. This expected value can be calculated as,

$$\mathbb{E}[M(T)] = \int_0^\infty a \, \frac{dF_{M(T)}(a)}{da} \, da. \tag{4.41}$$

It is now convenient to introduce a new integration variable η defined by the equation $F_{M(T)}(a) = e^{-\eta}$, which implicitly defines a as a function of η. Note that $a = 0\,(\infty)$ corresponds to $\eta = \infty$ (0). It is obtained that,

$$\frac{dF_{M(T)}(a)}{da}\, da = \frac{dF_{M(T)}(a)}{d\eta}\,\frac{d\eta}{da}\, da = \frac{dF_{M(T)}(a)}{d\eta}\, d\eta = -e^{-\eta}\, d\eta\,. \tag{4.42}$$

Substituted into Eq. (4.41), this leads to the equation,

$$\mathbb{E}[M(T)] = \int_0^\infty a(\eta)\, e^{-\eta}\, d\eta\,, \tag{4.43}$$

where $a = a(\eta)$ is a function of η. The way η is defined, $\eta = \nu_X^+(0)\,T \exp\{-a^2/(2\sigma_X^2)\}$. Our focus is on large a-values, that is, small η-values, and it is obtained by solving with respect to a:

$$\begin{aligned} a &= \sigma_X \sqrt{2\ln(\nu_X^+(0)\,T) - 2\ln\eta} \\ &= \sigma_X \sqrt{2\ln(\nu_X^+(0)\,T)}\left\{1 - \frac{\ln\eta}{\ln(\nu_X^+(0)\,T)}\right\}^{1/2} \\ &= \sigma_X \sqrt{2\ln(\nu_X^+(0)\,T)}\left\{1 - \frac{\ln\eta}{2\ln(\nu_X^+(0)\,T)} - \frac{(\ln\eta)^2}{8(\ln(\nu_X^+(0)\,T))^2} + \ldots\right\}. \end{aligned} \tag{4.44}$$

It can be shown that the main contribution to the integral in Eq. (4.43) comes from small η-values, and it is found that,

$$\begin{aligned} \mathbb{E}[M(T)] &\approx \sigma_X \sqrt{2\ln(\nu_X^+(0)\,T)}\left\{1 - \frac{\int_0^\infty \ln\eta\, e^{-\eta}\,d\eta}{2\ln(\nu_X^+(0)\,T)} - \frac{\int_0^\infty (\ln\eta)^2\, e^{-\eta}\,d\eta}{8(\ln(\nu_X^+(0)\,T))^2} + \ldots\right\} \\ &= \sigma_X \sqrt{2\ln(\nu_X^+(0)\,T)}\left\{1 + \frac{\lambda_E}{2\ln(\nu_X^+(0)\,T)} - \frac{\frac{\pi^2}{6} + \lambda_E^2}{8(\ln(\nu_X^+(0)\,T))^2} + \ldots\right\} \end{aligned} \tag{4.45}$$

because $\int_0^\infty \ln\eta\, e^{-\eta}\,d\eta = -\lambda_E$ and $\int_0^\infty (\ln\eta)^2\, e^{-\eta}\,d\eta = \pi^2/6 + \lambda_E^2$, where $\lambda_E = 0.5772\ldots$ denotes Euler's constant. Usually, $\ln(\nu_X^+(0)\,T)$ is sufficiently large to warrant the following approximation

$$\mathbb{E}[M(T)] \approx \sigma_X \left\{\sqrt{2\ln(\nu_X^+(0)\,T)} + \frac{\lambda_E}{\sqrt{2\ln(\nu_X^+(0)\,T)}}\right\}. \tag{4.46}$$

Similarly, it is found that,

$$\mathbb{E}[M(T)^2] = \int_0^\infty a(\eta)^2\, e^{-\eta}\,d\eta \approx 2\sigma_X^2\left\{\ln(\nu_X^+(0)\,T) + \lambda_E\right\}. \tag{4.47}$$

From Eqs. (4.47) and (4.46), the expression for the variance of $M(T)$ may be derived. It is obtained that,

$$\sigma_{M(T)}^2 = \mathbb{E}[M(T)^2] - \mathbb{E}[M(T)]^2 \approx \frac{\pi^2\sigma_X^2}{12\ln(\nu_X^+(0)\,T)}. \tag{4.48}$$

One may note that $\mathbb{E}[M(T)] \to \infty$, while $\sigma_{M(T)} \to 0$ when $T \to \infty$. In Fig. 4.5 it is clearly seen how the mean value of $M(T)$ increases, while the standard deviation decreases with increasing values of T.

Let us conclude this discussion of the extreme value distribution of a stationary Gaussian process by showing that it approaches one of the three asymptotic extreme value distributions discussed in Chap. 2, viz. the Gumbel distribution. The expression for $F_{M(T)}(a)$ given by Eq. (4.34) can be written as,

$$F_{M(T)}(a) = \exp\Big\{ - \exp\Big(- h(a)\Big)\Big\}, \quad (a \to \infty), \tag{4.49}$$

where $h(a) = \frac{a^2}{2\sigma_X^2} + \ln(\nu_X^+(0)\, T)$. Let a_0 denote the solution of the equation $h(a) = 0$. This gives $a_0 = \sigma_X \sqrt{2\ln(\nu_X^+(0)\, T)}$. It can be verified that the range of values of a where most of the extreme value distribution "lives," cf. Fig. 4.5, satisfies $|a - a_0| << a_0$ for $a_0 \to \infty$. Then $h(a) = h(a) - h(a_0) \approx h'(a_0)\,(a - a_0)$ for large values of a_0. Hence, it follows that asymptotically,

$$\begin{aligned} F_{M(T)}(a) &\approx \exp\{ - \exp(- h'(a_0)\,(a - a_0))\} \\ &= \exp\Big\{ - \exp\Big(- \frac{a - a_0}{\sigma_0}\Big)\Big\}, \quad (a_0 \to \infty)\,, \end{aligned} \tag{4.50}$$

where $\sigma_0 = \sigma_X / (\sqrt{2\ln(\nu_X^+(0)\, T)})$. This is clearly an extreme value distribution of the Gumbel type. It may be noted that the mean value of this Gumbel distribution is $a_0 + 0.5772\sigma_0$, cf. Sect. 2.2, which agrees with the mean value given in Eq. (4.46). Similarly, the variance of this Gumbel distribution is $\pi^2\sigma_0^2/6$, which coincides with the variance derived in Eq. (4.48).

The Gumbel distribution limit will follow for a large class of extreme value models of the form given in Eq. (4.49) with a differentiable function h with the properties that $h(a) \to \infty$ when $a \to \infty$, and $h'(a) > 0$ for $a > a_l$ for some value a_l.

4.7 The Crossing Rate of Transformed Processes

Assume that two stationary and differentiable processes $X(t)$ and $Y(t)$ satisfy the equation

$$Y(t) = h\big[X(t)\big], \tag{4.51}$$

where $h(\cdot)$ is a given differentiable function. The following useful result can then be shown: the upcrossing rate $\nu_Y^+(b)$ of $Y(t)$ is determined by the upcrossing rate $\nu_X^+(a)$ of $X(t)$ by the relation (Naess, 1983; Grigoriu, 1984),

$$\nu_Y^+(b) = \sum_{j=1}^{n} \nu_X^+(a_j), \tag{4.52}$$

where $a_1, \ldots, a_n$ denote all possible x-solutions of the equation $b = h(x)$.

Let us apply this result on the example $Y(t) = X(t)^2$, where $X(t)$ is a stationary Gaussian process with mean value zero. It can be shown that $Y(t)$ is also stationary. In this particular case, $h(x) = x^2$ so that the equation $b = h(x)$ has the solutions $a_1 = \sqrt{b}$ and $a_2 = -\sqrt{b}$ $(b \geq 0)$. According to Eqs. (4.52) and (4.22), it is obtained that,

$$\nu_Y^+(b) = \frac{1}{\pi}\frac{\sigma_{\dot{X}}}{\sigma_X} \exp\Big\{-\frac{b}{2\sigma_X^2}\Big\}, \quad b \geq 0. \tag{4.53}$$

The extreme value distribution for the $Y(t)$ process then becomes

$$F_{M_Y(T)}(b) = \exp\{-\nu_Y^+(b)\,T\} = \exp\Big\{-\nu_Y^+(0)\,T\,\exp\Big(-\frac{b}{2\sigma_X^2}\Big)\Big\}, \tag{4.54}$$

where

$$\nu_Y^+(0) = \frac{1}{\pi}\frac{\sigma_{\dot{X}}}{\sigma_X}\;(= 2\,\nu_X^+(0)). \tag{4.55}$$

Note that $F_{M_Y(T)}(b)$ is in fact a Gumbel distribution. Analogously to the Gaussian case, one will find that, for example,

$$\mathbb{E}[M_Y(T)] = 2\sigma_X^2\,\big\{\ln(\nu_Y^+(0)\,T) + \lambda_E\big\}. \tag{4.56}$$

For the similar case of $Z(t) = X(t)\,|X(t)|$, then $\nu_Z^+(b) = \nu_Y^+(|b|)/2$ because in this case the equation $b = h(x) = x|\,x|$ has only one solution, viz. $a = \text{sign}(b)\sqrt{|b|}$.

4.8 Hermite Moment Models

When only statistical moments of a response process are available, it has been proposed to use Hermite moments to capture non-Gaussian behavior and its effect on extreme response statistics (Winterstein, 1985, 1988). Assuming that a stationary response process $X(t)$ can be related to a stationary standard Gaussian process $U(t)$ by a strictly increasing function g by $X = g(U)$, an approximation to g is sought in terms of Hermite polynomials. Specifically,

$$\frac{X - \mu_X}{\sigma_X} = X_0 = g(U) \approx \kappa\Big(U + \Sigma_{n=3}^{N}\, g_n\, He_{n-1}(U)\Big)$$
$$= \kappa\Big(U + g_3\,(U^2 - 1) + g_4\,(U^3 - 3U) + \dots\Big). \tag{4.57}$$

The expansion coefficients g_n control the shape of the standardized distribution, while the κ parameter is a scaling factor ensuring that $X_0(t)$ has unit variance. For $N = 4$, the g_n can be expressed in terms of the central moments $\alpha_n = \mathbb{E}(X_0^n)$, assuming that $\alpha_4 > 3$ (Winterstein, 1988),

$$g_3 = \frac{\alpha_3}{4 + 2\sqrt{1 + 1.5(\alpha_4 - 3)}}, \tag{4.58}$$

$$g_4 = \frac{\sqrt{1 + 1.5(\alpha_4 - 3)} - 1}{18}, \tag{4.59}$$

and

$$\kappa = \big(1 + 2\,g_3^2 + 6\,g_4^2\big)^{-1/2}. \tag{4.60}$$

The condition $\alpha_4 > 3$ corresponds to a “softening” response, signifying that the tails of the distribution is wider than the Gaussian. The opposite case, $\alpha_4 < 3$, is discussed by Winterstein (1988). For some recent work on this method, cf. Zhang et al. (2019). Having obtained an approximation of the function g in terms of statistical moments estimated from the recorded or simulated response time series, the result from the previous section can now be applied to determine the crossing rate of the response process to estimate extreme value statistics by the point process method.

Previously, a common model adopted in ocean engineering was that the ocean surface elevation could be regarded as a Gaussian random field. However, in recent years there has been a development toward implementing also non-Gaussian, second-order wave field models. This topic is briefly discussed in Sect. 11.7 on extremes of non-Gaussian random fields. With a Gaussian model of the random wave field, the dynamic response of marine structures to the ocean waves would then appear to be represented as a transformation of a Gaussian input process. There will be several examples of such modeling in this book, and it will be clear that, in general, a simple, marginal transformation between the input and output processes will not be adequate. However, it is a rather interesting observation that the upcrossing rate for a wide range of nonlinear dynamical models is largely determined by the probability density (Naess and Gaidai, 2008). This indicates that in such cases a marginal transformation may give reasonably good results also for the upcrossing rate.

4.9 Return Period

Let Z be a random variable, and let

$$p = \text{Prob}\{Z > z\} = 1 - F_Z(z). \tag{4.61}$$

Assume that a series of independent observations of Z can be made. The mean number of observations to the first time the observed (measured) value of Z exceeds z is called the ***return period*** for exceedance of z, and it is denoted by $\tilde{R}(z)$. It can be shown that,

$$\tilde{R}(z) = \frac{1}{p} = \frac{1}{1 - F_Z(z)}. \tag{4.62}$$

This equation can be explained by recognizing that, on average, $1/p$ trials must be conducted before an event of probability p occurs.

Note that $\tilde{R}(z)$ refers to the number of observations and that these are assumed to be statistically independent. To express the return period in terms of time, knowledge about the time interval between the observations is needed. If the observation interval is Δt, the return period specified in terms of time, which is denoted by $R(z)$, will be given as,

$$R(z) = \Delta t \, \tilde{R}(z). \tag{4.63}$$

The observation interval Δt must be chosen sufficiently long such that the individual observations become approximately independent. Note that $R(\xi_p(T)) = T/(1-p)$, where $\xi_p(T)$ is given by Eq. (4.39) and T is the "observation" interval.

A design load with a probability of 10^{-2} of being exceeded during one year is often used in connection with the design of offshore structures. If $X(t)$ denotes a relevant load process considered for such a design provision, and ξ denotes the corresponding load level, then $\text{Prob}\{Z > \xi\} = 0.01$, where $Z = \max(X(t); 0 \leq t \leq 1\,\text{year})$. The return period for exceedance of ξ then becomes,

$$\tilde{R}(\xi) = \frac{1}{\text{Prob}\{Z > \xi\}} = \frac{1}{0.01} = 100. \tag{4.64}$$

The reference period in this case is one year; therefore, $R(\xi) = 100$ years.

It should be mentioned that time varying loads caused by, for example, ocean waves cannot generally be considered as stationary over an extended period of time. This implies that quantities such as yearly maxima must be calculated by using so-called long-term statistics. This is discussed in the next section.

4.10 Long-Term Extreme Value Distributions

Clearly, the estimation of the extreme loads or load effects on, e.g., a marine structure subjected to the ocean environment over the design life of the structure must take into account the changing weather conditions. This is done in a consistent manner by invoking an appropriate long-term statistical method.

There are basically three different approaches to estimating characteristic long-term extreme values. These methods are based on (1) all peak values, (2) all short-term extremes, or (3) the long-term extreme value. A more detailed description follows, where $X(t)$ denotes a zero mean stochastic process, for example the wave elevation or a corresponding load effect, that reflects the changing environmental conditions. Therefore, $X(t)$ is a non-stationary process. Let T denote the long-term time duration, e.g., 1 year, or a service life of, e.g., 30 years, and let $\tilde{T}$ denote the duration of each short-term weather condition, assuming that $T = K\tilde{T}$, where K is a large integer. The long-term situation is considered to be a sequence of K short-term conditions, where each short-term condition is assumed to be stationary. Significantly, around the world there are different kinds of weather conditions, also at sea. A coarse characterization of sea states:

- Extratropical, with slowly varying wave conditions
- Tropical, with rare hurricanes that represent very rapidly changing weather conditions

In this section, the discussions and derivations are limited to extratropical conditions.

Let W denote the vector of parameters that describes the short-term environmental condition. W can be considered as a random vector variable. For simplicity, let us assume that $W = (H_s, T_s)$, where H_s is the significant wave height and T_s a suitable spectral period (generic notation). For example, T_s may represent the spectral peak period T_p or the mean zero-crossing period T_z. In principle, the analysis is entirely similar if W contains more parameters, e.g., dominant wave direction, wind speed, etc.

4.10.1 All Peak Values

A peak value of $X(t)$, denoted generically by X_p, is defined here as the maximum value of $X(t)$ between two consecutive zero-upcrossings. For each short-term condition, let $F_{X_p|H_sT_s}(\xi|h_s, t_s)$ denote the conditional distribution of the peak value. Battjes (1970) showed that the long-term distribution $F_{X_p}(\xi)$ of the peak value X_p is given as follows:

$$F_{X_p}(\xi) = \frac{1}{\nu_X^+(0)} \int_{h_s} \int_{t_s} \nu_X^+(0|h_s, t_s)\, F_{X_p|H_sT_s}(\xi|h_s, t_s) f_{H_sT_s}(h_s, t_s)\, dh_s dt_s\,, \tag{4.65}$$

where $\overline{\nu_X^+(0)}$ denotes the long-term average zero-upcrossing rate given by

$$\overline{\nu_X^+(0)} = \int_{h_s}\int_{t_s} \nu_X^+(0|h_s, t_s)\, f_{H_sT_s}(h_s, t_s)\, dh_s dt_s\,. \tag{4.66}$$

Here, $\nu_X^+(0|h_s, t_s)$ denotes the average zero-upcrossing rate for the short-term stationary condition characterized by $H_s = h_s$ and $T_s = t_s$.

In practical applications, a commonly adopted statistical distribution for the peak values in a short-term condition is the Rayleigh distribution, that is,

$$F_{X_p|H_sT_s}(\xi|h_s, t_s) = 1 - \exp\Big(-\frac{\xi^2}{2\sigma_X(h_s, t_s)^2}\Big)\,. \tag{4.67}$$

In the modeling of ocean waves, it is sometimes appropriate to use a more accurate distribution of the peak values, or wave crest heights. A distribution that is frequently used, is one proposed by Forristall (2000).

Under the assumption that all peak values can be considered as statistically independent, which may not always be very accurate, the peak value ξ_q with a probability q of being exceeded per year is found by solving the following equation:

$$F_{X_p}(\xi_q) = 1 - \frac{q}{S^{(1y)} \cdot \overline{\nu_X^+(0)}}\,, \tag{4.68}$$

where $S^{(1y)} = 365{\cdot}24{\cdot}3600$ denotes the number of seconds in a year. The short-term duration $\tilde{T}$ does not enter into this analysis. In the long run, the relative frequency of the various sea states is reflected in the joint density $f_{H_sT_s}(h_s, t_s)$, which can be approximated by using an appropriate scatter diagram, if that is available. An example of such a scatter diagram is shown in Table 4.2.

Let the scatter diagram be divided into m intervals for the h_s-values and n intervals for the t_s-values. It may often be an acceptable approximation to assume that $\nu_X^+(0|h_s, t_s) = T_z^{-1} \approx c\,T_s^{-1}$ for a fixed constant c, cf. Eq. (4.24). Equation (4.65) with the Rayleigh approximation for $F_{X_p|H_sT_s}(\xi|h_s, t_s)$ can then be approximately expressed in the following form:

$$F_{X_p}(\xi) \approx \frac{1}{\overline{t_s^{-1}}}\sum_{i=1}^{m}\sum_{j=1}^{n}\Big\{1 - \exp\Big(-\frac{\xi^2}{2\sigma_X(h_i, t_j)^2}\Big)\Big\}\frac{K_{ij}}{t_j\,K}\,, \tag{4.69}$$

where

$$\overline{t_s^{-1}} = \sum_{i=1}^{m}\sum_{j=1}^{n}\frac{K_{ij}}{t_j\,K}\,. \tag{4.70}$$

Table 4.2 Scatter diagram northern North Sea, 1973–2001. Values given for h_s and t_p are upper-class limits

	t_p(s)																	
h_s (m)	3	4	5	6	7	8	9	10	11	12	13	14	15	16	17	18	19	20	>20
0.5	18	15	123	113	110	390	260	91	38	42	32	3	19	13	9	1	3	2	7
1.0	16	49	675	433	589	1442	1802	959	273	344	125	33	64	29	13	1	7	1	6
1.5	5	32	417	893	1107	1486	2757	1786	636	731	299	121	92	43	18	10	5	2	13
2.0	1	0	102	741	1290	1496	2575	1968	780	868	492	200	116	51	31	8	4	4	8
2.5	0	0	9	256	969	1303	2045	1892	803	941	484	181	157	58	23	19	5	1	8
3.0	0	0	1	45	438	1029	1702	1898	705	957	560	218	196	92	40	11	4	2	5
3.5	0	0	1	4	124	650	1169	1701	647	865	456	237	162	100	36	12	6	1	5
4.0	0	0	2	0	33	270	780	1369	573	868	427	193	157	91	51	13	3	0	1
4.5	0	0	0	0	3	90	459	1017	466	761	380	127	137	86	31	23	6	5	0
5.0	0	0	0	0	0	15	228	647	408	737	354	119	96	50	32	18	2	4	1
5.5	0	0	0	0	0	2	68	337	363	580	283	94	92	31	24	10	6	2	0
6.0	0	0	0	0	0	1	20	166	221	418	307	63	76	24	13	9	4	0	0
6.5	0	0	0	0	0	0	5	50	140	260	257	59	49	20	12	4	2	2	2
7.0	0	0	0	0	0	0	0	23	90	180	193	41	53	20	5	3	3	0	0
7.5	0	0	0	0	0	0	0	6	25	93	121	45	46	17	5	5	0	1	0
8.0	0	0	0	0	0	0	0	3	14	50	84	26	47	11	6	0	1	0	0
8.5	0	0	0	0	0	0	0	0	7	25	45	23	25	20	8	0	0	0	0
9.0	0	0	0	0	0	0	0	1	2	12	30	22	20	19	0	0	0	0	0
9.5	0	0	0	0	0	0	0	0	1	2	20	21	14	7	1	1	0	1	0
10.0	0	0	0	0	0	0	0	0	0	2	5	4	21	6	2	0	0	0	0
10.5	0	0	0	0	0	0	0	0	0	3	4	8	9	12	2	0	0	0	0
11.0	0	0	0	0	0	0	0	0	0	0	2	0	4	3	1	0	1	0	0
11.5	0	0	0	0	0	0	0	0	0	0	2	1	2	3	0	0	0	0	0
12.0	0	0	0	0	0	0	0	0	0	0	0	0	1	2	1	0	0	0	0
12.5	0	0	0	0	0	0	0	0	0	0	0	0	0	1	0	0	0	0	0
13.0	0	0	0	0	0	0	0	0	0	0	0	0	0	0	0	1	0	0	0

Here, K_{ij} equals the number of observations in condition (i, j), that is, in the h_s-interval $(h_i - \Delta h/2, h_i + \Delta h/2)$ and the t_s-interval $(t_j - \Delta t/2, t_j + \Delta t/2)$; $i = 1, \ldots, m$, $j = 1, \ldots, n$. $K = \sum_{i=1}^{m} \sum_{j=1}^{n} K_{ij}$ is the total number of observations, or sea states. Also note that the values for h_s and t_s included on the scatter diagram in Table 4.2 are upper-class limits, that is, $h_i + \Delta h/2$ and $t_j + \Delta t/2$.

4.10.2 All Short-Term Extremes

The conditional distribution $F_{\tilde{X}|H_sT_s}(\xi|h_s, t_s)$ of the largest peak value, $\tilde{X}$, during a short-term condition is clearly given by the expression

$$F_{\tilde{X}|H_sT_s}(\xi|h_s, t_s) = \left(F_{X_p|H_sT_s}(\xi|h_s, t_s) \right)^{k^{(st)}}, \tag{4.71}$$

where $k^{(st)} = \nu_X^+(0|h_s, t_s)\,\tilde{T}$ is the number of peak values during the short-term condition specified by $H_s = h_s$ and $T_s = t_s$. The validity of Eq. (4.71) is again based on the assumption that all peak values are independent.

The long-term distribution of the short-term extreme peak values is often approximated by the expression,

$$F_{\tilde{X}}(\xi) = \int_{h_s} \int_{t_s} F_{\tilde{X}|H_sT_s}(\xi|h_s, t_s)\, f_{H_sT_s}(h_s, t_s)\, dh_s dt_s\,. \tag{4.72}$$

Although the error is usually not significant, the averaging done in Eq. (4.72) is not quite correct in the sense that it is not a so-called ergodic average (Naess, 1984), which would be the correct approach. To achieve this, Eq. (4.72) has to be modified to read (Krogstad, 1985),

$$F_{\tilde{X}}(\xi) = \exp\left\{ \int_{h_s} \int_{t_s} \ln F_{\tilde{X}|H_sT_s}(\xi|h_s, t_s)\, f_{H_sT_s}(h_s, t_s)\, dh_s dt_s \right\}. \tag{4.73}$$

The root of this problem of averaging resides in the very notion of long-term statistics, and how it has to be interpreted. By its very definition, it is a notion built on evolution in time. And all information that is extracted about its properties is obtained by taking time averages along the observed time histories. That is precisely what is wrong with Eq. (4.72), because it applies a simple ensemble average. On the other hand, Eq. (4.73), which is an ergodic average, results from taking the appropriate time averages.

Assuming for illustration that $\tilde{T}$ = 3 h, an estimate of the value ξ_q, which has a probability q of being exceeded per year, is in this case determined by the equation

$$F_{\tilde{X}}(\xi_q) = 1 - \frac{q}{365 \cdot 8}. \tag{4.74}$$

If Eq. (4.72) is used, a relation analogous to Eq. (4.69) would be (with $\tilde{T} = 3$ h and $T_z^{-1} \approx c\, T_s^{-1}$)

$$F_{\tilde{X}}(\xi) \approx \sum_{i=1}^{m} \sum_{j=1}^{n} \Big\{ 1 - \exp\Big(- \frac{\xi^2}{2\,\sigma_X(h_i, t_j)^2} \Big) \Big\}^{\frac{60^2 \times 3 \times c}{t_j}} \frac{K_{ij}}{K} . \tag{4.75}$$

4.10.3 The Long-Term Extreme Value

The distribution of the extreme value $\hat{X} = \hat{X}(T)$, that is, the global extreme value over a long-term period T, can be expressed as follows (Naess, 1984),

$$F_{\hat{X}}(\xi) = \exp\Big\{ - T \int_{h_s} \int_{t_s} \nu_X^+(\xi\,|h_s, t_s)\, f_{H_s T_s}(h_s, t_s)\, dh_s dt_s \Big\} , \tag{4.76}$$

where $\nu_X^+(\xi\,|h_s, t_s)$ denotes the average ξ-upcrossing rate for the short-term stationary situation characterized by $H_s = h_s$ and $T_s = t_s$.

From Eqs. (4.34) and (4.35), it follows that for the case of a zero mean Gaussian process, Eq. (4.76) would read,

$$F_{\hat{X}}(\xi) =$$
$$\exp\Big\{ - T \int_{h_s} \int_{t_s} \frac{\sigma_{\dot{X}}(h_s, t_s)}{2\pi\sigma_X(h_s, t_s)} \exp\Big(- \frac{\xi^2}{2\sigma_X(h_s, t_s)^2} \Big)\, f_{H_s T_s}(h_s, t_s)\, dh_s dt_s \Big\} , \tag{4.77}$$

where the standard deviations σ_X and $\sigma_{\dot{X}}$ in the long-term situation become functions of the environmental parameters h_s and t_s, as indicated.

With $T = 1$ year $= S^{(1y)}$ seconds, the value ξ_q, which has a probability q of being exceeded per year, is now calculated from the equation,

$$F_{\hat{X}}(\xi_q) = 1 - q . \tag{4.78}$$

With reference to Table 4.2, Eq. (4.77) can then be expressed as a relation analogous to Eq. (4.69) in the following way (with $T = 1$ year and $T_z^{-1} \approx c\, T_s^{-1}$):

$$F_{\hat{X}}(\xi) \approx \exp\Big\{ - \sum_{i=1}^{m} \sum_{j=1}^{n} \frac{S^{(1y)} \times c}{t_j} \exp\Big(- \frac{\xi^2}{2\,\sigma_X(h_i, t_j)^2} \Big) \frac{K_{ij}}{K} \Big\} . \tag{4.79}$$

For the purpose of estimating extreme load effects, the use of scatter diagrams calls for a certain amount of caution. If the scatter diagram is too coarse, leading to poor resolution in the tail regions, the long-term extreme value estimates may become inaccurate. In such cases, it is recommended to use a properly adapted smooth joint density of the parameters characterizing the short-term sea states. For our purposes, the joint density of $W = (H_s, T_s)$ is needed. For North Sea applications, the spectral period T_s is often the spectral peak period T_p due to the fact that a commonly adopted spectral model is the JONSWAP spectrum, which is usually parameterized by the significant wave height and the spectral peak period. The marginal distribution of H_s is now often modeled as one of the following two alternatives:

- A three-parameter Weibull distribution
- A combination of a lognormal and a Weibull distribution

The following probabilistic model given by Haver (1980) and Haver and Nyhus (1986) has been frequently adopted as the latter model. Expressed in terms of probability densities, it assumes the form,

$$f_{H_s}(h_s) = \frac{1}{\sqrt{2\pi}\alpha h_s} \exp\left\{ -\frac{(\ln h_s - \theta)^2}{2\alpha^2} \right\}, \qquad h_s \leq \eta, \tag{4.80}$$

and

$$f_{H_s}(h_s) = \frac{\beta}{\rho}\left(\frac{h_s}{\rho}\right)^{\beta-1} \exp\left\{ -\left(\frac{h_s}{\rho}\right)^{\beta} \right\}, \qquad h_s > \eta, \tag{4.81}$$

where the value of the transition parameter η separating the lognormal model for the smaller values of H_s from the Weibull model for the larger values, will depend on the geographic location. A requirement is that $\lim_{h_s \uparrow \eta} f_{H_s}(h_s) = \lim_{h_s \downarrow \eta} f_{H_s}(h_s)$, that is, $f_{H_s}(h_s)$ is continuous at η.

This marginal density for the significant wave height is complemented by the conditional density of the spectral peak period T_p given the value of H_s using a lognormal model:

$$f_{T_p|H_s}(t_p|h_s) = \frac{1}{\sqrt{2\pi}\sigma t_p} \exp\left\{ -\frac{(\ln t_p - \mu)^2}{2\sigma^2} \right\}, \tag{4.82}$$

where the parameters μ and σ are assumed to depend on the significant wave height h_s in the following manner:

$$\mu = a_1 + a_2 h_s^{a_3}, \tag{4.83}$$

$$\sigma^2 = b_1 + b_2 \exp(-b_3 h_s), \tag{4.84}$$

for suitably chosen constants a_i and b_i, $i = 1, 2, 3$.

The joint density for the environmental parameters is then obtained by multiplying the marginal density for the significant wave height with the conditional density for the spectral peak period, that is,

$$f_W(w) = f_{H_sT_p}(h_s, t_p) = f_{H_s}(h_s)\, f_{T_p|H_s}(t_p|h_s)\,. \tag{4.85}$$

The following set of parameter values was cited by Haver (2002) for locations in the northern North Sea (Statfjord area): $\alpha = 0.6565$, $\theta = 0.77$, $\eta = 2.90$, $\beta = 2.691$, $\rho = 1.503$, $a_1 = 1.134$, $a_2 = 0.892$, $a_3 = 0.225$, $b_1 = 0.005$, $b_2 = 0.120$, $b_3 = 0.455$.

4.10.4 Simplified Methods

In this chapter three alternative long-term statistical approaches to estimate the extreme wave-induced response for ULS design checks at a given annual probability of exceedance (or return period) have been outlined. It is emphasized that a long-term approach is the most accurate approach, if it can be achieved at all. If computer models can be used, a long-term analysis is possible with the computational power that is accessible today. However, such analyses may still become a challenge in terms of required computer time. Therefore, simplified design methods are still very popular, but such methods would always need to be validated against the full long-term approach.

The applicability of simplified approaches depends on the character of the response, especially whether it can be considered quasistatic or dynamic, which response values are relevant and also which accuracy is required; that is, whether the analysis is carried out in pre-engineering or the detailed design phase. In this connection, the fact that nonlinear hydrodynamic effects might be present and cause sum- or difference-frequency excitation, respectively, should be considered.

A bottom fixed structure with a natural period below 3 seconds, say, could be considered to have a quasistatic behavior under steady wave loading and a design approach with appropriately chosen wave height and period would be relevant. For structures with natural periods above 3 seconds, the quasistatic approach might still be used in combination with the use of a dynamic amplification factor (DAF) determined by a stochastic analysis for relevant sea states, if the DAF is limited, say, to less than 1.5. The simplified methods would be relevant for early design phases while a stochastic dynamic approach should be used in the detailed design phase.

Method of Equivalent Storms

Based on earlier work by Jahns and Wheeler (1972) and Haring and Heideman (1978), Tromans and Vanderschuren (1995) proposed an alternative approach to the calculation of the long-term extreme load or load effect. In their approach, the focus is on storm events, similar to what is done in a peaks-over-threshold analysis. This approach is particularly relevant for tropical areas with rare hurricanes. Hence, the long-term situation is considered as a sequence of storm events. The method is based on the assumption that the distribution of the storm extreme response value can be approximated by a Gumbel extreme value distribution conditional on the most probable extreme response for that storm. The distribution of the most probable extreme value itself is assumed to follow a generalized Pareto distribution, which is determined by fitting to data. By invoking the rule of total probability, as exemplified by Eq. (4.72), the long-term extreme response value distribution can be calculated.

Contour Line Method

In recent years, the environmental contour line approach (Winterstein et al., 1993; Haver and Winterstein, 2008) has been advocated as a rational basis for choosing the appropriate short-term design storms leading to load and response extremes corresponding to a prescribed return period, e.g., 100 years, or equivalently, a prescribed annual probability of exceedance, which otherwise has to be obtained from a long-term analysis.

Environmental contour line plots are convenient tools for complicated structural dynamic systems where a full long-term response analysis is extremely time consuming. Environmental contour lines make it possible to obtain reasonable long-term extremes by concentrating the short-term considerations to a few sea states in the scatter diagram.

The contour line approach can be applied for an offshore site if the joint probability density for the significant wave height and the spectral peak period is available in the form of a joint model as described by Eqs. (4.80)–(4.82). This joint model must be fitted to the available data given in the form of a scatter diagram such as the one in Table 4.2. As demonstrated by Moan et al. (2005), prediction of extreme values is very sensitive to the amount of environmental data available to represent the long-term variability of the sea states. The fitting of appropriate analytical densities ensures a smoothing and can facilitate a reasonably accurate representation of the long-term extreme response.

Contour lines corresponding to a constant annual exceedance probability can be obtained by transforming the joint model to a space consisting of independent, standard Gaussian variables and then using the inverse first-order reliability method (IFORM), (see, e.g., Winterstein et al. 1993). In the standard Gaussian space, the contour line corresponding to an annual exceedance probability of q will be circles with radius $r = \Phi^{-1}(1 - q/2920)$, where Φ denotes the distribution of a standard Gaussian variable, and $2920 = 365 \times 8$ is the number of 3-hour sea states per year. Transforming these circles back to the physical parameter space provides the q-probability contour lines. Approximate contour lines can be obtained by determining the probability density for the point defined by the marginal q-probability significant wave height and the conditional median spectral peak period, and then estimating the q-probability contour line by the line of constant probability

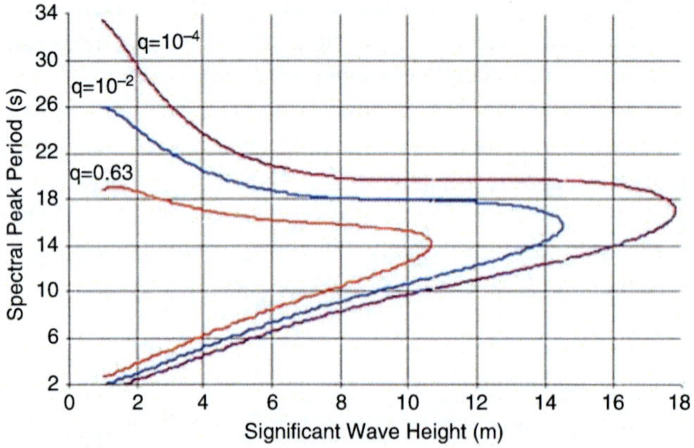

Fig. 4.6 Environmental contour line plot for the wave conditions in the Statfjord area

density. Contour lines based on the joint model discussed in Sect. 4.10.3 are plotted in Fig. 4.6, cf. Haver (2002). Even for the most complicated systems, simple methods may often be used to identify the most critical range of the q-probability contour line regarding a prediction of the q-probability response extreme.

The advantage of this method is that analyses of only a few sea states are required. As the most unfavorable sea state along the q-probability contour line is identified, a proper estimate for the q-probability response is taken as the p-fractile of the distribution of the 3-hour extreme response value. It is important to note that the median 3-hour extreme value for this sea state, i.e., $p = 0.50$, will not represent a proper estimate for the q-probability extreme value because this characteristic value will not account for the inherent randomness of the 3-hour extreme value. The fractile level, p, will depend on the aimed exceedance probability target, q, and the degree of nonlinearity of the system. For most practical systems, $p = 0.90$ seems reasonable for $q = 10^{-2}$, while $p = 0.95$ may be more adequate for $q = 10^{-4}$. As an alternative to using a p-fractile above 0.5, the desired load effect may be obtained by multiplying the median or expected maximum value with a factor of 1.2–1.3. Anyway, this simplified contour line method ideally needs to be validated by a full long-term analysis for the relevant type of environmental conditions and load effects.

Appendix

In this appendix the validity of Eq. (4.30) is demonstrated. The derivation of the formula for the mean rate of upcrossings of a given level, as it is expressed in Eq. (4.9), was based on assumptions about the probability of the number of upcrossings during short time intervals. Using similar notation as in Sect. 4.2, that is, $p_n = p_n(\Delta t) = \text{Prob}\{N^+(a, \Delta t) = n\}$, it was assumed in particular that

$p_n/p_1 \to 0$, when $\Delta t \to 0$ for $n \geq 2$. In other words, the probability of occurrence of two or more upcrossings during a short time interval can be neglected compared with the probability of one upcrossing during the same time interval. To simplify the notation somewhat, $\lambda = \nu_X^+(a)$ is used. According to Eq. (4.9), it can therefore be assumed that for sufficiently small Δt,

$$p_1(\Delta t) = \lambda \Delta t \tag{4.86}$$

and

$$p_n(\Delta t) = 0 \, , \quad n \geq 2. \tag{4.87}$$

Because $\sum_{n=0}^{\infty} p_n = 1$, it follows that,

$$p_0(\Delta t) = 1 - \lambda \Delta t. \tag{4.88}$$

To proceed, it is now required to use the assumption that the upcrossings of the level a are independent events. This implies that the number of upcrossings in a given time interval is statistically independent of the number of upcrossings in another, nonoverlapping time interval. This may be used to derive the following equation:

$$\begin{aligned} p_0(t + \Delta t) &= \text{Prob}\{\text{No upcrossings in } (0, t + \Delta t)\} \\ &= \text{Prob}\{[\text{No upcrossings in } (0, t)] \text{ and} \\ &\quad \times [\text{No upcrossings in } (t, t + \Delta t)]\} \\ &= p_0(t)\, p_0(\Delta t). \end{aligned} \tag{4.89}$$

Equation (4.89) together with Eq. (4.88) gives,

$$\frac{p_0(t + \Delta t) - p_0(t)}{\Delta t} = -\lambda \, p_0(t). \tag{4.90}$$

Strictly speaking, this equation is only approximately correct. However, on the basis of the assumptions made, it is realized that the approximation becomes better the smaller Δt becomes. This leads to the differential equation,

$$\frac{dp_0(t)}{dt} = -\lambda \, p_0(t), \tag{4.91}$$

which has the solution $p_0(t) = C \, \exp(-\lambda t)$, where C is a constant. Clearly $p_0(0) = 1$. This gives $C = 1$. The solution is therefore,

$$p_0(t) = e^{-\lambda t}, \tag{4.92}$$

which corresponds to Eq. (4.30).

While in the process, the expression for $p_n(t)$ will also be derived. It is realized that $n \geq 1$ upcrossings in the interval $(0, t + \Delta t)$ can occur as follows: { n upcrossings in $(0, t)$ and 0 upcrossings in $(t, t + \Delta t)$ } or { $n - 1$ upcrossings in $(0, t)$ and 1 upcrossing in $(t, t + \Delta t)$ }, etc. This can be expressed by the equation,

$$p_n(t + \Delta t) = \sum_{i=0}^{n} p_{n-i}(t)\, p_i(\Delta t) = p_n(t)\, p_0(\Delta t) + p_{n-1}(t)\, p_1(\Delta t). \tag{4.93}$$

The last equality follows from Eq. (4.87).

Similarly to the preceding derivation, this leads to the differential equations,

$$\frac{dp_n(t)}{dt} = -\lambda\, p_n(t) + \lambda\, p_{n-1}(t)\,, \quad n = 1, 2, \ldots \tag{4.94}$$

with the initial conditions $p_n(0) = 0$ $(n \geq 1)$. These equations can be solved in several ways. One way is to introduce the auxiliary functions $u_n(t)$, $n = 0, 1, \ldots$ defined by $p_n(t) = e^{-\lambda t}\, u_n(t)$. When this is substituted into Eq. (4.94), it leads to the equations,

$$\frac{du_n(t)}{dt} = \lambda\, u_{n-1}(t)\,, \quad n = 1, 2, \ldots\,, \tag{4.95}$$

with initial conditions $u_n(0) = 0$, $n = 1, 2, \ldots$. In particular,

$$\frac{du_1(t)}{dt} = \lambda\, u_0(t) = \lambda, \tag{4.96}$$

which gives $u_1(t) = \lambda\, t$, because $u_1(0) = 0$. Successive solution of Eq. (4.95) gives,

$$u_n(t) = \frac{(\lambda\, t)^n}{n!}, \tag{4.97}$$

and thereby,

$$p_n(t) = \frac{(\lambda\, t)^n}{n!}\, e^{-\lambda t}, \tag{4.98}$$

which holds for $n = 0, 1, 2, \ldots$ $(0! = 1)$.

The Poisson process is a frequently used model for phenomena characterized by events that occur approximately independent of each other. Subject to certain conditions, the Poisson process can be used to model the stream of telephone calls through a telephone exchange or the stream of cars passing through a road crossing.

Chapter 5
The ACER Method

5.1 Introduction

Extreme value statistics, even in applications, is generally based on asymptotic results. This is done either by assuming that the epochal extremes, for example, yearly extreme wind speeds at a given location, are distributed according to the so-called generalized (asymptotic) extreme value distribution with unknown parameters to be estimated on the basis of the observed data, cf. Chap. 2 or Coles (2001) and Beirlant et al. (2004), or by assuming that the exceedances above high thresholds follow a generalized (asymptotic) Pareto distribution with parameters that are estimated from the data, cf. Chap. 3 or Coles (2001), Beirlant et al. (2004), Davison and Smith (1990), and Reiss and Thomas (2007). As was discussed in Chap. 1, the major problem with both of these approaches is that the asymptotic extreme value theory itself cannot be used in practice to decide to what extent it is applicable for the observed data. And since statistical tests to decide this issue are rarely precise enough to settle this problem, the assumption that a specific asymptotic extreme value distribution is the appropriate distribution for the observed data is based more or less on faith or convenience.

On the other hand, one can reasonably assume that in many cases long time series obtained from practical measurements do contain values that are large enough to provide useful information about extreme events that are truly asymptotic. This cannot be strictly proved in general, of course, but the accumulated experience indicates that asymptotic extreme value distributions do provide reasonable, if not always very accurate, predictions when based on measured data. This is amply documented in the vast literature on the subject, and good references to this literature are Beirlant et al. (2004), Embrechts et al. (1997), and Falk et al. (2004).

However, even if the situation might be tolerable, it is clearly not satisfactory. In an effort to improve on the described state of affairs, an approach to the extreme value prediction problem has been developed that is less restrictive and more flexible than the ones using only asymptotic theory (Naess and Gaidai 2009; Naess et al.

A. Naess, *Applied Extreme Value Statistics*,
https://doi.org/10.1007/978-3-031-60769-1_5

2013). The approach is based on two separate components that are designed to improve on two important aspects of extreme value prediction based on observed data.

⇒ The first component provides a nonparametric empirical representation of the extreme value distribution inherent in the data. It also has the capability to accurately capture and display the effect of statistical dependence in the data on the extreme value distribution, which opens for the opportunity of using all the available data in the analysis.

⇒ The second component is then constructed based on the realization that all, or at least almost all, the available data are subasymptotic. Hence, it is important to set up the method of analysis to also incorporate subasymptotic data into the estimation of extreme value distributions, which will be shown to have some importance for accurate prediction.

The proposed method has been used on a wide variety of estimation problems, and the experience is that it represents a very powerful addition to the toolbox of methods for extreme value estimation and prediction. Several such examples will be presented in this book, cf. Chaps. 6, 9, and 12. A recent and interesting example of its practical use on the distribution of defects in metallic materials is given by Cetin and Naess (2012). Fatigue failures in such materials are intimately related to the distribution of defects, cf. Murakami (2002). Another recent example is its application to sea ice dynamics proposed by Sinsabvarodom et al. (2022). Needless to say, what is presented in this chapter is by no means considered the end of the development of the ACER method. It is a novel method, and it is to be expected that several aspects of the proposed approach will see improvements.

5.2 A Sequence of Conditioning Approximations

In this section a sequence of nonparametric distribution functions will be constructed that converges to the exact extreme value distribution for the time series considered. This constitutes the core of the proposed approach.

Consider a stochastic process $Z(t)$, which has been observed over a time interval, $(0, T)$ say. Assume that values $X_1, \ldots, X_N$, which have been derived from the observed process, are allocated to the discrete times $t_1, \ldots, t_N$ in $(0, T)$. This could be simply the observed values of $Z(t)$ at each t_j, $j = 1, \ldots, N$, or it could be average values or peak values over smaller time intervals centered at the t_j's. Our goal is to accurately determine the distribution function of the extreme value $M_N = \max\{X_j\,; j = 1, \ldots, N\}$. Specifically, we want to estimate $P(\eta) = \text{Prob}(M_N \leq \eta)$ accurately for large values of η without asymptotics. Clearly, $P(\eta) = \text{Prob}(X_1 \leq \eta, X_2 \leq \eta, \ldots, X_N \leq \eta)$. Since N will typically be a large number, direct estimation of this joint distribution function is not a practical option. Hence, we need to develop another approach.

An underlying premise for the development in this chapter is that a rational approach to the study of the extreme values of the sampled time series is to consider exceedances of the individual random variables X_j above given thresholds, as in classical extreme value theory. The alternative approach of considering the exceedances by upcrossing of given thresholds by a continuous stochastic process has already been discussed in Chap. 4, see also Naess and Gaidai (2008) and Naess et al. (2007). The approach taken in the present chapter seems to be the appropriate way to deal with the recorded data time series of, for example, the hourly or daily largest wind speeds observed at a given location, just to cite a concrete example.

The following basic rule from probability theory: $\text{Prob}(\mathscr{A} \cap \mathscr{B}) = \text{Prob}(\mathscr{A}|\mathscr{B})\text{Prob}(\mathscr{B})$ for two events $\mathscr{A}$ and $\mathscr{B}$ turns out to be an important key to estimating $P(\eta)$. Using this basic rule repeatedly, it is obtained that

$$
\begin{aligned}
P(\eta) &= \text{Prob}(M_N \le \eta) = \text{Prob}\{X_N \le \eta, \ldots, X_1 \le \eta\} \\
&= \text{Prob}\{X_N \le \eta | X_{N-1} \le \eta, \ldots, X_1 \le \eta\}\text{Prob}\{X_{N-1} \le \eta, \ldots, X_1 \le \eta\} \\
&= \text{Prob}\{X_N \le \eta | X_{N-1} \le \eta, \ldots, X_1 \le \eta\} \\
&\quad \cdot \text{Prob}\{X_{N-1} \le \eta | X_{N-2} \le \eta, \ldots, X_1 \le \eta\}\text{Prob}\{X_{N-2} \le \eta, \ldots, X_1 \le \eta\} \\
&\;\vdots \\
&= \prod_{j=2}^{N} \text{Prob}\{X_j \le \eta | X_{j-1} \le \eta, \ldots, X_1 \le \eta\} \cdot \text{Prob}(X_1 \le \eta).
\end{aligned}
\tag{5.1}
$$

In general, the variables X_j are statistically dependent. Hence, instead of assuming that all the X_j are statistically independent, which leads to the classical approximation,

$$
P(\eta) \approx P_1(\eta) := \prod_{j=1}^{N} \text{Prob}(X_j \le \eta), \tag{5.2}
$$

where := means "by definition," the following one-step memory approximation will to some extent account for the dependence between the X_j's

$$
\text{Prob}\{X_j \le \eta | X_{j-1} \le \eta, \ldots, X_1 \le \eta\} \approx \text{Prob}\{X_j \le \eta | X_{j-1} \le \eta\}, \tag{5.3}
$$

for $2 \le j \le N$. With this approximation, it is obtained that

$$
P(\eta) \approx P_2(\eta) := \prod_{j=2}^{N} \text{Prob}\{X_j \le \eta | X_{j-1} \le \eta\}\text{Prob}(X_1 \le \eta). \tag{5.4}
$$

By conditioning on one more data point, the one-step memory approximation is extended to

$$\text{Prob}\{X_j \le \eta | X_{j-1} \le \eta, \ldots, X_1 \le \eta\} \approx \text{Prob}\{X_j \le \eta | X_{j-1} \le \eta, X_{j-2} \le \eta\}, \tag{5.5}$$

where $3 \le j \le N$, which leads to the approximation,

$$P(\eta) \approx P_3(\eta) := \prod_{j=3}^{N} \text{Prob}\{X_j \le \eta | X_{j-1} \le \eta, X_{j-2} \le \eta\} \cdot \text{Prob}\{X_2 \le \eta | X_1 \le \eta\} \text{Prob}(X_1 \le \eta). \tag{5.6}$$

For a general k, $2 \le k \le N$, it is obtained that

$$P(\eta) \approx P_k(\eta) := \prod_{j=k}^{N} \text{Prob}\{X_j \le \eta | X_{j-1} \le \eta, \ldots, X_{j-k+1} \le \eta\} \cdot \prod_{j=2}^{k-1} \text{Prob}\{X_j \le \eta | X_{j-1} \le \eta \ldots, X_1 \le \eta\} \cdot \text{Prob}(X_1 \le \eta), \tag{5.7}$$

where $P(\eta) = P_N(\eta)$. It follows that the sequence of approximations $P_1(\eta), P_2(\eta), \ldots$ constitutes a sequence of increasingly accurate representations of the exact extreme value distribution $P(\eta)$.

It should be noted that the one-step memory approximation adopted above is not a Markov chain approximation, as being discussed in Smith (1992), Coles (1994), Smith et al. (1997), nor do the k-step memory approximations lead to kth-order Markov chains, which are proposed in Yun (1998, 2000). An effort to relinquish the Markov chain assumption to obtain an approximate distribution of clusters of extremes is reported by Segers (2005).

It is now necessary to have a closer look at the values for $P(\eta)$ obtained by using Eq. (5.7) as compared to Eq. (5.2). Equation (5.2) can be rewritten in the form

$$P(\eta) \approx P_1(\eta) = \prod_{j=1}^{N} \big(1 - \alpha_{1j}(\eta)\big), \tag{5.8}$$

where $\alpha_{1j}(\eta) = \text{Prob}\{X_j > \eta\}$, $j = 1, \ldots, N$. Then the approximation based on assuming independent data can be written as

$$P(\eta) \approx F_1(\eta) := \exp\Big(-\sum_{j=1}^{N} \alpha_{1j}(\eta)\Big), \tag{5.9}$$

for large values of η, noting that $1 - x \approx \exp(-x)$ with high accuracy for small x. The relative error of this approximation is less than 0.5% for values $|x| < 0.1$, and it decreases rapidly for decreasing values of $|x|$.

Similarly, Eq. (5.7) can be expressed as

$$P(\eta) \approx P_k(\eta) = \prod_{j=k}^{N} \left(1 - \alpha_{kj}(\eta)\right) \prod_{j=1}^{k-1} \left(1 - \alpha_{jj}(\eta)\right), \tag{5.10}$$

where $\alpha_{kj}(\eta) = \text{Prob}\{X_j > \eta \,|\, X_{j-1} \leq \eta, \ldots, X_{j-k+1} \leq \eta\}$, for $j \geq k \geq 2$, denotes the exceedance probability conditional on $k - 1$ previous non-exceedances. From Eq. (5.10), it is obtained that, for large values of η,

$$P(\eta) \approx F_k(\eta) := \exp\Big(- \sum_{j=k}^{N} \alpha_{kj}(\eta) - \sum_{j=1}^{k-1} \alpha_{jj}(\eta)\Big), \quad \eta \to \infty, \tag{5.11}$$

and $F_k(\eta) \to P(\eta)$ as $k \to N$ with $F_N(\eta) = P(\eta)$.

For the sequence of approximations $F_k(\eta)$ to have practical significance, it is implicitly assumed that there is a cutoff value k_c satisfying $k_c \ll N$ such that effectively $F_{k_c}(\eta) = F_N(\eta)$. It may be noted that for k-dependent stationary data sequences, that is, for data where X_i and X_j are independent whenever $|j - i| > k$, then $P(\eta) = P_{k+1}(\eta)$ exactly, and, under rather mild conditions on the joint distributions of the data, $\lim_{N\to\infty} P_1(\eta) = \lim_{N\to\infty} P(\eta)$ (Watson 1954). In fact, it can be shown that $\lim_{N\to\infty} P_1(\eta) = \lim_{N\to\infty} P(\eta)$ is true for weaker conditions than k-dependence (Leadbetter et al. 1983). However, for finite values of N, the picture is much more complex, and purely asymptotic results should be used with some caution.

Returning to Eq. (5.11), extreme value prediction by the conditioning approach described above reduces to estimation of (combinations) of the $\alpha_{kj}(\eta)$ functions. In accordance with the previous assumption about a cutoff value k_c, for all k-values of interest, $k \ll N$, so that $\sum_{j=1}^{k-1} \alpha_{jj}(\eta)$ is effectively negligible compared to $\sum_{j=k}^{N} \alpha_{kj}(\eta)$. Hence, for simplicity, the following approximation is adopted, which is applicable to both stationary and nonstationary data

$$F_k(\eta) = \exp\Big(- \sum_{j=k}^{N} \alpha_{kj}(\eta)\Big), \quad k \geq 1. \tag{5.12}$$

Going back to the definition of $\alpha_{1j}(\eta)$, it follows that $\sum_{j=1}^{N} \alpha_{1j}(\eta)$ is equal to the expected number of exceedances of the threshold η during the time interval $(0, T)$. Equation (5.9) therefore expresses the approximation that the stream of exceedance events constitute a (nonstationary) Poisson process. This opens for an understanding of Eq. (5.12) by interpreting the expressions $\sum_{j=k}^{N} \alpha_{kj}(\eta)$ as the expected effective number of (assumed) independent exceedance events provided by conditioning on $k-1$ previous observations.

5.3 Empirical Estimation of the Average Conditional Exceedance Rates

The concept of average conditional exceedance rate (ACER) of order k is now introduced as follows:

$$\varepsilon_k(\eta) = \frac{1}{N-k+1} \sum_{j=k}^{N} \alpha_{kj}(\eta)\,,\; k = 1, 2, \ldots \tag{5.13}$$

It is noted that this empirical ACER function also depends on the number of data points N. In contrast to the average upcrossing rate of Chap. 4, which would typically express the average number of upcrossings per time unit, the ACER functions are exceedance rates per data point.

The behavior and diagnostic power of the ACER functions will be demonstrated for several examples in Chap. 6. In terms of the ACER function, we may now write

$$F_k(\eta) = \exp\Big(- (N-k+1)\varepsilon_k(\eta)\Big)\,,\; k \geq 1\,. \tag{5.14}$$

In practice, there are typically two scenarios for the underlying process $Z(t)$. We may consider it to be either a stationary process or, in fact, even an ergodic process, which allows the replacement of ensemble averages with time averages, cf. Doob (1953), Cramer and Leadbetter (1967), and Wong and Hajek (1985). The alternative is to view $Z(t)$ as a process that depends on certain parameters whose variation in time may be modelled as an ergodic process in its own right. For each set of values of the parameters, the premise is that $Z(t)$ can then be modelled as an ergodic process. This would be the scenario that can be used to model long-term statistics (Naess 1984; Schall et al. 1991).

For both these scenarios, the empirical estimation of the ACER function $\varepsilon_k(\eta)$ proceeds in a completely analogous way by counting the total number of favorable incidents, that is, exceedances combined with the requisite number of preceding non-exceedances, for the total data time series and then finally dividing by $N-k+1 \approx N$. This can be shown to apply for the long-term situation, as briefly discussed below.

A few more details on the numerical estimation of $\varepsilon_k(\eta)$ for $k \geq 2$ may be appropriate. Initially, the following random functions are introduced:

$$A_{kj}(\eta) = \mathbf{1}\{X_j > \eta, X_{j-1} \leq \eta, \ldots, X_{j-k+1} \leq \eta\}\,,\; j = k, \ldots, N,\; k = 2, 3, \ldots \tag{5.15}$$

and

$$B_{kj}(\eta) = \mathbf{1}\{X_{j-1} \le \eta, \ldots, X_{j-k+1} \le \eta\}, \; j = k, \ldots, N, \; k = 2, \ldots, \tag{5.16}$$

where $\mathbf{1}\{\mathscr{A}\}$ denotes the indicator function of some event $\mathscr{A}$, that is, $\mathbf{1}\{\mathscr{A}\} = 1$ if the event occurs, and $\mathbf{1}\{\mathscr{A}\} = 0$ if not. Then, since $\mathbb{E}[\mathbf{1}\{\mathscr{A}\}] = \text{Prob}\{\mathscr{A}\}$, where $\mathbb{E}[\cdot]$ denotes the expectation operator, it follows that

$$\begin{aligned}\alpha_{kj}(\eta) &= \frac{\text{Prob}\{X_j > \eta, X_{j-1} \le \eta, \ldots, X_{j-k+1} \le \eta\}}{\text{Prob}\{X_{j-1} \le \eta, \ldots, X_{j-k+1} \le \eta\}} \\ &= \frac{\mathbb{E}[A_{kj}(\eta)]}{\mathbb{E}[B_{kj}(\eta)]}, \; j = k, \ldots, N, \; k = 2, \ldots .\end{aligned} \tag{5.17}$$

Assuming an ergodic process, which implies stationarity, then obviously $\varepsilon_k(\eta) = \alpha_{kk}(\eta) = \ldots = \alpha_{kN}(\eta)$, and by ergodicity, replacing ensemble means with corresponding time averages, it may be assumed that for the time series at hand

$$\varepsilon_k(\eta) = \lim_{N\to\infty} \frac{\sum_{j=k}^N a_{kj}(\eta)}{\sum_{j=k}^N b_{kj}(\eta)}, \tag{5.18}$$

where $a_{kj}(\eta)$ and $b_{kj}(\eta)$ are the realized values of $A_{kj}(\eta)$ and $B_{kj}(\eta)$, respectively, for the observed time series. Denoting a realization of the time series $X_1, X_2, \ldots$ by $x_1, x_2, \ldots$, then $a_{kj}(\eta) = 1$ if $x_j > \eta, x_{j-1} \le \eta, \ldots, x_{j-k+1} \le \eta$; otherwise $a_{kj}(\eta) = 0$. Similarly, $b_{kj}(\eta) = 1$ if $x_{j-1} \le \eta, \ldots, x_{j-k+1} \le \eta$; otherwise $b_{kj}(\eta) = 0$.

Clearly, $\lim_{\eta\to\infty} \mathbb{E}[B_{kj}(\eta)] = 1$. Hence, $\lim_{\eta\to\infty} \tilde{\varepsilon}_k(\eta)/\varepsilon_k(\eta) = 1$, where

$$\tilde{\varepsilon}_k(\eta) = \frac{\sum_{j=k}^N \mathbb{E}[A_{kj}(\eta)]}{N-k+1}. \tag{5.19}$$

The advantage of using the modified ACER function $\tilde{\varepsilon}_k(\eta)$ for $k \ge 2$ is that it is easier to use for non-stationary or long-term statistics than $\varepsilon_k(\eta)$. Since our focus is on the values of the ACER functions at the extreme levels, we may use any function that provides correct predictions of the appropriate ACER function at these extreme levels.

To see why Eq. (5.19) may be applicable for nonstationary time series, it is recognized that, for large values of η,

$$\begin{aligned}P(\eta) &\approx \exp\Big(-\sum_{j=k}^N \alpha_{kj}(\eta)\Big) = \exp\Big(-\sum_{j=k}^N \frac{\mathbb{E}[A_{kj}(\eta)]}{\mathbb{E}[B_{kj}(\eta)]}\Big) \\ &\approx \exp\Big(-\sum_{j=k}^N \mathbb{E}[A_{kj}(\eta)]\Big).\end{aligned} \tag{5.20}$$

If the time series can be segmented into K blocks such that $\mathbb{E}[A_{kj}(\eta)]$ remains approximately constant within each block and such that $\sum_{j \in C_i} \mathbb{E}[A_{kj}(\eta)] \approx \sum_{j \in C_i} a_{kj}(\eta)$ for a sufficient range of η-values, where C_i denotes the set of indices for block no. i, $i = 1, \ldots, K$, then $\sum_{j=k}^{N} \mathbb{E}[A_{kj}(\eta)] \approx \sum_{j=k}^{N} a_{kj}(\eta)$. Hence,

$$P(\eta) \approx \exp\big(- (N - k + 1)\hat{\varepsilon}_k(\eta)\big) , \quad \eta \to \infty, \tag{5.21}$$

where, in this case,

$$\hat{\varepsilon}_k(\eta) = \frac{1}{N - k + 1} \sum_{j=k}^{N} a_{kj}(\eta) , \tag{5.22}$$

which is the empirical counterpart of Eq. (5.19).

It is of interest to note what events are actually counted for the estimation of the various $\varepsilon_k(\eta)$, $k \geq 2$. Let us start with $\varepsilon_2(\eta)$. It follows from the definition of $\varepsilon_2(\eta)$ that $\varepsilon_2(\eta)\,(N - 1)$ can be interpreted as the expected number of exceedances above the level η satisfying the condition that an exceedance is counted only if it is immediately preceded by a non-exceedance. Hence, if a realization of a stochastic process is sampled with a sufficiently high frequency to approximate the realization with good accuracy, then $\varepsilon_2(\eta)\,(N - 1)$ is approximately the expected number of upcrossings of η among the $N - 1$ data points. This implies that $\varepsilon_2(\eta)$ is approximately the average upcrossing rate per data point. Therefore, if the $N - 1$ data points correspond to a time interval of length T, then $\varepsilon_2(\eta)\,(N-1)/T$ expresses the average upcrossing rate per time unit.

Assuming that we are studying a narrow-band stochastic response process, sampling a realization by extracting the local peak values would be a typical procedure for studying the extremes. A reinterpretation of $\varepsilon_2(\eta)\,(N - 1)$ for this case would lead us to the conclusion that it equals the average number of clumps of exceedances above η for the realizations considered, where a clump of exceedances is defined as a maximum number of consecutive exceedances of the peak values above η. Figure 5.1 illustrates how clumps of exceedances would be counted for the estimation of $\varepsilon_2(\eta)$. A scrutiny of the figure shows that four clumps can be identified.

In general, $\varepsilon_k(\eta)\,(N - k + 1)$ then equals the average number of clumps of exceedances above η separated by at least $k - 1$ non-exceedances. Hence, for the illustrating example of Fig. 5.1, it follows that for the estimation of $\varepsilon_3(\eta)$, the two arrows show that what were three clumps for the estimation of $\varepsilon_2(\eta)$ now become one clump for the estimation of $\varepsilon_3(\eta)$. Hence, in Fig. 5.1, we can now identify only two clumps for the estimation of $\varepsilon_3(\eta)$.

If the time series analyzed is obtained by extracting local peak values from a narrow-band response process, it is interesting to note that there is a certain similarity between the ACER approximations and the envelope approximations for extreme value prediction (Naess and Gaidai 2008; Vanmarcke 1975). For alternative statistical approaches to account for the effect of clustering on the extreme value

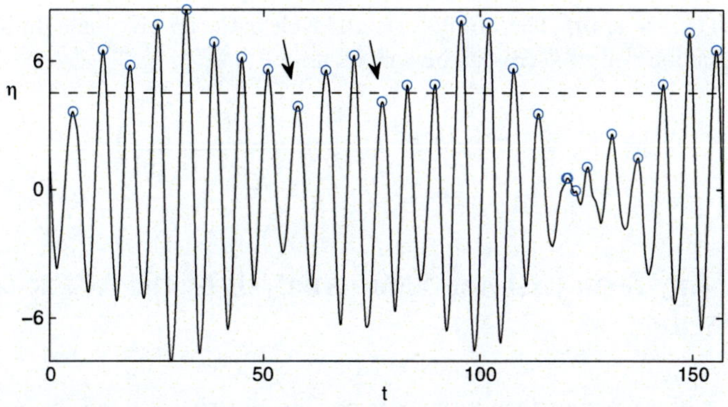

Fig. 5.1 An illustration of clumps of exceedances counted for the estimation of $\varepsilon_2(\eta)$

distribution, the reader may consult Leadbetter (1983, 1995), Hsing (1987, 1991), Ferro and Segers (2003), and Robert (2009). In these works the emphasis is on the notion of the extremal index, which characterizes the clumping or clustering tendency of the data and its effect on the extreme value distribution, cf. Sect. 2.9. In the ACER functions, these effects are automatically accounted for.

Now, let us look at the problem of estimating a confidence interval for $\varepsilon_k(\eta)$, assuming a stationary time series. If R realizations of the requisite length of the time series are available or if one long realization can be segmented into R subseries, then the sample standard deviation $\hat{s}_k(\eta)$ can be estimated by the standard formula

$$\hat{s}_k(\eta)^2 = \frac{1}{R-1}\sum_{r=1}^{R}\left(\hat{\varepsilon}_k^{(r)}(\eta) - \hat{\varepsilon}_k(\eta)\right)^2, \tag{5.23}$$

where $\hat{\varepsilon}_k^{(r)}(\eta)$ denotes the ACER function estimate from realization no. r, and $\hat{\varepsilon}_k(\eta) = \sum_{r=1}^{R}\hat{\varepsilon}_k^{(r)}(\eta)/R$.

Assuming that realizations are independent, for a suitable number R, e.g., $R \geq 20$, Eq. (5.23) leads to a good approximation of the 95% confidence interval CI = $\big(C^-(\eta), C^+(\eta)\big)$ for the value $\varepsilon_k(\eta)$, where

$$C^{\pm}(\eta) = \hat{\varepsilon}_k(\eta) \pm 1.96\,\hat{s}_k(\eta)/\sqrt{R}\,. \tag{5.24}$$

Alternatively, and which also applies to the non-stationary case, it is consistent with the adopted approach to assume that the stream of conditional exceedances over a threshold η constitute a Poisson process, possibly non-homogeneous. Hence, the variance of the estimator $\hat{E}_k(\eta)$ of $\tilde{\varepsilon}_k(\eta)$, where

$$\hat{E}_k(\eta) = \frac{\sum_{j=k}^{N} A_{kj}(\eta)}{N-k+1}, \tag{5.25}$$

is $\mathrm{Var}[\hat{E}_k(\eta)] = \tilde{\varepsilon}_k(\eta)$. Therefore, for high levels η, the approximate limits of a 95% confidence interval of $\varepsilon_k(\eta)$ can be written as

$$C^{\pm}(\eta) = \hat{\varepsilon}_k(\eta)\Big(1 \pm \frac{1.96}{\sqrt{(N-k+1)\hat{\varepsilon}_k(\eta)}}\Big). \tag{5.26}$$

5.4 Long-Term Extreme Value Analysis by the ACER Method

In Sect. 4.10 we studied long-term extreme value distributions for the point process method. This was related to a scatter diagram of short-term environmental conditions. If the whole time series over a long-term scenario is available, we have shown in the previous section that the long-term statistics using ACER functions may be estimated directly from the time series, cf. Eq. 5.21. However, in many cases it would be more practical to analyze each short-term condition separately and combine the obtained ACER functions after that. This would, e.g., be the typical approach in a simulation based long-term statistical analysis where the short-term response time series would be simulated and the resulting time series subjected to an ACER analysis. The obtained ACER functions would then be combined to form the long-term ACER function. One advantage of doing a long-term analysis this way would be the opportunity to apply the ACER function definition of Eq. (5.18) to each short-term condition.

Adapting Eq. (4.76) to the discrete scenario for the ACER analysis over the scatter diagram of Table 4.2, an estimated long-term ACER function would be expressed as follows:

$$\hat{\varepsilon}_k(\eta) = \sum_{h_i}\sum_{t_j} \hat{\varepsilon}_k(\eta|h_i, t_j)\, f_{H_s T_s}(h_i, t_j)\, \Delta h_i \Delta t_j\,, \tag{5.27}$$

where the ACER functions $\hat{\varepsilon}_k(\eta|h_i, t_j)$ have been estimated for each separate short-term condition (h_i, t_j), or (i, j) for short.

An alternative equivalent formulation is obtained as in Sect. 4.10.3. Assume that the number of sea states in condition (i, j) is K_{ij}, $i = 1, \ldots, m$ and $j = 1, \ldots, n$, and $K = \sum_{i=1}^{m}\sum_{j=1}^{n} K_{ij}$, cf. Sect. 4.10.1. We may then write

$$\hat{\varepsilon}_k(\eta) = \sum_{i=1}^{m}\sum_{j=1}^{n} \hat{\varepsilon}_k^{(ij)}(\eta)\, \frac{K_{ij}}{K}\,, \tag{5.28}$$

where the ACER function $\hat{\varepsilon}_k^{(ij)}(\eta)$ is estimated for condition (i, j). So, again we obtain the long-term extreme value distribution as

$$P(\eta) \approx \exp\big(- (N - k + 1)\hat{\varepsilon}_k(\eta)\big) , \tag{5.29}$$

using either Eq.(5.27) or (5.28), where N is the number of data over the long-term period. However, please note the cautionary comments on the use of a scatter diagram for long-term analysis in Sect. 4.10.3.

5.5 Estimation of Extremes for the Asymptotic Gumbel Case

The second component of the approach to extreme value estimation presented in this chapter was originally derived for a time series with an asymptotic extreme value distribution of the Gumbel type, cf. Naess and Gaidai (2009). This case is therefore presented first, also because the extension of the asymptotic distribution to a parametric class of extreme value distribution tails that are capable of capturing to some extent subasymptotic behavior is more transparent, and perhaps more obvious, for the Gumbel case. The reason behind the efforts to extend the extreme value distributions to the subasymptotic range is the fact that the ACER functions allow us to use not only asymptotic data, which is clearly an advantage since proving that observed extremes are asymptotic is not possible.

The effect of the asymptotic distribution being of the Gumbel type on the possible subasymptotic functional forms of $\varepsilon_k(\eta)$ cannot easily be decided in any detail. However, using the asymptotic form as a guide, it is assumed that the behavior of the ACER in the tail is dominated by a function of the form $\exp\{-a(\eta - b)^c\}$ $(\eta \geq \eta_1 \geq b)$ where a, b, and c are suitable constants, and η_1 is an appropriately chosen tail marker. Hence, it will be assumed that

$$\varepsilon_k(\eta) = q_k(\eta) \exp\{-a_k(\eta - b_k)^{c_k}\} , \quad \eta \geq \eta_1 , \tag{5.30}$$

where the function $q_k(\eta)$ is slowly varying compared with the exponential function $\exp\{-a_k(\eta - b_k)^{c_k}\}$ and a_k, b_k, and c_k are suitable constants, which in general will be dependent on k. Note that the value $c_k = q_k(\eta) = 1$ corresponds to the asymptotic Gumbel distribution, which is then a special case of the assumed tail behavior. And, of course, any extreme value distribution with an ACER function of the form given by Eq. (5.30) is asymptotically Gumbel, cf. comments at the end of Sect. 4.6.

Note that under the assumptions made, a plot of $-\log\big|\log\big(\varepsilon_k(\eta)/q_k(\eta)\big)\big|$ versus $\log(\eta - b_k)$ will exhibit a perfectly linear tail behavior. This will be illustrated in Chap. 9.

It is realized that if the function $q_k(\eta)$ could be replaced by a constant value, q_k say, one would immediately be in a position to apply a linear extrapolation strategy for deep tail prediction problems. In general, $q_k(\eta)$ is not constant, but its variation in the tail region is often sufficiently slow to allow for its replacement by a constant, possibly by adjusting the tail marker η_1. The proposed statistical approach to the prediction of extreme values is therefore based on the assumption that we can write

$$\varepsilon_k(\eta) = q_k \exp\{-a_k(\eta - b_k)^{c_k}\}, \quad \eta \geq \eta_1, \tag{5.31}$$

where a_k, b_k, c_k, and q_k are appropriately chosen constants. In a certain sense, this is a minimal class of parametric functions that can be used for this purpose, which makes it possible to achieve three important goals. Firstly, the parametric class contains the asymptotic form given by $c_k = q_k = 1$ as a special case. Secondly, the class is flexible enough to capture to a certain extent subasymptotic behavior of any extreme value distribution that is asymptotically Gumbel. Thirdly, the parametric functions agree with a wide range of known special cases, of which a very important example is the extreme value distribution for a regular stationary Gaussian process, which has $c_k = 2$.

The viability of this approach has been successfully demonstrated for extreme value statistics of the response processes related to a wide range of different dynamical systems, cf. Naess and Gaidai (2008) and Naess et al. (2007). In Chap. 6, the ACER method will be applied to the problem of predicting extreme wind speeds. The performance of the annual maxima method and the POT method for this purpose will also be discussed.

As to the question of finding the parameters a, b, c, q (the subscript k, if it applies, is suppressed), the adopted approach is to determine these parameters by minimizing the following mean square error function with respect to all four arguments:

$$F(a, b, c, q) = \sum_{j=1}^{J} w_j \left| \log \hat{\varepsilon}(\eta_j) - \log q + a(\eta_j - b)^c \right|^2, \tag{5.32}$$

where $\eta_1 < \ldots < \eta_J$ denotes the levels where the ACER function has been estimated, and w_j denotes a weight factor that puts more emphasis on the more reliably estimated $\hat{\varepsilon}(\eta_j)$, cf. the discussion about the BLUE below. However, the choice of weight factor is to some extent arbitrary. With $\big(C^-(\eta), C^+(\eta)\big)$ denoting the 95% confidence interval for the value $\varepsilon_k(\eta)$, we have previously used $w_j = \big(\log C^+(\eta_j) - \log C^-(\eta_j)\big)^{-\theta}$ with $\theta = 1$ and 2, combined with a Levenberg–Marquardt least squares optimization method (Gill et al. 1981). This has usually worked well, provided reasonable initial values for the parameters were chosen. Note that the form of w_j puts some restriction on the use of the data. Usually, there is a level η_j beyond which w_j is no longer defined, that is, $C^-(\eta_j)$ becomes negative. Hence, the summation in Eq. (5.32) has to stop before that happens. Also, the data should be preconditioned by establishing the tail marker η_1 based on inspection of the empirical ACER functions.

In general, to improve robustness of results, it is recommended to apply a nonlinearly constrained optimization (Forst and Hoffmann 2010). The set of constraints is written as

$$\begin{cases} \log q - a(\eta_i - b)^c < 0\,, \\ 0 < q < +\infty\,, \\ \min\limits_j X_j < b \le \eta_1\,, \\ 0 < a < +\infty\,, \\ 0 < c < 5\,. \end{cases} \tag{5.33}$$

Here, the first nonlinear inequality constraint is evident, since under our assumption we have $\hat{\varepsilon}(\eta_i) = q \exp\{-a(\eta_i - b)^c\}$, and $\hat{\varepsilon}(\eta_i) < 1$ by definition.

A note of caution: When the parameter c is equal to 1.0 or close to it, that is, the distribution is close to the Gumbel distribution, the optimization problem becomes ill-defined or close to ill-defined. It is seen that when $c = 1.0$, there is an infinity of (b, q) values that gives exactly the same value of $F(a, b, c, q)$. Hence, there is no well-defined optimum in parameter space. There are simply too many parameters. This problem is alleviated by fixing the q-value, and the obvious choice is $q = 1$. The restriction $c < 5$ is a practical one and has no real significance beyond limiting the range of c-values. A c-value larger than 5 should cause an inspection of the choice of tail marker. For c-values larger than 1.0, the c-value would typically decrease with increasing tail marker, indicating a deep tail approach to the Gumbel distribution, which is the asymptotic limit.

Although the Levenberg–Marquardt method generally works well with four or, when appropriate, three parameters, a more direct and transparent optimization method has also been developed for the problem at hand. It is realized by scrutinizing Eq. (5.32) that if b and c are fixed, the optimization problem reduces to a standard weighted linear regression problem. That is, with both b and c fixed, the optimal values of a and $\log q$ are found using closed-form weighted linear regression formulas in terms of w_j, $y_j = \log \varepsilon(\eta_j)$ and $x_j = (\eta_j - b)^c$. In that light, it can also be concluded that the best linear unbiased estimators (BLUEs) are obtained for $w_j = \sigma_{yj}^{-2}$, where $\sigma_{yj}^2 = \mathrm{Var}[y_j]$ (empirical) (Draper and Smith 1998; Montgomery et al. 2002). Unfortunately, this is not a very practical weight factor for the kind of problem we have here because the summation in Eq. (5.32) then typically would have to stop at undesirably small values of η_j.

It is obtained that the optimal values of a and q are given by the relations

$$a^*(b, c) = -\frac{\sum_{j=1}^{N} w_j (x_j - \overline{x})(y_j - \overline{y})}{\sum_{j=1}^{N} w_j (x_j - \overline{x})^2}\,, \tag{5.34}$$

and

$$\log q^*(b, c) = \overline{y} + a^*(b, c)\overline{x}\,, \tag{5.35}$$

where $\overline{x} = \sum_{j=1}^{N} w_j x_j / \sum_{j=1}^{N} w_j$, with a similar definition of $\overline{y}$.

To calculate the final optimal set of parameters, one may use the Levenberg–Marquardt method on the function $\tilde{F}(b, c) = F(a^*(b, c), b, c, q^*(b, c))$ to find the optimal values b^* and c^* and then use Eqs. (5.34) and (5.35) to calculate the corresponding a^* and q^*.

For a simple construction of a confidence interval for the target deep tail extreme value provided by a particular ACER function as given by the fitted parametric curve, the empirical confidence band is reanchored to the fitted curve by centering the individual confidence intervals $\mathrm{CI}_{0.95}$ for the point estimates of the ACER function on the fitted curve. Under the premise that the specified class of parametric curves fully describes the behavior of the ACER functions in the tail, parametric curves are fitted as described above to the boundaries of the reanchored confidence band. These curves are used to determine a first estimate of a 95% confidence interval for the target extreme value. To obtain a more precise estimate of the confidence interval, a bootstrapping method would be recommended. A comparison of estimated confidence intervals by both these methods will be presented in Sect. 6.2 on extreme value prediction for synthetic data.

As a final point, it has been observed that the predicted value is not very sensitive to the choice of η_1, provided it is chosen with some care. This property is easily recognized by looking at the way the optimized fitting is done. If the tail marker is in the appropriate domain of the ACER function, the optimal fitted curve does not change appreciably by moving the tail marker.

5.6 Estimation of Extremes for the General Case

For independent data in the general case, the ACER function $\varepsilon_1(\eta)$ can be expressed asymptotically as

$$\varepsilon_1(\eta) \underset{\eta \to \infty}{\simeq} \left[1 + \gamma\big(a(\eta - b)\big)\right]^{-\frac{1}{\gamma}} , \tag{5.36}$$

where $a > 0$, b, γ are constants. This follows from the explicit form of the generalized extreme value (GEV) distribution, which has been discussed in Chap. 2.

Again, the implication of this assumption on the possible subasymptotic functional forms of $\varepsilon_k(\eta)$ in the general case is not a trivial matter. The approach we have chosen is to assume that the class of parametric functions needed for the prediction of extreme values for the general case can be modelled on the relation between the Gumbel distribution and the GEV distribution. While the extension of the asymptotic Gumbel case to the proposed class of subasymptotic distributions was fairly transparent, this is not equally so for the general case. However, using a similar kind of approximation, the behavior of the mean exceedance rate in the subasymptotic part of the tail is assumed to follow a function largely of the form

$\left[1+\gamma\big(a(\eta-b)^c\big)\right]^{-\frac{1}{\gamma}}$ $(\eta \ge \eta_1 \ge b)$ where $a > 0, b, c > 0$, and $\gamma > 0$ are suitable constants, and η_1 is an appropriately chosen tail level. Hence, it will be assumed that (Naess 2010)

$$\varepsilon_k(\eta) = q_k(\eta)\,\left[1+\gamma_k\big(a_k(\eta-b_k)^{c_k}\big)\right]^{-\frac{1}{\gamma_k}}\ ,\ \ \eta \ge \eta_1, \tag{5.37}$$

where the function $q_k(\eta)$ is weakly varying compared with the function $\left[1+\gamma_k\big(a_k(\eta-b_k)^{c_k}\big)\right]^{-\frac{1}{\gamma_k}}$ and $a_k > 0$, b_k, $c_k > 0$, and $\gamma_k > 0$ are suitable constants that in general will be dependent on k. Note that the values $c_k = 1$ and $q_k(\eta) = 1$ correspond to the asymptotic limit, which is then a special case of the general expression given in Eq. (5.37). Another method to account for subasymptotic effects has been proposed by Eastoe and Tawn (2012), building on ideas developed by Tawn (1990), Ledford and Tawn (1996), and Heffernan and Tawn (2004). In this approach, the asymptotic form of the marginal distribution of exceedances is kept, but it is modified by a multiplicative factor accounting for the dependence structure of exceedances within a cluster.

An alternative form to Eq. (5.37) would be to assume that

$$\varepsilon_k(\eta) = \left[1+\gamma_k\big(a_k(\eta-b_k)^{c_k}+d_k(\eta)\big)\right]^{-\frac{1}{\gamma_k}}\ ,\ \ \eta \ge \eta_1, \tag{5.38}$$

where the function $d_k(\eta)$ is weakly varying compared with the function $a_k(\eta - b_k)^{c_k}$. However, for estimation purposes, the form given by Eq. (5.37) appears to be preferable as it leads to somewhat simpler estimation procedures.

For practical identification of the ACER functions given by Eq. (5.37), it is expedient to assume that the unknown function $q_k(\eta)$ varies sufficiently slowly to be replaced by a constant. In general, $q_k(\eta)$ is not constant, but its variation in the tail region is assumed to be sufficiently slow to allow for its replacement by a constant. Hence, as in the Gumbel case, it is in effect assumed that $q_k(\eta)$ can be replaced by a constant for $\eta \ge \eta_1$, for an appropriate choice of tail marker η_1. For simplicity of notation, in the following we shall suppress the index k on the ACER functions, which will then be written as

$$\varepsilon(\eta) = q\,\left[1+\tilde{a}\,(\eta-b)^c\right]^{-\xi}\ ,\ \ \eta \ge \eta_1, \tag{5.39}$$

where $\xi = 1/\gamma, \tilde{a} = a\gamma$.

For the analysis of data, first the tail marker η_1 is provisionally identified from visual inspection of the log plot $(\eta, \ln\hat{\varepsilon}(\eta))$. The value chosen for η_1 corresponds to the beginning of regular tail behavior in a sense to be discussed below.

The optimization process to estimate the parameters is done relative to the log plot, as for the Gumbel case. The mean square error function to be minimized is in the general case written as

$$F(\tilde{a}, b, c, q, \xi) = \sum_{j=1}^{N} w_j \left| \log \hat{\varepsilon}(\eta_j) - \log q + \xi \log\left[1 + \tilde{a}(\eta_j - b)^c\right] \right|^2, \qquad (5.40)$$

where w_j is a weight factor as defined previously.

An option for estimating the five parameters $\tilde{a}, b, c, q, \xi$ is again to use the Levenberg–Marquardt least squares optimization method, which can be simplified also in this case by observing that if $\tilde{a}$, b, and c are fixed in Eq. (5.40), the optimization problem reduces to a standard weighted linear regression problem. That is, with $\tilde{a}$, b, and c fixed, the optimal values of ξ and $\log q$ are found using closed-form weighted linear regression formulas in terms of w_j, $y_j = \log \hat{\varepsilon}(\eta_j)$ and $x_j = \log(1 + \tilde{a}(\eta_j - b)^c)$.

It is obtained that the optimal values of ξ and $\log q$ are given by relations similar to Eqs. (5.34) and (5.35). To calculate the final optimal set of parameters, the Levenberg–Marquardt method may then be used on the function $\tilde{F}(\tilde{a}, b, c) = F(\tilde{a}, b, c, q^*(\tilde{a}, b, c), \xi^*(\tilde{a}, b, c))$ to find the optimal values $\tilde{a}^*$, b^*, and c^*, and then the corresponding ξ^* and q^* can be calculated. The optimal values of the parameters may, e.g., also be found by a sequential quadratic programming (SQP) method (Numerical Algorithms Group 2010). This general case formulation of the ACER method used on financial risk issues, which often requires heavy tailed distributions, is discussed in Chap. 7.

For the construction of confidence intervals, the procedure discussed at the end of the previous section can easily be adapted to the present case.

Chapter 6
Some Practical Aspects of Extreme Value Analyses

6.1 Introduction

As stated in Chaps. 1 and 5, extreme value statistics, even in applications, is generally based on asymptotic results. This is done either by assuming that the epochal extremes, for example, yearly extreme wind speeds at a given location, are distributed according to the generalized (asymptotic) extreme value distribution with unknown parameters to be estimated on the basis of the observed data, cf. Chap. 2, or by assuming that the exceedances above high thresholds follow a generalized (asymptotic) Pareto distribution with parameters that are estimated from the data, cf. Chap. 3. With the ACER method now available, the performance of these three methods on simulated or measured data may be compared. Note that all calculations of the empirical ACER functions in this chapter were performed using the ACER program package for Matlab (Karpa 2012). This package also allows for optimized fitting of parametric functions for prediction of long return period extreme values for the case of asymptotic Gumbel distributions, cf. Sect. 5.5, which totally dominates engineering applications.

The first example in this chapter deals with synthetic data, which allows us to control the exact result to be predicted by the different methods. The second example looks at measured wind speed data at three locations along the coast of Norway. The measurement periods are ranging from 12 to 16 years. The goal would then, e.g., be to predict a 100-year return period value for the wind speed on the basis of these data. This represents a classical problem in wind engineering. For some other problems of this kind of engineering challenges where the ACER method has been applied, cf., e.g., Gaidai et al. (2016, 2018), Yu et al. (2020), and Sinsabvarodom et al. (2022).

A. Naess, *Applied Extreme Value Statistics*,
https://doi.org/10.1007/978-3-031-60769-1_6

6.2 Extreme Value Prediction for Synthetic Data

In this section the performance of the ACER method and also the 95% CI estimation will be illustrated. And 20 years of synthetic wind speed data are subjected to analysis, amounting to 2000 data points, which is not much for detailed statistics. However, this case may represent a real situation when nothing but a limited data sample is available. In this case it would appear crucial to provide extreme value estimates utilizing all data available. As will be demonstrated, the tail extrapolation technique proposed for the ACER method performs on the average better than the asymptotic POT or Gumbel methods.

The extreme value statistics will be analyzed by application to synthetic data for which the exact extreme value statistics can be calculated (Naess and Clausen 2001). In particular, it is assumed that the underlying (normalized) stochastic process $Z(t)$ is stationary and Gaussian with mean value zero and standard deviation equal to one. It is also assumed that the mean zero upcrossing rate $\nu^+(0)$ is such that the product $\nu^+(0)T = 10^3$ where $T = 1$ year, which seems to be typical for the wind speed process. Using the Poisson assumption, the distribution of the yearly extreme value of $Z(t)$ is then calculated by the formula

$$F^{1\text{yr}}(\eta) = \exp\left\{-\nu^+(\eta)T\right\} = \exp\left\{-10^3 \exp\left(-\frac{\eta^2}{2}\right)\right\}, \tag{6.1}$$

where $T = 1$ year and $\nu^+(\eta)$ is the mean upcrossing rate per year, and η is the scaled wind speed. The 100-year return period value $\eta^{100\text{yr}}$ is then calculated from the relation $F^{1\text{yr}}(\eta^{100\text{yr}}) = 1 - 1/100$, which gives $\eta^{100\text{yr}} = 4.80$.

The Monte Carlo simulated data to be used for the synthetic example are generated based on the observation that the peak events extracted from measurements of the wind speed process are usually separated by 3–4 days. This is done to obtain approximately independent data, as required by the POT method. In accordance with this, peak event data are generated from the extreme value distribution

$$F^{3\text{d}}(\eta) = \exp\left\{-q \exp\left(-\frac{\eta^2}{2}\right)\right\}, \tag{6.2}$$

where $q = \nu^+(0)T = 10$, which corresponds to $T = 3.65$ days, and $F^{1\text{yr}}(\eta) = \left(F^{3\text{d}}(\eta)\right)^{100}$.

Since the data points (i.e., $T = 3.65$ days maxima) are independent, $\varepsilon_k(\eta)$ is independent of k. Therefore, $k = 1$ is chosen. Since there are 100 data from 1 year, the data amounts to 2000 data points. For estimation of a 95% confidence interval for each value of the ACER function $\varepsilon_1(\eta)$ for the chosen range of η-values, the required standard deviation in Eq. (11.37) was based on 20 estimates of the ACER function using the yearly data. This provided a 95% confidence band on the ACER function based on 2000 data. From these data, the predicted 100 year return level is obtained from $\hat{\varepsilon}_1(\eta^{100\text{yr}}) = 10^{-4}$. A nonparametric bootstrapping method was also used to estimate a 95% confidence interval based on 1000 resamples of size 2000.

The POT prediction of the 100 year return level was based on using maximum likelihood estimates (MLEs) of the parameters in Eq. (3.3) for a specific choice of threshold. The 95% confidence interval was obtained from the parametrically bootstrapped density of the POT estimate for the given threshold. A sample of 1000 data sets was used. One of the unfortunate features of the POT method is that the estimated 100 year value may vary significantly with the choice of threshold, so also for the synthetic data. We have followed the standard recommended procedures for identifying a suitable threshold (Coles 2001).

Note that in spite of the fact that the true asymptotic distribution of exceedances is the exponential distribution Eq. (3.4), the POT method used here is based on adopting Eq. (3.3). The reason is simply that this is the recommended procedure (Coles 2001), which is somewhat unfortunate, but understandable, the reason being that the GP distribution provides greater flexibility in terms of curve fitting. If the correct asymptotic distribution of exceedances had been used in this example, poor results for the estimated return period values would be obtained. The price to pay for using the GP distribution is that the estimated parameters may easily lead to an asymptotically inconsistent distribution.

The 100 year return level predicted by the Gumbel method was based on using the method of moments for parameter estimation on the sample of 20 yearly extremes. This choice of estimation method is due to the small sample of extreme values. The 95% confidence interval was obtained from the parametrically bootstrapped density of the Gumbel prediction. This was based on a sample of size of 10,000 data sets of 20 yearly extremes. The results obtained by the method of moments were compared with the corresponding results obtained by using the maximum likelihood method. While there were individual differences, the overall picture was one of very good agreement.

In order to get an idea about the performance of the ACER, POT, and Gumbel methods, 100 independent 20-year MC simulations as discussed above were done. Table 6.1 compares predicted values and confidence intervals for a selection of 10 cases together with average values over the 100 simulated cases. It is seen that the average of the 100 predicted 100 year return levels is slightly better for the ACER method than for both the POT and the Gumbel methods. But more significantly, the range of predicted 100 year return levels by the ACER method is 4.34–5.36, while the same for the POT method is 4.19–5.87 and for the Gumbel method 4.41–5.71. Hence, in this case the ACER method performs consistently better than both these methods. It is also observed from the estimated 95% confidence intervals that the ACER method, as implemented in this book, provides slightly higher accuracy than the other two methods. Lastly, it is pointed out that the confidence intervals of the 100 year return level values by the ACER method obtained by either the simplified extrapolated confidence band approach or nonparametric bootstrapping are very similar except for a slight mean shift. As a final comparison, the 100 bootstrapped

Table 6.1 100-year return level estimates and 95% CI (BCI = CI by bootstrap) for A=ACER, G=Gumbel, and P=POT. Exact value = 4.80

Sim.No.	A $\hat{\eta}^{100yr}$	A CI	A BCI	G $\hat{\eta}^{100yr}$	G BCI	P $\hat{\eta}^{100yr}$	P BCI
1	5.07	(4.67, 5.21)	(4.69, 5.42)	4.41	(4.14, 4.73)	4.29	(4.13, 4.52)
10	4.65	(4.27, 4.94)	(4.37, 5.03)	4.92	(4.40, 5.58)	4.88	(4.42, 5.40)
20	4.86	(4.49, 5.06)	(4.44, 5.19)	5.04	(4.54, 5.63)	5.04	(4.48, 5.74)
30	4.75	(4.22, 5.01)	(4.33, 5.02)	4.75	(4.27, 5.32)	4.69	(4.24, 5.26)
40	4.54	(4.20, 4.74)	(4.27, 4.88)	4.80	(4.31, 5.39)	4.73	(4.19, 5.31)
50	4.80	(4.35, 5.05)	(4.42, 5.14)	4.91	(4.41, 5.50)	4.79	(4.31, 5.34)
60	4.84	(4.36, 5.20)	(4.48, 5.19)	4.85	(4.36, 5.43)	4.71	(4.32, 5.23)
70	5.02	(4.47, 5.31)	(4.62, 5.36)	4.96	(4.47, 5.53)	4.97	(4.47, 5.71)
80	4.59	(4.33, 4.81)	(4.38, 4.98)	4.76	(4.31, 5.31)	4.68	(4.15, 5.27)
90	4.84	(4.49, 5.11)	(4.60, 5.30)	4.77	(4.34, 5.32)	4.41	(4.23, 4.64)
100	4.62	(4.29, 5.05)	(4.45, 5.09)	4.79	(4.31, 5.41)	4.53	(4.05, 4.88)
Av. 100	4.82	(4.41, 5.09)	(4.48, 5.18)	4.84	(4.37, 5.40)	4.72	(4.27, 5.23)

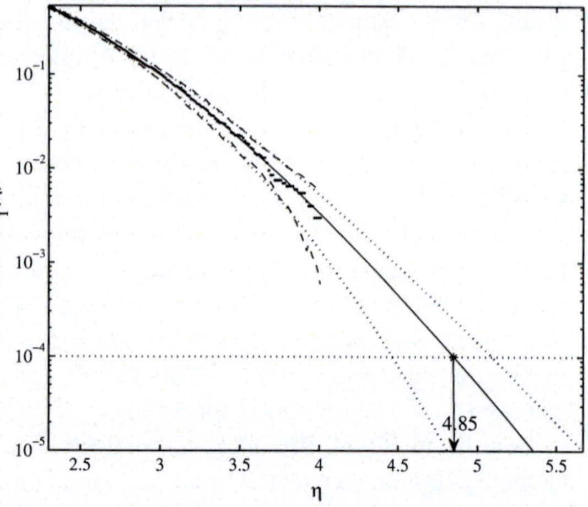

Fig. 6.1 Synthetic data ACER $\hat{\varepsilon}_1$: Monte Carlo simulation (*); optimized curve fit (—); empirical 95% confidence band (- -); and optimized confidence band (· · ··). Tail marker $\eta_1 = 2.3$

confidence intervals obtained for the ACER and Gumbel methods missed the target value three times, while for the POT method this number was 18.

An example of the ACER plot and results obtained for one set of data is presented in Fig. 6.1. The predicted 100-year value is 4.85 with a predicted 95% confidence interval (4.45, 5.09). Figure 6.2 presents POT predictions based on MLE for different thresholds in terms of the number n of data points above the threshold. The predicted value is 4.7 at $n = 204$, while the 95% confidence interval is (4.25, 5.28). The same data set as in Fig. 6.1 was used. This was also used for the Gumbel plot shown in Fig. 6.3. In this case the predicted value based on the method of moments (MMs) is $\hat{\eta}_{\mathrm{MM}}^{100\mathrm{yr}} = 4.75$ with a parametric bootstrapped 95% confidence interval of (4.34, 5.27). Prediction based on the Gumbel–Lieblein BLUE method

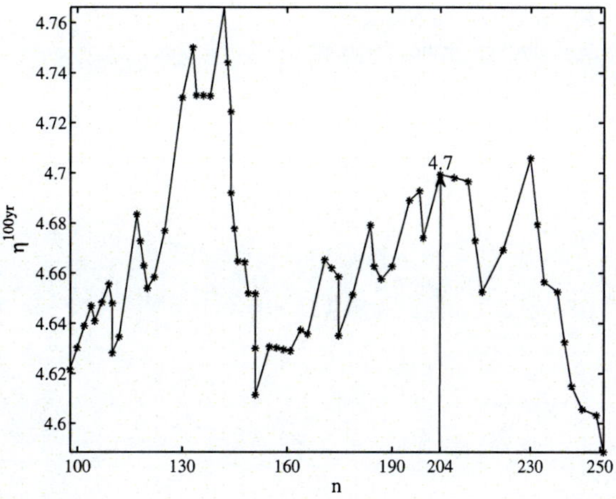

Fig. 6.2 The point estimate $\hat{\eta}^{100\text{yr}}$ of the 100-year return period value based on 20-year synthetic data as a function of the number n of data points above the threshold. The return level estimate = 4.7 at $n = 204$

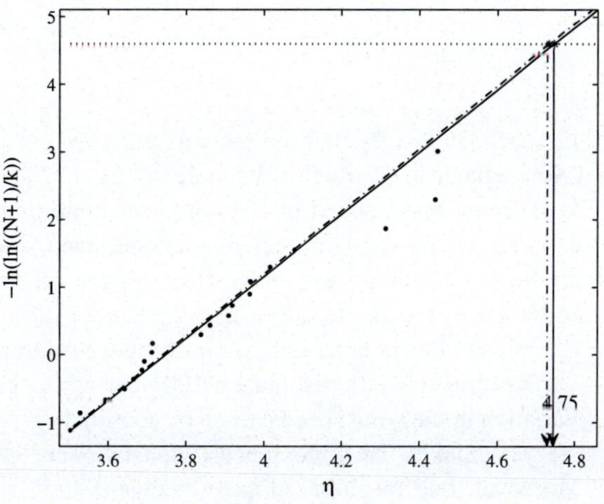

Fig. 6.3 The point estimate $\hat{\eta}^{100\text{yr}}$ of the 100-year return period value based on 20-year synthetic data. Lines are fitted by the method of moments—solid line (—) and the Gumbel–Lieblein BLUE method—dash-dotted lite (−·−). The return level estimate by the method of moments is 4.75 and by the Gumbel–Lieblein BLUE method is 4.73

(GL), cf., e.g., Cook (1985), is $\hat{\eta}_{GL}^{100\text{yr}} = 4.73$ with a parametric bootstrapped 95% confidence interval equal to (4.35, 5.14).

6.3 Measured Wind Speed Data

In this section we analyze real wind speed data, measured at three weather stations off the coast of Norway: at Torsvåg Fyr weather station (station number 90800), Sula weather station (station number 65940), and Obrestad Fyr weather station (station number 44080), cf. Karpa and Naess (2013). Figure 6.4 shows the geographical position of each station. The hourly maximum of the 3-second wind gust (10 meters above the ground) was recorded during 13 years (1997–2010) at the first station,

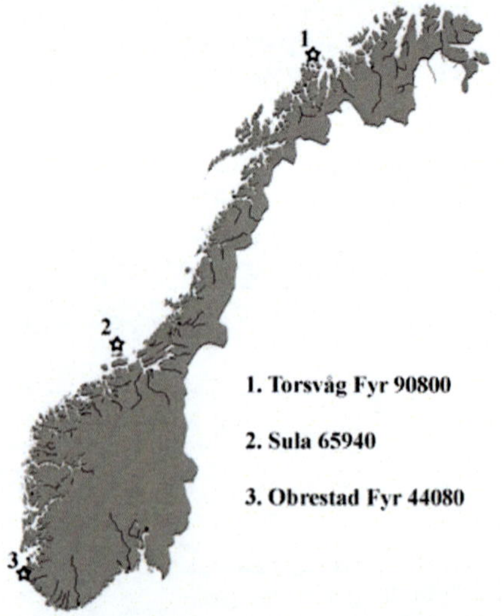

Fig. 6.4 Map of Norway with marked weather stations

12 years (1998–2010) at the second, and 16 years (1994–2010) at the third station (Norwegian Meteorological Institute 2012).

Extreme wind speed prediction is an important issue for design of structures exposed to the weather variations. Significant efforts have been devoted to the problem of predicting extreme wind speeds on the basis of measured data by various authors over several decades; see, e.g., Cook (1982), Naess (1998a), Palutikof et al. (1999), and Perrin et al. (2006) for extensive references to previous work.

The objective is to estimate a 100-year wind speed for each of these locations. Variation in the wind speed caused by seasonal variations in the wind climate during the year makes the wind speed a nonstationary process on the scale of months. Moreover, due to global climate change, yearly statistics may vary on the scale of years. The latter is, however, a slow process, and for the purpose of long-term prediction, it is assumed here that within a time span of 100 years, a quasi-stationary model of the wind speeds applies. This may not be entirely true, of course, but for the three stations under study, no apparent trend in the wind speed was detected over the period of registration.

Figures 6.5, 6.6, and 6.7 demonstrate the plots of the time series observed from each station. A conspicuous feature of the time series is the clear seasonal variation of the wind speeds. Note the paucity of data at certain times at the Sula station. The practical consequence of this is to shorten the effective length of the time series. It is of some importance to note that the samples from Torsvåg Fyr and Obrestad Fyr stations contain outlying observations, such as 45.3 m/s in June 06, 1997, 43.7 m/s in May 10, 2001 and 60.8 m/s in September 09, 2008 for Obrestad Fyr station, and 45.3 m/s in July 12, 1998 and July 31, 1999 for Torsvåg Fyr station. Such wind speeds are most likely spurious for the corresponding time periods and latitudes. Moreover, observations from the weather stations in the close

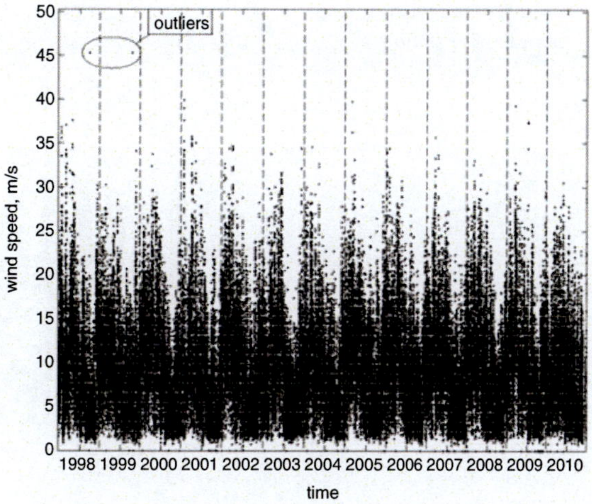

Fig. 6.5 Observations from Torsvåg Fyr station

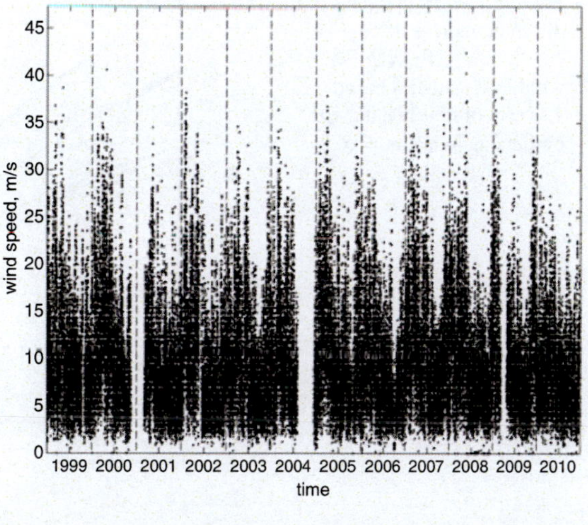

Fig. 6.6 Observations from Sula station

neighborhood of Obrestad Fyr confirm that no heavy storm has occurred during the period in question, while no information from the stations in the neighborhood of Torsvåg Fyr is available. Therefore, the outliers from Obrestad Fyr station have to be rejected, while the outliers from Torsvåg Fyr are kept, mainly in order to show that the ACER method is largely insensitive to observations of this kind.

In Figs. 6.8, 6.9, and 6.10, $\hat{\varepsilon}_k(\eta)$ is plotted versus wind speeds, η/σ for different values of k for the three stations. The figures reveal that there is significant dependence between the data, but that this is largely accounted for by $k = 48$ since there is a marked degree of convergence of the $\hat{\varepsilon}_k(\eta)$ for $k \geq 48$. It should be noted that $k = 48$ obviously corresponds to 2 days separated exceedances for hourly observations. For $k \geq 96$, that is, 4 days declustered data, full convergence has been achieved for all practical purposes. However, Figs. 6.8, 6.9, and 6.10 also

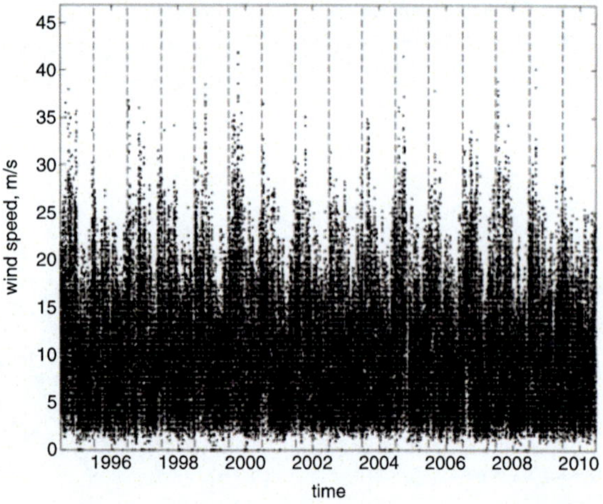

Fig. 6.7 Observations from Obrestad Fyr station

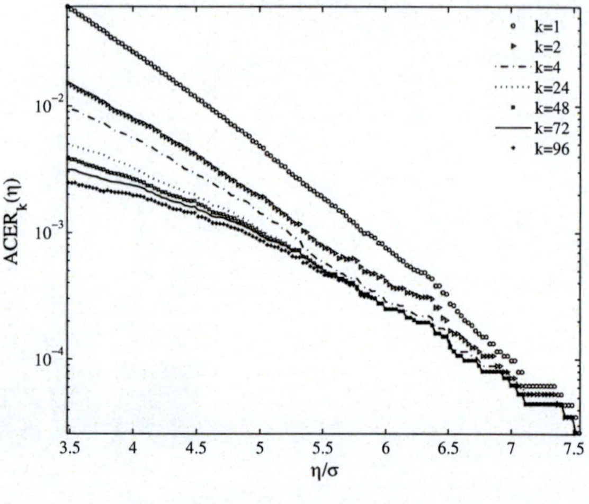

Fig. 6.8 ACER estimates for different degrees of conditioning. Torsvåg Fyr wind gust statistics based on 13 years of hourly data with outliers included, $\sigma = 5.30$

reveal that for extreme value estimation $\hat{\varepsilon}_1(\eta)$ can be used since the ACER functions all converge in the far tail. This clearly demonstrates the power of an ACER function plot as a diagnostic tool to decide on the value of k needed for extreme value estimation in a particular case. In spite of significant dependence effects for the lower wind speeds, for the extreme wind speeds this is largely absent. This makes it possible to choose $k = 1$, which makes much more data available for estimation, with a possible reduction of uncertainty in estimation as a result.

Figures 6.11, 6.12, and 6.13 show the plots of the optimized fit to the data for $\hat{\varepsilon}_1(\eta)$ for each station. The 100-year return level value η^{100yr} and its 95% CI are estimated parametrically. For the data with outliers from Torsvåg Fyr weather station, $\hat{\eta}^{100yr} = 47.46$ m/s and the 95% CI = (42.11, 50.71) with parameters of the optimal curve: $q = 0.44$, $b = 9.02$, $a = 0.1$, $c = 1.33$. The predicted 100-year return wind speed and 95% CI for the data without outliers are $\hat{\eta}^{100yr} = 47.21$ m/s; CI =

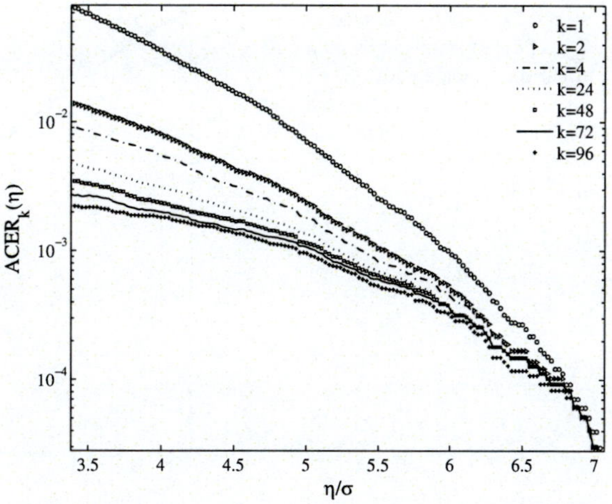

Fig. 6.9 ACER estimates for different degrees of conditioning. Sula wind gust statistics based on 12 years of hourly data, $\sigma = 5.49$

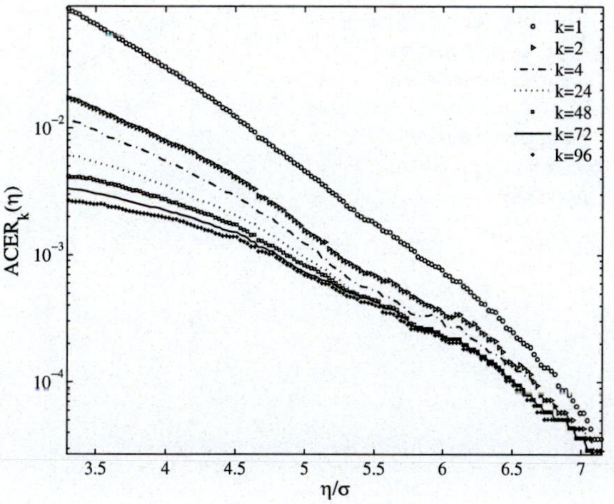

Fig. 6.10 ACER estimates for different degrees of conditioning. Obrestad Fyr wind gust statistics based on 16 years of hourly data, $\sigma = 5.47$

(39.94, 50.60) with optimal parameters $q = 0.47$, $b = 8.49$, $a = 0.09$, $c = 1.36$. In the case of the Sula wind station, $\hat{\eta}^{100yr} = 46.33$ m/s; CI = (43.41, 47.77), where the parameters of the optimal curve are $q = 0.58$, $b = 0$, $a = 0.005$, $c = 2.07$. Finally, for the Obrestad Fyr station, the 100-year return level value is $\hat{\eta}^{100yr} = 48.38$ m/s with confidence interval (43.18, 50.74) and optimal parameters $q = 0.29$, $b = 12.34$, $a = 0.13$, $c = 1.27$.

The annual maxima method is applied to the wind gust data to compare the estimated 100-year return level values. The application of this method to the wind data is based on the premise that the time series of the yearly maxima can be modeled as a sequence of iid random variables, which would seem to be a reasonable assumption to make as a first approximation. The Gumbel estimate $\hat{\eta}^{100yr}$ is based on the method of moments (MMs) and the Gumbel–Lieblein BLUE method (GL), cf., e.g., Cook (1985). A computer program has been written in the Matlab language

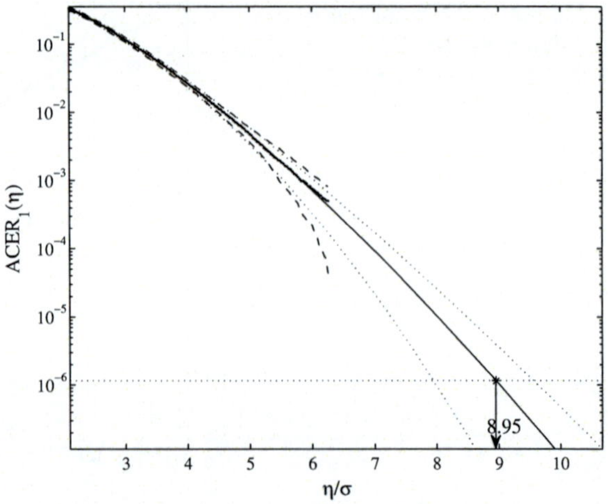

Fig. 6.11 Plot for Torsvåg Fyr of $\hat{\varepsilon}_1(\eta)$ against η/σ on a logarithmic scale with $\eta_1 = 2.07\sigma$ for the optimized parameter values with the 95% confidence band, $\sigma = 5.30$

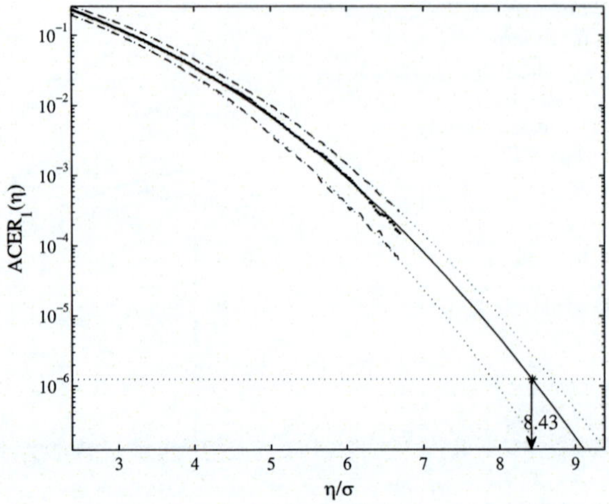

Fig. 6.12 Plot for Sula of $\hat{\varepsilon}_1(\eta)$ against η/σ on a logarithmic scale with $\eta_1 = 2.36\sigma$ for the optimized parameter values with the 95% confidence band, $\sigma = 5.49$

to implement both methods. Figures 6.14, 6.15, and 6.16 present observed yearly extremes extracted from the hourly data together with fitted straight lines on Gumbel probability paper. Hereby, the 100-year return level values for the first station with outliers included are $\hat{\eta}_{MM}^{100yr} = 51.33$ m/s and $\hat{\eta}_{GL}^{100yr} = 51.57$ m/s, while in the case of rejected outlying observations, $\hat{\eta}_{MM}^{100yr} = 44.31$ m/s and $\hat{\eta}_{GL}^{100yr} = 45.84$ m/s. For Sula and Obrestad Fyr stations, the 100-year return level values are $\hat{\eta}_{MM}^{100yr} = 48.66$ m/s with $\hat{\eta}_{GL}^{100yr} = 52.9$ m/s and $\hat{\eta}_{MM}^{100yr} = 48.59$ m/s with $\hat{\eta}_{GL}^{100yr} = 53.79$ m/s, respectively.

Despite the fact that the Gumbel–Lieblein BLUE method is considered as one of the best available conventional Gumbel methods, the application of the GL method requires tables of the BLUE coefficients which are not easily available for annual data with sample size $N > 24$. Observed results reveal sensitivity of this method

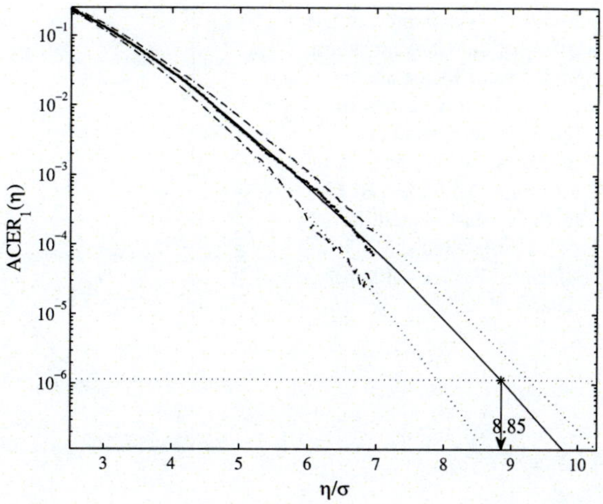

Fig. 6.13 Plot for Obrestad Fyr of $\hat{\varepsilon}_1(\eta)$ against η/σ on a logarithmic scale with $\eta_1 = 2.5\sigma$ for the optimized parameter values with the 95% confidence band, $\sigma = 5.47$

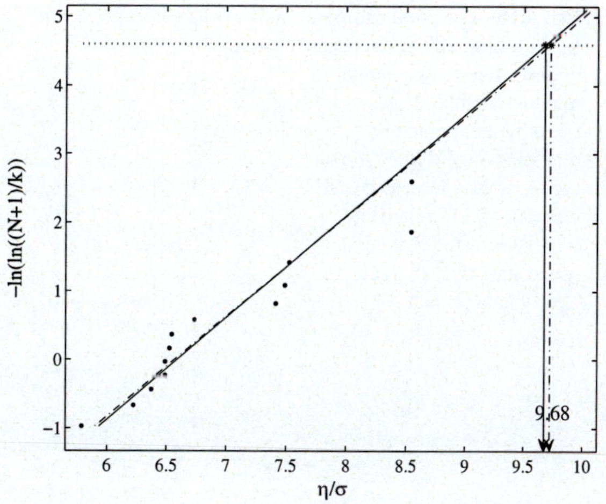

Fig. 6.14 The point estimate $\hat{\eta}^{100yr}$ of the 100-year return period value by the annual maxima method. Lines are fitted by the method of moments—solid line (—) and the Gumbel–Lieblein BLUE method—dash-dotted lite (–·–). Torsvåg Fyr wind speed statistics, 13 years of hourly data, $\sigma = 5.30$

to outliers, which also applies for the method of moments. It is also noted that the Gumbel–Lieblein BLUE method seems to have a tendency to overestimate predicted return level values, while the method of moments seems to be reasonably stable for the studied sets of data.

The POT method is also applied to the wind gust time series. Immediately, this would seem to be an unwarranted approach since the time series of wind speeds is conspicuously nonstationary. Efforts have been made to account for the seasonal variations by explicit modelling of the parameters of the GP distribution, cf. Coles (2001). However, this does not seem to be a widely adopted procedure. Instead, the POT method is applied directly to the full time series, recognizing that the extracted relevant data comes from a period of 3 or 4 months, which may be considered to represent a more or less stationary period. And by manipulating time frames, this is

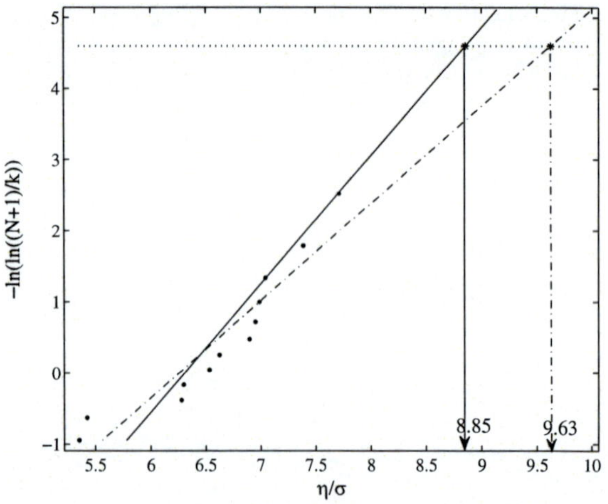

Fig. 6.15 The point estimate $\hat{\eta}^{100yr}$ of the 100-year return period value by the annual maxima method. Lines are fitted by the method of moments—solid line (—) and the Gumbel–Lieblein BLUE method—dash-dotted lite (– · · –). Sula wind speed statistics, 12 years of hourly data, $\sigma = 5.49$

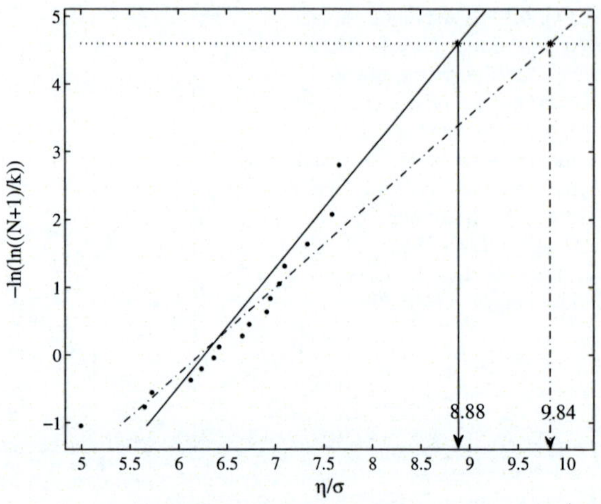

Fig. 6.16 The point estimate $\hat{\eta}^{100yr}$ of the 100-year return period value by the annual maxima method. Lines are fitted by the method of moments—solid line (—) and the Gumbel–Lieblein BLUE method—dash-dotted lite (– · · –). Obrestad Fyr wind speed statistics, 16 years of hourly data, $\sigma = 5.47$

considered to represent 1 year. This trick approximately circumvents the problem of nonstationarity. This will be our approach here.

Following the WAFO-group (2000), data were declustered beforehand. It was done in such a way that peak events separated by 3.5 days or more were extracted from the measured data and selected for the analysis to achieve approximate independence of the exceedances (Naess 1998b). Figures 6.17, 6.18, and 6.19 present POT estimates of η^{100yr} for different threshold numbers based on maximum likelihood estimation, cf. Chap. 3. Estimates were obtained by using the MATLAB (2009) Statistics Toolbox routine gpfit. It is interesting to observe the unstable characteristics of the estimates over a range of threshold values. While it is clearly of interest to discuss methods for stabilizing the POT estimates, this issue is considered to be outside the scope of this book.

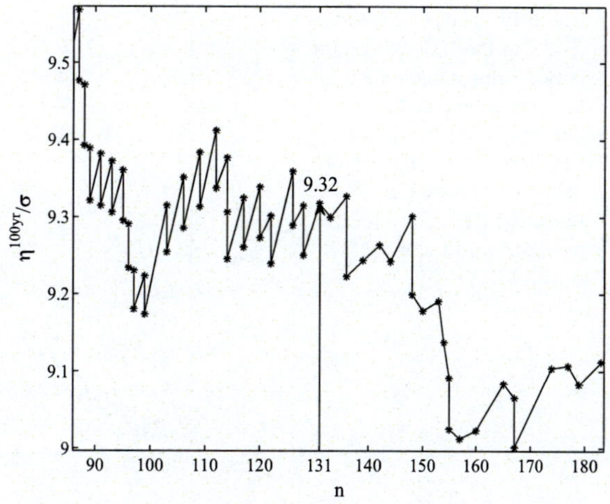

Fig. 6.17 The point estimate $\hat{\eta}^{100yr}$ of the 100-year return period value based on 13 years of wind data for Torsvåg Fyr station as a function of the number n of data points above the threshold. The return level estimate = 49.41 at n = 131, $\sigma = 5.30$

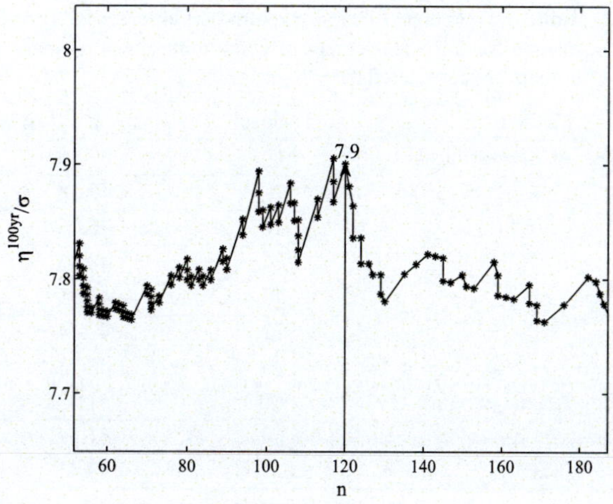

Fig. 6.18 The point estimate $\hat{\eta}^{100yr}$ of the 100-year return period value based on 12 years of wind data for Sula station as a function of the number n of data points above the threshold. The return level estimate = 43.42 at n = 120, $\sigma = 5.49$

In Tables 6.2, 6.3, 6.4, and 6.5, the 100-year return period values are listed together with the predicted 95% confidence intervals for all methods and each station. In case of the annual maxima method, the 95% confidence intervals are estimated from a parametric bootstrapping of the Gumbel estimates based on a sample of size 10,000 data sets of 13, 12, and 16 yearly extremes. For the POT method, the bootstrapped 95% confidence intervals were estimated by using the MATLAB (2009) Statistics Toolbox routine `bootstrp`. A total of 10,000 samples are generated by sampling with replacement from the observed exceedances above high thresholds.

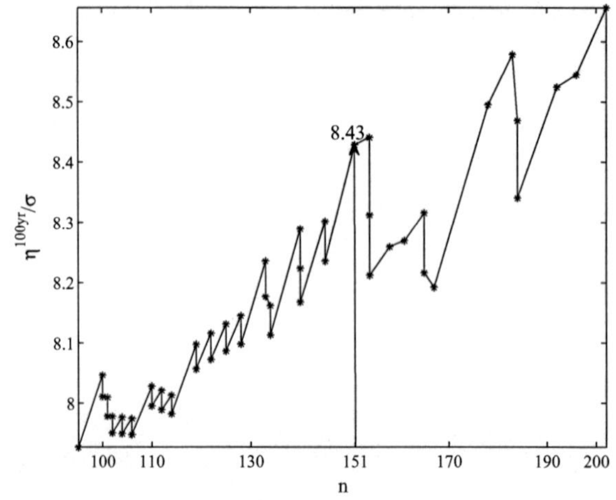

Fig. 6.19 The point estimate $\hat{\eta}^{100yr}$ of the 100-year return period value based on 12 years of wind data for Obrestad Fyr station as a function of the number n of data points above the threshold. The return level estimate = 46.1 at n = 151, σ = 5.47

Table 6.2 Predicted 100-year return period levels for Torsvåg Fyr weather station by the ACER method for different degrees of conditioning, annual maxima, and POT methods, respectively; outliers are considered true observations

Method	Spec	$\hat{\eta}^{100yr}$ m/s	95% CI (η^{100yr}) m/s
ACER, various k	1	47.46	(42.11, 50.71)
	2	48.18	(41.48, 51.31)
	4	46.96	(42.25, 49.63)
	24	48.36	(43.44, 51.63)
	48	47.54	(43.46, 49.75)
	72	47.44	(44.39, 48.79)
	96	48.78	(44.53, 51.61)
Annual maxima	MM	51.33	(43.08, 61.57)
	GL	51.57	(44.24, 60.67)
POT	–	49.41	(40.95, 59.42)

6.4 Extreme Value Prediction for a Narrow-Band Process

In engineering mechanics, a classical extreme response prediction problem is the case of a lightly damped mechanical oscillator subjected to random forces. To illustrate this prediction problem, the response process of a linear mechanical oscillator driven by a Gaussian white noise will be investigated. Let $X(t)$ denote the displacement response. The dynamic model can then be expressed as $\ddot{X}(t) + 2\zeta\omega_e\dot{X}(t) + \omega_e^2 X(t) = W(t)$, where ζ = relative damping, ω_e = undamped eigenfrequency, and $W(t)$ = a stationary Gaussian white noise (of suitable intensity). By choosing a small value for ζ, the response time series will exhibit narrow-band characteristics, that is, the spectral density of the response process $X(t)$ will assume significant values only over a narrow range of frequencies. This manifests itself by producing a strong beating of the response time series, which means that the size

Table 6.3 Predicted 100-year return period levels for Torsvåg Fyr weather station by the ACER method for different degrees of conditioning, annual maxima, and POT methods, respectively; outliers are rejected

Method	Spec	$\hat{\eta}^{100yr}$ m/s	95% CI (η^{100yr}) m/s
ACER, various k	1	47.21	(39.94, 50.60)
	2	47.79	(41.13, 50.93)
	4	46.32	(42.00, 49.04)
	24	47.22	(43.26, 50.04)
	48	46.38	(43.60, 48.19)
	72	46.32	(44.24, 47.37)
	96	47.80	(44.45, 49.95)
Annual maxima	MM	44.31	(39.36, 50.39)
	GL	45.84	(40.72, 52.41)
POT	–	42.62	(39.01, 47.31)

Table 6.4 Predicted 100-year return period levels for Sula weather station by the ACER method for different degrees of conditioning, annual maxima, and POT methods, respectively

Method	Spec	$\hat{\eta}^{100yr}$ m/s	95% CI (η^{100yr}) m/s
ACER, various k	1	46.33	(43.41, 47.77)
	2	46.81	(44.08, 49.04)
	4	47.99	(44.80, 50.57)
	24	46.65	(44.10, 48.07)
	48	46.83	(44.28, 48.03)
	72	45.80	(43.01, 46.96)
	96	45.69	(42.32, 47.01)
Annual maxima	MM	48.66	(41.58, 57.58)
	GL	52.90	(44.29, 63.39)
POT	–	43.42	(39.07, 47.80)

of the response peaks will change slowly in time, see Fig. 6.20. A consequence of this is that neighbouring peaks are strongly correlated, and there is a conspicuous clumping of the peak values. This gives rise to the problem of accurate prediction, since the usual assumption of independent peak values is then violated.

Many approximations have been proposed to deal with this correlation problem, but no completely satisfactory solution has been presented. In this section it will be shown that the ACER method solves this problem efficiently and elegantly in a statistical sense. In Fig. 6.21 are shown some of the ACER functions for the example time series. It may be verified from Fig. 6.20 that there are approximately 32 sample points between two neighbouring peaks in the time series. To illustrate a point, the time series consisting of all sample points has been analyzed. Usually, in practice, only the time series obtained by extracting the peak values would be used for the ACER analysis of a narrow-band process. In the present case, the first ACER function is then based on assuming that all the sampled data points are independent, which is obviously completely wrong. The second ACER function, which is based

Table 6.5 Predicted 100-year return period levels for Obrestad Fyr weather station by the ACER method for different degrees of conditioning, annual maxima, and POT methods, respectively

Method	Spec	$\hat{\eta}^{100yr}$ m/s	95% CI (η^{100yr}) m/s
ACER, various k	1	48.38	(43.18, 50.74)
	2	48.11	(42.38, 50.69)
	4	48.81	(42.34, 51.59)
	24	47.90	(42.87, 50.53)
	48	48.90	(43.82, 50.72)
	72	49.47	(44.06, 51.52)
	96	48.55	(43.46, 49.96)
Annual maxima	MM	48.59	(42.10, 56.84)
	GL	53.79	(46.16, 63.53)
POT	–	46.10	(41.00, 55.00)

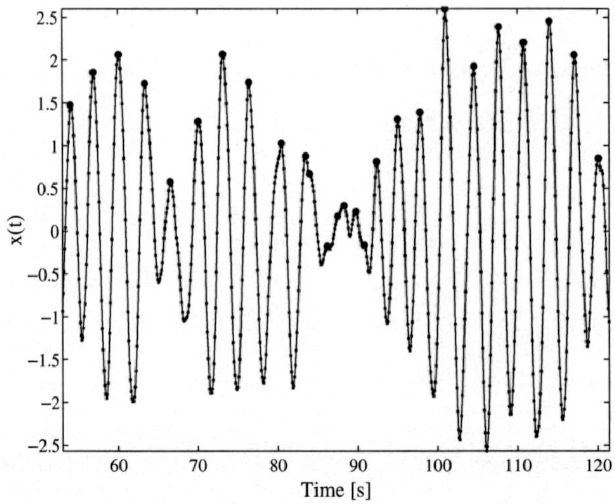

Fig. 6.20 Part of the narrow-band response time series of the linear oscillator with fully sampled and peak values indicated

on counting each exceedance with an immediately preceding non-exceedance, is nothing but an upcrossing rate. Using this ACER function is largely equivalent to assuming independent peak values. It is now interesting to observe that, e.g., the 25th ACER function can hardly be distinguished from the second ACER function. In fact, the ACER functions after the second do not change appreciably until one starts to approach the 32nd, which corresponds to hitting the previous peak value in the conditioning process. So, the important information concerning the dependence structure in the present time series seems to reside in the peak values, which may not be very surprising. It is seen that the ACER functions show a significant change in value as a result of accounting for the correlation effects in the time series. To verify the full dependence structure in the time series, it is necessary to continue the

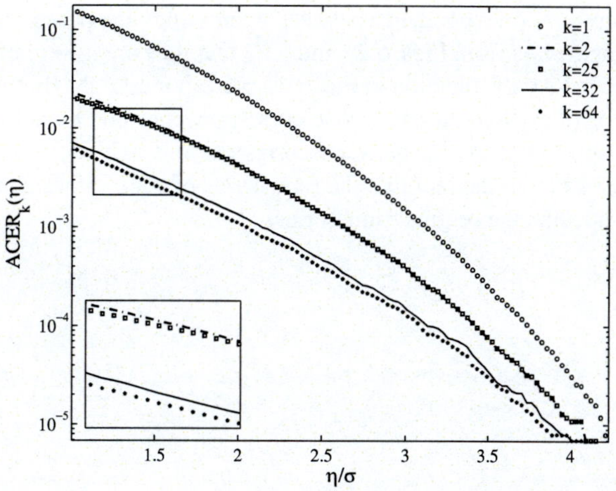

Fig. 6.21 Comparison between ACER estimates for different degrees of conditioning for the narrow-band time series

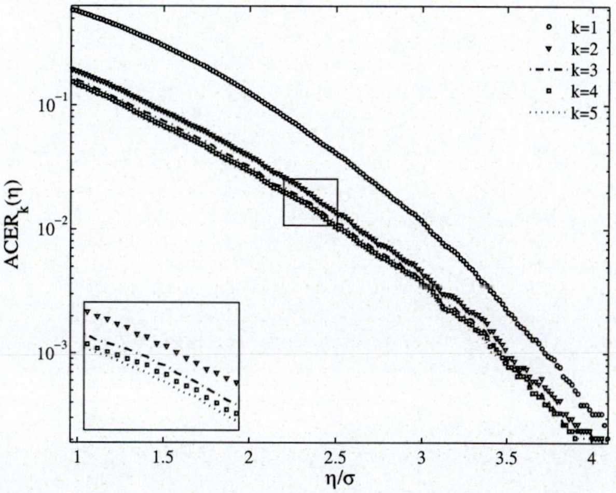

Fig. 6.22 Comparison between ACER estimates for different degrees of conditioning based on the time series of the peak values, cf. Fig. 6.20

conditioning process down to at least the 64th ACER function. In the present case there is virtually no difference between the 32nd and the 64th, which shows that the dependence structure in this particular time series is captured almost completely by conditioning on the previous peak value. It is interesting to contrast the method of dealing with the effect of sampling frequency discussed here with that of Robinson and Tawn (2000).

To illustrate the results obtained by extracting only the peak values from the time series, which would be the approach typically chosen in an engineering analysis, the ACER plots for this case are shown in Fig. 6.22. By recognizing that there is an almost one-to-one correspondence between upcrossings and peaks, it can be verified by comparing results from Figs. 6.21 and 6.22 that they are in very close agreement since the second ACER function in Fig. 6.21 corresponds to the first ACER function in Fig. 6.22 by the observed one-to-one correspondence and by noting that there is a factor of approximately 32 between corresponding ACER functions in the two figures. This is due to the fact that the time series of peak values contains about 32 times less data than the original time series.

Chapter 7
Estimation of Extreme Values for Financial Risk Assessment

7.1 Introduction

In 1991 the Norwegian government decided that the power market should be deregulated allowing for power trading. Following this, the Nord Pool market was established in 1996 as a common electricity market for Norway and Sweden. Finland followed into the Nord Pool market area in 1998, and western and eastern Denmark joined in 1999 and 2000, respectively. In 2002 Nord Pool spot was established and today runs the spot (1 day ahead) market for electricity in Norway, Sweden, Denmark, Finland, and Estonia. Today, a large part of the consumption of electricity in this market is traded through Nord Pool spot, and the spot market is the most important.

Due to the difficulties and costs of storing electricity (basically it cannot be stored), the observed spot price is highly volatile and the changes of these spot prices are often very large. Some stylized facts for the price changes in these spot data are that they display very heavy tails, significant serial correlation, seasonality, and volatility clustering (Weron 2006). The seasonality comes from the fact that the electricity cannot be stored, so the price during the high demand periods (during the day, weekdays, and the winter) tends to be higher. Most of the electricity in the market is produced by hydropower plants, so prices are also highly influenced by precipitation, giving higher electricity prices in dry years. Extreme price changes or spikes are observed in electricity markets around the world (Escribano et al. 2002) and have been studied extensively over the last years. Earlier work on modelling these spikes as an error process (Contreras et al. 2003; Garcia et al. 2005; Swider and Weber 2007) applying extreme value theory will be used here (Byström 2005; Chan and Gray 2006). This chapter largely follows the work reported by Dahlen et al. (2015).

In the examples presented, a conditional extreme value approach will be used, as suggested by McNeil and Frey (2000), to estimate tail quantiles for the return distribution and thus get a measure of the Value-at-Risk (VaR). They suggested

A. Naess, *Applied Extreme Value Statistics*,
https://doi.org/10.1007/978-3-031-60769-1_7

that a Generalized Pareto (GP) distribution is to be fitted, with the use of the POT method, to a data set of residuals from an AR-GARCH (Brockwell and Davis 2002) filtering process. Here, the ACER method will also be used to estimate the tails of these residuals, and the obtained results will be compared to those obtained with the use of the POT method. The methods will be compared in both in sample and out of sample performances. A reason for using the ACER method over the POT method is that there are essentially no asymptotic arguments and no requirements about independent data in the derivation of the ACER method, which in turn leads to better use of the data.

7.2 Value-at-Risk

To quantify the risk associated with the models applied, it was decided to use the Value-at-Risk (VaR) metric. The VaR metric, for the next day, is defined for a probability α as the value the loss will not exceed with probability α. For a series of stochastic variables X_t, which would here be the future price of electricity, the VaR will be defined as

$$\text{Prob}(X_t \leq \text{VaR}_\alpha) = \alpha. \tag{7.1}$$

So, the VaR is directly related to the quantiles of the distribution of electricity prices. For the models used in this chapter, the VaR is straightforward to calculate. For the POT method, the VaR is simply obtained by inverting the GP distribution for a given probability. This gives the VaR as, cf. Eq. (3.3),

$$\text{VaR}_\alpha = u + \frac{\sigma}{\gamma}\big((1-\alpha)^{-\gamma} - 1\big), \tag{7.2}$$

where u denotes the chosen threshold value, and σ and γ are the scale and shape parameters of the GP distribution of exceedances of u, respectively. For the ACER method, the VaR is found just as easily by inverting the estimated ACER function for the desired exceedance level α, cf. Eq. (5.39). The VaR then becomes

$$\text{VaR}_\alpha = b + \left(\frac{1}{\tilde{a}}\left(\left(\frac{q}{1-\alpha}\right)^{\xi} - 1\right)\right)^{1/c}, \tag{7.3}$$

where $\tilde{a}$, b, c, q, and ξ are the parameters estimated by fitting the assumed parametric ACER function to the observed data. In the derivation of Eq. (7.3), the approximation $\exp(-x) \approx 1 - x$ for small x is used. It should be noted that when using the conditional approach, the VaR estimations need to be inserted into the equation for the conditional quantiles. As a last reminder, it should be noted that for the VaR to be interpreted as the unlikely loss, these two methods need to be fitted to the left tail of the return distribution. So, using the VaR, a simple and straightforward method to quantify the risk in the Nordic Power Exchange spot market is at hand.

7.3 Application to Simulated Time Series of Electricity Prices

Before using the method developed on real time series, it was decided to try to motivate the use of the ACER method instead of the POT method by applying the two to simulated time series with known properties. By this approach, it should be possible to get an idea of the difference in performance for the two methods, and if the ACER method performs better, argue that it should be used when estimating extreme quantiles. The initial step was simply simulating iid innovations from a Student's t distribution and comparing the estimated quantiles with the ones from the actual distribution. Data sets were simulated consisting of 10 time series with 3650 independent outcomes of the t variable in each, which then provides us with data sets of 36,500 observations. This was repeated five times, resulting in five data sets for comparing the performance of the ACER and POT methods. For the data sets simulated, a t variable with $\nu = 4$ degrees of freedom was used. Applying the ACER and POT methods to the simulated data sets and estimating the 100 time series return level, which is the level that is expected to be exceeded once per 100 time series, the results presented in Table 7.1 were obtained. In this table, the estimated 100 time series return level is presented along with the percentage deviation from the real return level obtained from the Student's t distribution and 95% confidence intervals for the estimated return level. From this table it is seen that while the ACER method estimates a return level closer to the real return level four out of five times, the width of the confidence intervals is far greater when using the POT method. The reason for the difference in the confidence intervals is likely due to the fact that the ACER method uses approximately 48% of all observations, compared to approximately 10% of all observations used by the POT method.

Table 7.1 The 100 time series return level estimated with ACER and POT for the t model with $\nu = 4$

η_{ACER}^{100ts}	CI_{ACER}	η_{POT}^{100ts}	CI_{POT}
32.99 (2.1%)	(29.32, 37.80)	33.82 (4.7%)	(23.75, 43.88)
27.78 (14.0%)	(24.17, 31.59)	24.56 (24.0%)	(18.35, 31.05)
31.68 (1.9%)	(28.43, 34.72)	30.24 (6.4%)	(21.87, 38.61)
29.76 (7.7%)	(26.88, 33.31)	31.83 (1.5%)	(22.46, 41.20)
31.80 (1.5%)	(26.29, 34.25)	29.05 (10.1%)	(21.15, 36.96)

7.4 Electricity Price Data

In this chapter, the daily spot price at the Nordic Power Exchange, Nord Pool, is studied. It will be the price at 9 a.m. for a 10-year period, from January 01, 2000 to December 31, 2009. The price region considered is the middle of Norway. It should be noted that this price region was established on November 20, 2006. Before that time, it was integrated with the rest of Norway. It is also worth mentioning that the spot price in this market is actually a 1-day futures price, where the price for the following day is set by an auction at noon.

Plotted in Fig. 7.1 are the daily prices observed over this 10-year period and in Fig. 7.2 are the daily price changes, presented as logarithmic returns. Observable from these figures are large spikes in both the price and price change processes. There also seems to be some volatility clustering in the return series. In Table 7.2, some descriptive statistics is presented along with test values for the Ljung–Box test (Brockwell and Davis 2002) for different lags on both the return and squared return series. $Q(h)$ denotes the test statistic at lag h for the return series, while $Q^2(h)$ is the same for the squared series. A large excess kurtosis and positive skewness

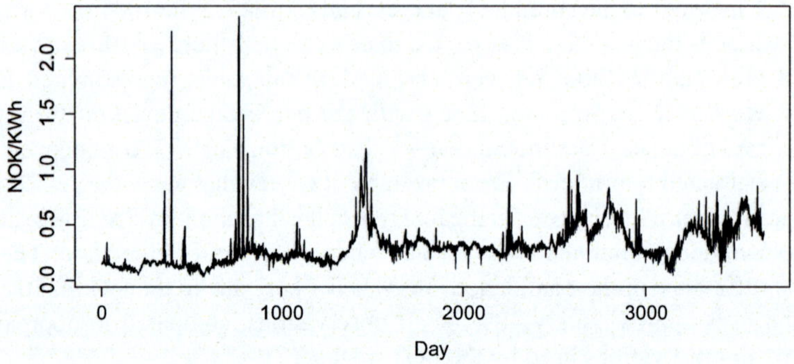

Fig. 7.1 Daily electricity spot prices on Nord Pool for the 10-year period

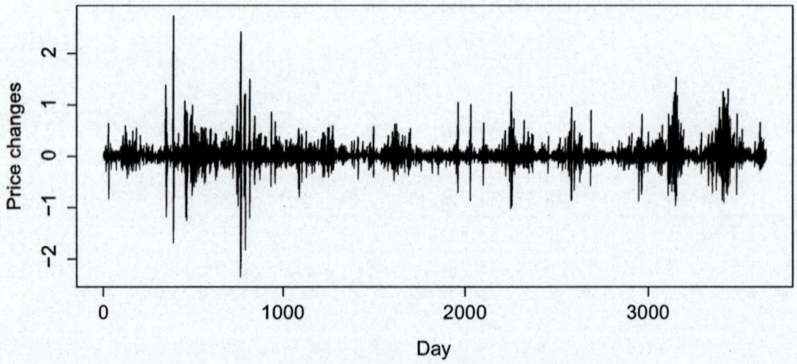

Fig. 7.2 Daily electricity price changes on Nord Pool for the 10-year period

Table 7.2 Descriptive statistics for the return series

Mean	Skew.	Ex.Kurt.	$Q(1)$	$Q(2)$	$Q(7)$	$Q^2(1)$	$Q^2(2)$	$Q^2(7)$
$3.00 \cdot 10^{-4}$	0.97	23.27	182.20	430.86	1505.67	265.50	606.49	863.32

are observed, meaning that both very small and very large price changes occur often compared to a normal distribution, and large positive price changes are more common than the large negative price changes. From the Ljung–Box test, significant serial correlation for all lags considered here is observed, which is expected. For the squared returns, there is significant serial correlation for both seven and ten lags, meaning that there is significant volatility clustering or a GARCH effect.

7.5 Conditional Approach

Following the conditional approach of McNeil and Frey (2000) and the use of a similar model for the electricity spot prices as given by Byström (2005), the data were modelled with the use of an AR-GARCH model, and the ACER method was then applied for estimation of the residual tail quantiles. To accommodate the heavy tailed data, which are typically observed in finance, the proposed procedure of Sect. 5.6 was used. For these returns, the goal was to fit an AR-GARCH model where the seasonality in the process was modeled by an AR process. The terms included in this AR process were the AR(1) and AR(7) terms because of the clear seasonality over the day and over the week. For modeling the volatility, a GARCH(1,1) process (Bollerslev 1986) was chosen. This provides a model that should be able to capture the serial correlation over the week and the observed heteroscedasticity. The model then assumes the form

$$R_t = a_0 + a_1 R_{t-1} + a_7 R_{t-7} + \sigma_t Z_t, \tag{7.4}$$

$$\sigma_t^2 = \alpha_0 + \alpha_1 \sigma_{t-1}^2 Z_{t-1}^2 + \beta_1 \sigma_{t-1}^2, \tag{7.5}$$

where R_t denotes daily log return rates of spot prices X_t, that is, $R_t = \log(X_t/X_{t-1})$. The Z_t are iid random variables of mean value zero and variance equal to 1.0, while $a_0, a_1, a_7, \alpha_0, \alpha_1, \beta_1$ are all nonnegative constants. In this chapter, it will be assumed that Z_t is Normal or Student's t distributed, scaled to unit variance. The conditional quantiles, for these two models, can then be calculated as

$$q_{t,\alpha}^* = a_0 + a_1 R_{t-1} + a_7 R_{t-7} + \sigma_t q_\alpha, \tag{7.6}$$

where q_α is the standard α-quantile of the Normal or Student's t-distribution.

For the heavy tails observed in this type of data, this standard AR-GARCH process will not be sufficient to model the tails of the return distribution accurately.

The error distribution simply cannot be assumed to be Normal or Student's t, as will be observed later. Introduced by McNeil and Frey (2000) and applied to Nord Pool spot prices by Byström (2005), the extreme value theory approach has proven itself superior to the use of a standard AR-GARCH approach. This approach was used here, but instead of using only a POT fitted GP distribution for estimation of the tail quantiles for the standardized residual distribution, the ACER method was also applied. The performance of this method was then compared to the performance of the conditional model where the POT method was used. After using the ACER and POT methods to estimate the tail quantiles of the residual distribution, the conditional quantiles for these models were calculated as

$$q^*_{t,\alpha} = a_0 + a_1 R_{t-1} + a_7 R_{t-7} + \sigma_t \hat{q}_\alpha , \tag{7.7}$$

where $\hat{q}_\alpha$ is the quantile of the residual distribution, associated with probability α, estimated by the POT or ACER method. The standardized residuals obtained by the Normal AR-GARCH filter were still slightly serially correlated, but most of the heteroscedacity was removed. These conditional quantiles may also be regarded as the VaR(α) estimate when considering the distribution of the lower tail. For the in sample performance test, the conditional tail quantiles were estimated for all observations, and then the number of empirical exceedances over these quantiles was compared to what was to be expected. When dealing with the out of sample performance, the initial step was estimating a model using the first 5 years, and then this model was used to predict the tail quantiles for the next day. The model was then re-estimated with the last 5 years of the data, or a rolling window of length 5 years, for each day, and the tail quantiles for the next day were predicted. As for the in sample performance test, the empirical exceedances over the predicted tail quantiles are compared to what is expected.

The reason for the prefiltering of the data set was the desire to use a GARCH process to model the volatility and to accommodate for sudden changes in the volatility. The use of an AR process for the autocorrelation is mainly because of the POT method's need of iid observations. There are no such iid requirements for the ACER method, so the use of an AR process is strictly not needed in this case.

7.6 Results

Using the models introduced, the goal is to analyze the 10 years of daily spot price data from the Nord Pool. As an introduction, let us start off with an unconditional tail fitting, that is, apply the POT and ACER methods directly to the return series. Then, the conditional approach will be introduced, as detailed in McNeil and Frey (2000), where both the ACER and POT methods will be used to fit the residuals produced by the AR-GARCH filtering.

7.6.1 Unconditional Approach

As mentioned, the unconditional approach will be to apply the POT and ACER method directly to the return series. To compare the performance of these two methods, both in sample and out of sample fit will be considered. For the in sample fit, the methods are simply used to estimate the tail quantiles of the return distribution, and these quantiles will be compared to what is actually observed. For the out of sample test, the first 5 years of the data will be used to estimate the model and then predict the tail quantiles for the following day. The rolling window is then moved to the next day, the model is re-estimated, and the quantiles for the next day are calculated from the estimated parameters. This is repeated for the last 5 years of the data set.

For the POT method, it is necessary for the data used to be iid, which from the Ljung–Box test results is clearly not the case. There are both significant serial correlation and volatility clustering. To deal with this problem, the data will be declustered by extracting peaks with enough lags in between that it is reasonable to assume independence between the observations. For the ACER method, there is no need to decluster the data as the correlation between lags is accounted for in the choice of ACER function.

In Table 7.3, the number of empirical exceedances over the estimated in sample quantiles is presented for both methods, along with the number of expected exceedances over these quantiles. It is observed from this table that there is not a great difference between the number of exceedance over the estimated quantiles for the two methods. For the out of sample performance the number of exceedances over the estimated quantiles can be observed in Table 7.4. Again, there is no great difference between the results of the two methods, but the ACER method is again slightly more accurate than the POT method. A problem with doing such out of sample prediction is that the last 5 years of the data are used with no more emphasis on what happened yesterday than 5 years ago. This means that the estimated quantiles, and thus risk measures, will need a long time to be able to incorporate a rise or fall in volatility. This again leads to the effect that most of the exceedances over the predicted quantiles will be observations from the periods with high volatility, while one would ideally like the exceedances to be uniformly distributed over the period in question.

Table 7.3 In sample performance of the unconditional methods

Probability	Expected	POT	ACER
0.95	183	175	179
0.99	37	38	38
0.995	18	19	19
0.999	4	3	4
0.9995	2	3	2
0.9999	0	1	0

Table 7.4 Out of sample performance of the unconditional methods

Probability	Expected	POT	ACER
0.95	91	99	99
0.99	18	29	24
0.995	9	17	11
0.999	2	2	2
0.9995	1	1	1
0.9999	0	0	0

Table 7.5 Estimated AR-GARCH parameter values

Parameter	Value-N	Value-t
μ	$2.997 \cdot 10^{-3}(1.69 \cdot 10^{-3})$	$-1.628 \cdot 10^{-3}(1.14 \cdot 10^{-3})$
a_1	$-0.338(2.95 \cdot 10^{-2})$	$-0.283(1.57 \cdot 10^{-2})$
a_7	$0.388(2.33 \cdot 10^{-2})$	$0.425(1.45 \cdot 10^{-2})$
α_0	$2.985 \cdot 10^{-4}(1.62 \cdot 10^{-5})$	$8.518 \cdot 10^{-4}(2.16 \cdot 10^{-4})$
α_1	$6.362 \cdot 10^{-2}(1.43 \cdot 10^{-2})$	$0.455(8.93 \cdot 10^{-2})$
β_1	$0.931(1.70 \cdot 10^{-2})$	$0.740(2.58 \cdot 10^{-2})$
ν	–	$2.582(0.14)$

Table 7.6 Descriptive statistics for residual series

Mean	Skew.	Ex.Kurt.	$Q(1)$	$Q(2)$	$Q(7)$	$Q^2(1)$	$Q^2(2)$	$Q^2(7)$
$2.90 \cdot 10^{-3}$	4.44	97.72	6.70	102.91	141.82	$7.20 \cdot 10^{-3}$	$8.60 \cdot 10^{-3}$	$3.67 \cdot 10^{-1}$

7.6.2 Conditional Approach

To be able to accommodate for sudden changes in volatility, at least to some extent, the conditional approach is used. The first step is to filter the data with an AR-GARCH process. For this, a GARCH(1,1) process with AR parameters for the first and seventh lag is applied before using the POT and ACER methods for fitting to the standardized residuals. The residuals are standardized with the current volatility. The parameters for the AR-GARCH model with both normal and t distributed errors are presented in Table 7.5, with the standard errors in brackets. For the model with normally distributed errors, all parameters are significant at a 0.01 significance level, except α_0 which is significant at a 0.1 significance level and μ which is nonsignificant. For the model with t distributed errors, all parameters are significant at 0.01 significance level, except μ which is nonsignificant.

Descriptive statistics and Ljung–Box test results for the residual series, that is, the residual series after pre-filtering with the normal AR-GARCH model, can be found in Table 7.6. For the residual series, there are still positive skewness and high excessive kurtosis. It is observed that while there is still significant serial correlation, it has been greatly reduced, and there are no significant GARCH effects in the residual series. It was observed that it is possible to remove slightly more of the serial correlation by including more AR-terms, but the difference is minimal, so the model with less parameters is preferred.

Table 7.7 Generalized Pareto distribution parameters from POT

Parameter	Value
σ	0.5318
ξ	0.3118
λ	$9.32 \cdot 10^{-2}$
u	1

Table 7.8 Parameters estimated for the ACER method

Parameter	Value
$\tilde{a}$	0.254
b	0.010
c	1.181
q	0.46
ξ	0.334

Table 7.9 Predicted in sample right quantiles for the different methods

Probability	Expected	AR-GARCH-N	AR-GARCH-t	C-POT	C-ACER
0.95	182	128	32	187	182
0.99	37	57	9	34	35
0.995	18	43	4	19	18
0.999	4	28	2	4	4
0.9995	2	23	1	3	1
0.9999	0	17	0	1	0

After filtering the return series, the POT and ACER methods are applied to the series of standardized residuals. As this series is much closer to iid than the return series and observations over the chosen threshold seem to be independent of each other, there is no need to decluster the data in the same way that was done with the unconditional method. Nevertheless, it should be noted that only the observations over the chosen threshold are used, which in this case will be less than 10% of the data. Using the POT method to fit a GP distribution to the data, with the threshold u selected from inspection (Coles 2001), the parameters presented in Table 7.7 are obtained. Here λ is the empirical estimate of $P(X > u)$. Inverting Eq. (3.3) for the desired probabilities gives us the POT estimated quantiles, which in turn is inserted into Eq. (7.7) to get the conditional quantiles. For the ACER method, the same procedure is used, and the parameters for the extrapolated ACER function can be found in Table 7.8.

For the in sample performance of these two methods, the same procedure as for the unconditional approach is used. In Table 7.9, the number of exceedances over a given quantile for the different methods is presented. It is seen that for the standard AR-GARCH model with standard normally distributed errors, the quantiles are severely underestimated for all quantile levels, and for the same model with t distributed errors the quantiles for the lower levels are severely overestimated. It is also observed that the AR-GARCH model, where the POT method has been applied to the standardized residuals, is clearly able to estimate the extreme quantiles much

Table 7.10 Exceedances over predicted out of sample quantiles. Right tail

Probability	Expected	AR-GARCH-N	AR-GARCH-*t*	C-POT	C-ACER
0.95	91	117	32	94	92
0.99	18	42	9	17	17
0.995	9	27	4	11	10
0.999	2	12	2	4	2
0.9995	1	11	1	0	1
0.9999	0	4	0	0	0

Table 7.11 Yearly distribution of exceedances for the conditional ACER and POT method. Out of sample

Year	C-POT	C-ACER
1	16	16
2	17	17
3	19	20
4	24	21
5	18	18

better than just a standard AR-GARCH model. This is the same as was found by Byström (2005).

To compare the performance of the POT and ACER methods, it is important to assess the out of sample fit for the two methods. To do this, a model is estimated using the first 5 years of the data, and then the conditional quantiles for the next day are estimated. The model is then re-estimated using what is now the last 5 years of the data, and again the next day conditional quantiles is predicted. This gives us a period of 5 years for the out of sample prediction. In Table 7.10, the number of exceedances over the predicted out of sample quantiles is presented. It is seen from this that the performance of the conditional POT and the conditional ACER method is quite similar, with the conditional ACER method giving a slightly better out of sample fit for the more extreme quantiles. In Table 7.11, the counted exceedances over the predicted 95% day ahead quantile are presented. It is seen that the distribution of the exceedances is fairly even over the years. The exceedances are well distributed and do not become more frequent in the high volatility periods.

Chapter 8
The Upcrossing Rate via the Characteristic Function

8.1 Introduction

As was detailed in Chap. 4, a key function for a practical assessment of the extreme value distribution of stochastic response processes is the average rate of upcrossings of high levels by the response. An important class of such response processes can be expressed as a second-order stochastic Volterra series, that is, a stochastic Volterra series that has been truncated after the second-order term (Schetzen 1980). A substantial amount of work has been done to derive methods for efficient analysis of this model, starting with the seminal paper by Kac and Siegert (1947). Later, with the development of the offshore industry, this paper had an impact on investigations of the response statistics of large floating structures. Early contributions in this field of research were made, among many others, by Neal (1974), Vinje (1983), Langley (1984), Naess (1985b, 1990b), and Donley and Spanos (1990).

The type of stochastic Volterra series models that will be studied here can be expressed as a sum of a linear and a nonlinear, quadratic transformation of a Gaussian process. Such a representation of the response process would apply to the standard model for expressing the total wave forces or horizontal excursion responses of, e.g., a tension leg platform in a random sea way. It also applies to the response of a linear structure to a quadratic wind loading where the wind speed is modeled as a Gaussian process. It would also apply to the representation of large deformations of flexible structures subjected to Gaussian loads in which the strains/stresses exhibit a strong quadratic effect. The problem of determining the marginal probability distribution function of such response processes has been solved in the sense that computer programs are available that allow very accurate numerical calculations of these functions, cf. Naess and Johnsen (1992).

The focus of this chapter is to show that it is also possible to accurately calculate the average upcrossing rate of second-order stochastic Volterra models. This allows the formulation of approximate extreme value distributions for such response processes. An interesting aspect of the development in this chapter is the

A. Naess, *Applied Extreme Value Statistics*,
https://doi.org/10.1007/978-3-031-60769-1_8

surprising structural complexity of the problem of calculating the upcrossing rate of a second-order Volterra model. It is also hoped that this chapter may serve to illustrate the power of the point process approach to practical extreme value analysis via the upcrossing rate function. The performance of the numerical method that has been developed will be illustrated by application to three specific examples.

8.2 The Response Process

As already stated, the object of study in this chapter is a stochastic response process $Z(t)$ modeled as a second-order stochastic Volterra series. Specifically, it is assumed that $Z(t)$ can be written as a sum of a linear, first-order response component $Z_1(t)$ and a nonlinear, second-order component $Z_2(t)$. That is, cf. Naess (1990b),

$$Z(t) = Z_1(t) + Z_2(t), \tag{8.1}$$

where

$$Z_1(t) = \int_0^\infty h_1(\tau)X(t-\tau)\,d\tau, \tag{8.2}$$

and

$$Z_2(t) = \int_0^\infty \int_0^\infty h_2(\tau_1, \tau_2)X(t-\tau_1)X(t-\tau_2)\,d\tau_1 d\tau_2. \tag{8.3}$$

In Eqs. (8.2) and (8.3), $X(t)$ denotes a stationary, real Gaussian process. $X(t)$ could represent a random wave elevation process or a stochastic wind velocity field. It was chosen here to limit the exposition to the case of a unidirectional situation. How to deal with the multidirectional case is explained in detail in Naess (1990b). Based on the results from this reference, it will be recognized that all results obtained in this chapter apply equally well to the multidirectional case.

The functions $h_1(\tau)$ and $h_2(\tau_1, \tau_2)$ characterize the physical system that is modeled. $h_1(\tau)$ is an ordinary impulse response function defining a linear dynamical system. $h_2(\tau_1, \tau_2)$, which is referred to as the quadratic impulse response function, characterizes the second-order properties of the physical system, but in contrast to the linear impulse response function, it does not have a direct physical interpretation.

To derive expressions suitable for practical numerical calculations, the input process $X(t)$ is represented as follows:

$$X(t) = \sum_{j=-N}^{N} \left[S_X(\omega_j)\Delta\omega\right]^{1/2} B_j\, e^{i\omega_j t}, \tag{8.4}$$

where $S_X(\omega)$ denotes the two-sided spectral density of $X(t)$. Throughout this chapter, when the summation index runs from negative to positive values, it invariably omits zero. $0 < \omega_1 < \cdots < \omega_N$ is an equidistant discretization of the pertinent part of the positive frequency axis. $\omega_{-i} = -\omega_i$, $\Delta\omega = \omega_{i+1} - \omega_i$. The assumption of an equidistant discretization is adopted for simplicity of presentation and is not necessary. In fact, often a non-equidistant version is used to avoid having too many frequencies, which is sometimes convenient. The formulas are easily adapted to cover the situation of non-equidistant discretization. $\{B_i\}$ is a set of independent, complex Gaussian $N(0, 1)$-variables with independent, identically distributed real and imaginary parts. These variables can be assumed to satisfy the relation $B_{-i} = B_i^*$, where * signifies complex conjugation. $\mathrm{i}^2 = -1$. Hence, $X(t)$ becomes a Gaussian process with zero mean value. The case of a nonzero mean value does not create any difficulties for the analysis, as is easily recognized.

By substituting Eq. (8.4) into Eqs. (8.2) and (8.3), the following expressions are obtained:

$$Z_1(t) = \sum_{i=-N}^{N} q_i B_i e^{\mathrm{i}\omega_i t}, \tag{8.5}$$

where

$$q_i = \hat{H}_1(\omega_i)[S_X(\omega_i)\Delta\omega]^{1/2}, \tag{8.6}$$

and

$$Z_2(t) = \sum_{i=-N}^{N} \sum_{j=-N}^{N} Q_{ij} B_i B_j^* e^{\mathrm{i}(\omega_i - \omega_j)t}, \tag{8.7}$$

where

$$Q_{ij} = \hat{H}_2(\omega_i, -\omega_j) \cdot [S_X(\omega_i) S_X(\omega_j)]^{1/2} \Delta\omega. \tag{8.8}$$

The function $\hat{H}_1(\omega)$ denotes the linear transfer function, corresponding to $h_1(\tau)$. Typically, the second order Volterra series is used for modeling the stochastic loading process. For the response process to be of the same type, it is necessary that the equations of motion of the dynamic system considered are linear and time invariant. Such a system is characterized by a linear transfer function, $\hat{L}(\omega)$ say. The implication of this is that $\hat{H}_1(\omega)$ can be expressed as follows:

$$\hat{H}_1(\omega) = \hat{L}(\omega)\hat{K}_1(\omega). \tag{8.9}$$

$\hat{L}(\omega)$ is assumed to be given as follows:

$$\hat{L}(\omega) = (-\omega^2 M + \mathrm{i}\omega C + K)^{-1}, \tag{8.10}$$

where $M = M(\omega)$, $C = C(\omega)$, and K are appropriate mass, damping, and stiffness parameters. The function $\hat{K}_1(\omega)$ is a linear transfer function associated with the loading process.

The function $\hat{H}_2(\omega, \omega')$ denotes the quadratic transfer function (QTF), which depends on two frequencies. It is obtained as a Fourier transform of $h_2(\tau_1, \tau_2)$. $\hat{H}_2(\omega, \omega')$ can be expressed as follows:

$$\hat{H}_2(\omega, \omega') = \hat{L}(\omega + \omega')\hat{K}_2(\omega, \omega'). \tag{8.11}$$

The function $\hat{K}_2(\omega, \omega')$ denotes a QTF for the quadratic forces on the structure.

It is shown by Naess (1987), see also Johnson and Kotz (1970), that by solving the eigenvalue problem (assumed nonsingular),

$$Q v_j = \lambda_j v_j; \tag{8.12}$$

to find the eigenvalues λ_j and orthonormal eigenvectors v_j, $j = -N, \ldots, -1, 1, \ldots, N$, the quadratic response process can be represented as

$$Z_2(t) = \sum_{j=-N}^{N} \lambda_j W_j(t)^2. \tag{8.13}$$

Here $W_j(t)$, $j = -N, \ldots, -1, 1, \ldots, N$, are real stationary Gaussian $N(0, 1)$-processes which can be represented as follows:

$$W_j(t) = \sum_{k=-N}^{N} v_j(\omega_k) B_k e^{\mathrm{i}\omega_k t}, \tag{8.14}$$

where $v_j(\omega_k)$ denotes the kth component of v_j. Note that the property $v_j(\omega_{-k}) = v_j(\omega_k)^*$ can be assumed, cf. Naess (1990b). For each fixed t, $\{W_j(t)\}$ becomes a set of independent Gaussian variables.

Having achieved the desired representation of the quadratic response $Z_2(t)$, it can then be shown that the first-order response can be expressed as

$$Z_1(t) = \sum_{j=-N}^{N} \beta_j W_j(t). \tag{8.15}$$

The (real) parameters β_j are given by the relations

$$\beta_j = \sum_{k=-N}^{N} \hat{H}_1^*(\omega_k) \sqrt{S_X(\omega_k) \Delta\omega}\, v_j(\omega_k). \tag{8.16}$$

This gives us the following representation of the total response process (Vinje 1983; Naess 1985b):

$$Z(t) = \sum_{j=-N}^{N} \{\lambda_j W_j(t)^2 + \beta_j W_j(t)\}. \tag{8.17}$$

Based on this representation, Naess (1987) describes how to calculate the statistical moments of the response process. Asymptotic upper and lower bounds on the mean level upcrossing rate are also given, but these bounds are valid only if specific conditions on the ratios β_j/λ_j are satisfied. It turns out that these conditions are very often violated in practical applications. Our goal is therefore to find a numerical method for accurate calculation of the mean level upcrossing rate of the total response process $Z(t)$. This is the topic of the next section.

8.3 The Average Crossing Rate

The calculation of the mean upcrossing rate $\nu_Z^+(\cdot; t)$ at time t of a stochastic process $Z(t)$ is usually based on the Rice formula. It states that

$$\nu_Z^+(\zeta; t) = \int_0^\infty \dot{z} f_{Z(t)\dot{Z}(t)}(\zeta, \dot{z})\, d\dot{z}, \tag{8.18}$$

where $f_{Z(t)\dot{Z}(t)}(\cdot, \cdot)$ denotes the joint probability density function of $Z(t)$ and $\dot{Z}(t) = dZ(t)/dt$. A direct application of Eq. (8.18) requires the calculation of $f_{Z(t)\dot{Z}(t)}$. An alternative approach, which turns out to be ideally suited for the problem at hand, is to express the mean crossing rate in terms of the characteristic function $M_{Z\dot{Z}}(u, v) = E[\exp\{i(u\, Z(t) + v\, \dot{Z}(t))\}]$. It has been shown by Naess (2000b) that in general

$$\begin{aligned}\nu_Z^+(\zeta, t) = &- \frac{1}{(2\pi)^2} \int_{-\infty}^{+\infty} \int_{-\infty}^{+\infty} \frac{1}{v} \frac{\partial M_{Z\dot{Z}}(u, v)}{\partial v} e^{-i\zeta u}\, du\, dv \\ &- \frac{i}{4\pi} \int_{-\infty}^{+\infty} \left(\frac{\partial M_{Z\dot{Z}}(u, v)}{\partial v} \right)_{v=0} e^{-i\zeta u}\, du,\end{aligned} \tag{8.19}$$

where the integral w.r.t. v is interpreted as a principal value integral in the following sense: $\int_{-\infty}^{\infty} = \lim_{\varepsilon \to 0}(\int_{-\infty}^{-\varepsilon} + \int_{\varepsilon}^{\infty})$. A heuristic derivation of this formula is given in Sect. 8.5.3.

For the stationary case, it can be shown that the last integral on the right hand side of Eq. (8.19) vanishes. Hence, for a stationary process, the following formula is obtained (Vinje 1983; Naess 2000b):

$$\nu_Z^+(\zeta) = -\frac{1}{(2\pi)^2}\int_{-\infty}^{+\infty}\int_{-\infty}^{+\infty}\frac{1}{v}\frac{\partial M_{Z\dot{Z}}(u,v)}{\partial v}\,e^{-\mathrm{i}\zeta\,u}\,du\,dv, \tag{8.20}$$

where the upcrossing rate is now independent of t, and again, the integral with respect to v is to be interpreted as a principal value integral as described above.

An alternative expression useful for numerical calculations of the upcrossing rate $\nu_Z^+(\cdot)$, whether the process is stationary or not, can be obtained by considering the characteristic function as a function of two complex variables. It can then often be shown that this new function becomes holomorphic in suitable regions of $\mathbf{C}^2$, where $\mathbf{C}$ denotes the complex plane. Under suitable conditions, the use of complex function theory allows the derivation of the following alternative expression for the crossing rate, cf. Naess and Karlsen (2004):

$$\nu_Z^+(\zeta) = -\frac{1}{(2\pi)^2}\int_{-\infty-ia}^{\infty-ia}\int_{-\infty-ib}^{\infty-ib}\frac{1}{w^2}\,M_{Z\dot{Z}}(z,w)\,e^{-\,iz\zeta}\,dz\,dw, \tag{8.21}$$

where $0 < a < a_1$ for some positive constant a_1, and $b_0 < b < b_1$ for some constants $b_0 < 0$ and $b_1 > 0$.

The calculation of the characteristic function $M_{Z\dot{Z}}$ is discussed in Grigoriu (1995), but no explicit expression is derived. Here, a different approach will be followed, which leads to a convenient explicit representation of the characteristic function suitable for calculation of the integrals appearing in Eq. (8.19), (8.20), or (8.21). To this end, consider the multidimensional Gaussian vectors $W = (W_{-N},\ldots,W_N)'$ and $\dot{W} = (\dot{W}_{-N},\ldots,\dot{W}_N)'$. It is obtained that the covariance matrix of $(W', \dot{W}')'$ is given by

$$\boldsymbol{\Sigma} = \begin{pmatrix}\boldsymbol{\Sigma}_{11} & \boldsymbol{\Sigma}_{12}\\ \boldsymbol{\Sigma}_{21} & \boldsymbol{\Sigma}_{22}\end{pmatrix}, \tag{8.22}$$

where $\boldsymbol{\Sigma}_{11} = I$ = the $2N \times 2N$ identity matrix, $\boldsymbol{\Sigma}_{12} = (r_{ij}) = (E[W_i\dot{W}_j])$, $\boldsymbol{\Sigma}_{21} = (E[\dot{W}_i W_j])$, and $\boldsymbol{\Sigma}_{22} = (s_{ij}) = (E[\dot{W}_i\dot{W}_j])$; $i, j = -N,\ldots,-1,1,\ldots,N$. $r_{ij} = -r_{ji}$ and $\boldsymbol{\Sigma}_{12} = \boldsymbol{\Sigma}_{21}'$. It follows from Eq. (8.14) that the entries of the covariance matrix $\boldsymbol{\Sigma}$ can be expressed in terms of the eigenvectors v_j, cf. Naess (1987). Let $\Lambda = \mathrm{diag}(\lambda_{-N},\ldots,\lambda_N)$ be the diagonal matrix with the parameters λ_j on the diagonal, and let $\beta = (\beta_{-N},\ldots,\beta_N)'$, cf. Eq. (8.16). It will be shown in Sect. 8.5.3 that (Naess 2000a)

$$M_{Z\dot{Z}}(u,v) = \frac{1}{\sqrt{\det(A)}}\exp\left\{-\frac{1}{2}v^2\,\beta'\,C\,\beta + \frac{1}{2}d'\,A^{-1}\,d\right\}, \tag{8.23}$$

where

$$A = A(u, v) = I - 2\mathrm{i}\, u\, \Lambda\; - 2\,\mathrm{i}\, v \left(\Lambda\, \Sigma_{21} + \Sigma_{12}\, \Lambda\right) + 4\, v^2 \Lambda\, C\, \Lambda, \tag{8.24}$$

$$C = \Sigma_{22} - \; \Sigma_{21}\, \Sigma_{12}, \tag{8.25}$$

$$d = d(u, v) = \left(\mathrm{i}\, u\, I + \mathrm{i}\, v\, \Sigma_{12} - 2\, v^2\, \Lambda\, C\right)\beta. \tag{8.26}$$

8.4 Numerical Calculation

Early efforts to carry out numerical calculation of the mean crossing rate using Eq. (8.21) have been reported in Naess and Karlsen (2004). These initial investigations indicated that the method had the potential to provide very accurate numerical results. Equation (8.21) will be rewritten as follows:

$$\nu_Z^+(\zeta) = -\frac{1}{(2\,\pi)^2} \int_{-\infty-ia}^{\infty-ia} \frac{1}{w^2}\, I(\zeta, w)\, dw, \tag{8.27}$$

where

$$\begin{aligned} I = I(\zeta, w) &= \int_{-\infty-ib}^{\infty-ib} M(z, w)\, e^{-\mathrm{i}z\zeta}\, dz \\ &= \int_{-\infty-ib}^{\infty-ib} \exp\{-\,\mathrm{i}z\zeta + \ln M(z, w)\}\, dz. \end{aligned} \tag{8.28}$$

A numerical calculation of the mean upcrossing rate can start by calculating the function $I(\zeta, w)$ for specified values of ζ and w. However, a direct numerical integration of Eq. (8.28) is made difficult by the oscillatory term $\exp\{-\mathrm{i}\Re(z)\zeta\}$, where $\Re(z)$ denotes the real part of the complex number z. This problem can be avoided by invoking the method of steepest descent, also called the saddle point method. For this purpose, write

$$\begin{aligned} g(z) = g(z; w) &= -\,\mathrm{i}z\zeta + \ln M(z, w) \\ &= \phi(x, y) + \mathrm{i}\, \psi(x, y), \end{aligned} \tag{8.29}$$

where $z = x + \mathrm{i}y$. $\phi(x, y)$ and $\psi(x, y)$ become real harmonic functions when $g(z)$ is holomorphic. The idea is to identify the saddle point of the surface $(x, y) \to \phi(x, y)$ closest to the integration line from $-\infty - ib$ to $\infty - ib$. By shifting this integration line to a new integration contour that passes through the saddle point and then follows the path of steepest descent away from the saddle point, it can be shown that the function $\psi(x, y)$ stays constant, and therefore the oscillatory term in the integral degenerates to a constant. This is a main advantage of the method of steepest descent for numerical calculations. It can be shown that the integral does not change its value as long as the function $g(z)$ is a holomorphic function in the region bounded by the two integration contours and if the integrals vanish along the contour segments required to close the region.

If z_s denotes the identified saddle point, where $g'(z_s) = 0$, the steepest descent path away from the saddle point will follow the direction given by $-g'(z)^*$, for $z \neq z_s$, cf. Henrici (1977). Typically, the singular points of the function g will be around the imaginary axis, which indicates that the direction of the paths of steepest descent emanating from the saddle point will typically not deviate substantially from a direction orthogonal to the imaginary axis. This provides a guide for setting up a numerical integration procedure based on the path of steepest descent. First the saddle point z_s is identified. Then the path of steepest descent starting at z_s and going "right" is approximated by the sequence of points $\{z_j\}_{j=0}^{\infty}$ calculated as follows:

$$z_0 = z_s, \qquad z_1 = z_s + h, \tag{8.30}$$

$$\Delta z_j = -\frac{g'(z_j)^*}{|g'(z_j)|}\, h, \quad j = 1, 2, \ldots \tag{8.31}$$

$$z_{j+1} = z_j + \Delta z_j, \quad j = 1, 2, \ldots, \tag{8.32}$$

where h is a small positive constant.

Similarly, the path of steepest descent going "left" is approximated by the sequence $\{z_j\}_{j=0}^{-\infty}$ calculated by

$$z_{-1} = z_s - h, \tag{8.33}$$

$$\Delta z_j = -\frac{g'(z_j)^*}{|g'(z_j)|}\, h, \quad j = -1, -2, \ldots \tag{8.34}$$

$$z_{j-1} = z_j + \Delta z_j, \quad j = -1, -2, \ldots \tag{8.35}$$

A numerical estimate $\hat{I}$ of I can be obtained as follows:

$$\hat{I} = \hat{I}^+ + \hat{I}^-, \tag{8.36}$$

where

$$\hat{I}^+ = \frac{h}{2} \exp\{g(z_s)\} + \sum_{j=1}^{K} \Delta z_j \exp\{g(z_j)\}, \tag{8.37}$$

and

$$\hat{I}^{-} = \frac{h}{2}\exp\{g(z_s)\} - \sum_{j=-1}^{-K} \Delta z_j \exp\{g(z_j)\}, \tag{8.38}$$

for a suitably large integer K.

A numerical estimate $\hat{\nu}_Z^+(\zeta)$ of the mean crossing rate can now be calculated by the sum

$$\hat{\nu}_Z^+(\zeta) = -\frac{1}{(2\pi)^2}\Re\Big\{\sum_{j=-L}^{L}\frac{1}{w_j^2}\hat{I}(\zeta, w_j)\,\Delta w_j\Big\}, \tag{8.39}$$

where the discretization points w_j are chosen to follow the negative real axis from a suitably large negative number up to a point at $-\varepsilon$, where $0 < \varepsilon \leq a$, then follow a semicircle in the lower half plane to ε on the positive real axis, and finally follow this axis to a suitably large positive number. Since the numerical estimate does not necessarily have an imaginary part that is exactly equal to zero, the real part operator $\Re$ has been applied.

Generally, the CPU time required to carry out the computations above can be quite long, depending on the size of the problem, which is related to the number N of eigenvalues. It is therefore of interest to see if approximating formulas are accurate enough. The first such approximation to have a look at is the Laplace approximation for the inner integral over the saddle point (Henrici 1977). The simplest version of this approximation, adapted to the situation at hand, leads to the result

$$I = I(\zeta, w) \approx \sqrt{\frac{2\pi}{-\frac{\partial^2 g(z_s; w)}{\partial x^2}}}\exp\{\,g(z_s; w)\}, \tag{8.40}$$

which can be substituted directly into Eq. (8.39), leading to an approximation of $\nu_Z^+(\zeta)$, which is denoted by $\tilde{\nu}_Z^+(\zeta)$.

This approximation can also be exploited in the following way: (1) The full method is used for an inner interval of w-values, which contribute significantly to the integral in Eq. (8.27). (2) A Laplace approximation is then used in an outer interval of w-values where the contribution is less than significant. Of course, the level of significance is chosen according to some suitable criterion. By this procedure, the CPU time was reduced by factor of about 3. This method will be referred to as the hybrid method, and the corresponding approximation of $\nu_Z^+(\zeta)$ is denoted by $\check{\nu}_Z^+(\zeta)$.

A simple approximation proposed in Teigen and Naess (1999, 2003) is worth a closer scrutiny. It is based on the widely adopted simplifying assumption that the displacement process is independent of the velocity process. This leads to an alternative approximation of $\nu_Z^+(\zeta)$, which is denoted by $\overline{\nu}_Z^+(\zeta)$. It is given by the formula

$$\overline{\nu}_Z^+(\zeta) = \nu_Z^+(\zeta_{ref}) \frac{f_Z(\zeta)}{f_Z(\zeta_{ref})}, \tag{8.41}$$

where f_Z denotes the marginal probability density of the surge response, and ζ_{ref} denotes a suitable reference level, typically the mean response. Here, ζ_{ref} has been chosen as the point where f_Z assumes its maximum, which corresponds well with the mean response level. A general approximation for $\nu_Z^+(\zeta_{ref})$ is given in Teigen and Naess (2003). If only slow-drift response is considered, a good approximation is obtained by putting $\nu_Z^+(\zeta_{ref}) \approx 1/T_0$, where $T_0 = 2\pi/\omega_0$ is the slow-drift period. The advantage of Eq. (8.41) is that the rhs is much faster to calculate than the exact formula.

A few comments on why the mean upcrossing rate has practical significance may seem appropriate. As is already known, the extreme value distribution is not completely determined by the mean upcrossing rate. This is true only when the upcrossing events of the high response levels can be assumed to be statistically independent. Usually that is a good approximation except when the total damping is very small. For such cases, Naess (1999) has developed a simple, but effective, method to account for the effect of low damping on the extreme value distribution. This method is based on the mean upcrossing rate and the appropriate damping parameter. This is often advantageous since no time series of response is required. However, when those are available, the key to an accurate estimation of the extreme value distribution would be the application of the ACER method.

8.5 Numerical Examples

8.5.1 *Slow-Drift Response*

To illustrate the accuracy of the numerical method, the first example concerns a simple model for the slow-drift response of a moored offshore structure, cf. Naess and Machado (2000). Specifically, the response process for this case may be written as ($\lambda = \lambda_1 = \lambda_2$)

$$Z_2(t) = \lambda \Big(W_1(t)^2 + W_2(t)^2 \Big). \tag{8.42}$$

For this process, the upcrossing rate $\nu_{Z_2}^+(z)$ is given by the relation

$$\nu_{Z_2}^+(\zeta) = \frac{\hat{\sigma}_1}{\sqrt{2\pi}} \exp\Big(-\frac{\zeta}{2\lambda} + \frac{1}{2}\ln\Big(\frac{\zeta}{\lambda}\Big)\Big), \tag{8.43}$$

Table 8.1 Comparison of analytical and numerical mean upcrossing rate

ζ	$\nu_{Z_2}^+(\zeta)$	$\tilde{\nu}_{Z_2}^+(\zeta)$
0.5	$1.37 \cdot 10^{-2}$	$1.37 \cdot 10^{-2}$
1.0	$4.31 \cdot 10^{-3}$	$4.30 \cdot 10^{-3}$
2.0	$2.995 \cdot 10^{-4}$	$2.995 \cdot 10^{-4}$
3.0	$1.802 \cdot 10^{-5}$	$1.793 \cdot 10^{-5}$
5.0	$5.618 \cdot 10^{-8}$	$5.618 \cdot 10^{-8}$
7.0	$1.605 \cdot 10^{-10}$	$1.592 \cdot 10^{-10}$

Table 8.2 Main particulars of the moored deep floater (MDF)

Draught (m)	80.0
Column diameter (m)	10.0
Natural period surge/sway (s)	133.5
Natural period yaw (s)	121

where $\hat{\sigma}_1 = \sqrt{s_{11} - (r_{12})^2}$. This special case provides a suitable test for the accuracy of the numerical method.

Let $\tilde{\nu}_{Z_2}^+(\zeta)$ denote the mean upcrossing rate of $Z_2(t)$ calculated by the numerical method. Table 8.1 compares the analytical with the numerical mean upcrossing rate for different levels, and it is seen that the agreement is very good indeed.

8.5.2 Moored Deep Floater

The numerical results presented in this example are based on a specific model structure (Naess et al. 2006). It is a moored deep floater (MDF), which is also called a spar-buoy, with main particulars as detailed in Table 8.2. Figure 8.1 shows the submerged part of the floater in the form of a computer mesh, which is used for the calculation of the hydrodynamic transfer functions. The total mass (including added mass) of the MDF is $M = 12.5 \cdot 10^6$ kg. The damping ratio is set equal to $\xi = 0.06$, and the natural frequency in surge or sway is $\omega_0 = 0.047$ rad/s. Note that the second-order theory is based on the assumption that the QTF $\hat{H}_2(\omega_i, -\omega_j) = \hat{L}(\omega_i - \omega_j)\,\hat{K}_2(\omega_i, -\omega_j)$, where $\hat{K}_2(\cdot, \cdot)$ is a QTF characterizing the slowly varying surge forces on the MDF, and $\hat{L}(\cdot)$ is a linear transfer function for the surge motion of the MDF, that is,

$$\hat{L}(\omega) = \frac{1}{M\,[-\omega^2 + 2\mathrm{i}\xi\,\omega_0\omega + \omega_0^2]}. \tag{8.44}$$

An example of the quadratic transfer function for a floating offshore structure is presented in Sect. 9.6.5.

The deep floater studied in this example could represent the supporting structure of a floating wind turbine. The Hywind turbine is a particular case of such a

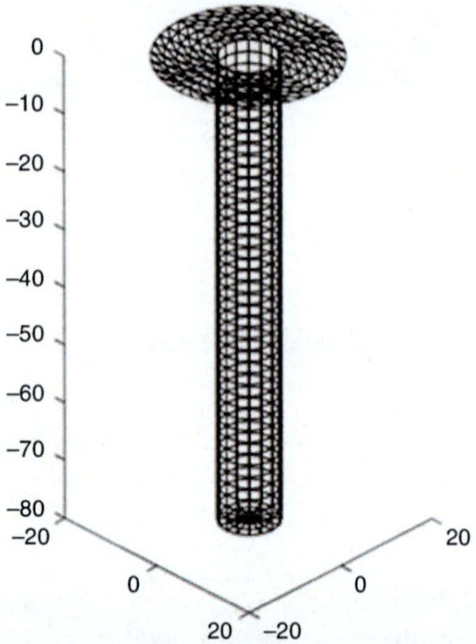

Fig. 8.1 Computer mesh of the submerged part of the moored deep floater and the near field of the sea surface

Fig. 8.2 A sketch of the Hywind floating wind turbine (©Equinor)

structure, where the concept of a moored deep floater is used as a supporting structure. A sketch of the Hywind turbine is presented in Fig. 8.2.

The random stationary sea state is specified by a JONSWAP spectrum, which is given as follows:

$$S_X(\omega) = \frac{\alpha g^2}{\omega^5} \exp\left\{ -\frac{5}{4}\left(\frac{\omega_p}{\omega}\right)^4 + \ln\gamma \, \exp\left[-\frac{1}{2\sigma^2}\left(\frac{\omega}{\omega_p} - 1\right)^2 \right] \right\}, \tag{8.45}$$

where $g = 9.81$ ms^{-2}, ω_p denotes the peak frequency in rad/s, and α, γ, and σ are parameters related to the spectral shape. $\sigma = 0.07$ when $\omega \leq \omega_p$, and $\sigma = 0.09$ when $\omega > \omega_p$. The parameter γ is chosen to be equal to 3.0. The parameter α is determined from the following empirical relationship:

$$\alpha = 5.06\Big(\frac{H_s}{T_p^2}\Big)^2 \big(1 - 0.287 \ln \gamma\big). \tag{8.46}$$

H_s = significant wave height and $T_p = 2\pi/\omega_p$ = spectral peak wave period. For the subsequent calculations, $H_s = 10.0$ m and $T_p = 12$ s. The natural frequency in surge is 0.047 rad/s, which is well below the range where the waves have noticeable energy. This is why the second-order, nonlinear term in the Volterra expansion is needed to capture the resonant motions in surge of the MDF.

To get an accurate representation of the response process, there is a specific requirement that must be observed. Since the damping ratio is only 6%, the resonance peak of the linear transfer function for the dynamics is quite narrow. Hence, to capture the dynamics correctly, the frequency resolution must secure a sufficient number of frequency values over the resonance peak. This usually leads to an eigenvalue problem with the Q-matrix of size of the order of magnitude 100×100. Using the full representation of this size in calculating the mean crossing rate by the general method described here would lead to very heavy calculations. In order to reduce this, the effect of restricting the calculations by retaining only some of the terms in Eq. (8.13) has been investigated.

For the specific example considered, where exactly 100 (positive) frequencies have been used, the values of the obtained eigenvalues λ_j have been plotted in Fig. 8.3. It is seen that a substantial portion of the response variance, which is given by $\mathrm{Var}[Z_2(t)] = 4\sum_{j=1}^{N}\lambda_j^2$, would be lost if only 10 or 20 eigenvalues were retained. This is also a factor to consider when deciding on the number of terms to retain.

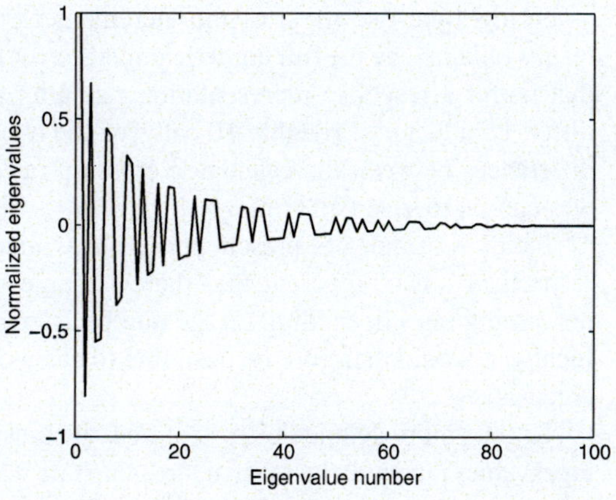

Fig. 8.3 The 100 normalized eigenvalues λ_j/λ_1

Table 8.3 Comparison of calculated mean upcrossing rate $\hat{\nu}^+_{Z_2}(\zeta)$ for different step lengths

$\eta = \zeta/\lambda_1$	$h = 1.0 \cdot 10^{-3}$	$h = 1.0 \cdot 10^{-2}$
2.0	$8.38 \cdot 10^{-3}$	$8.38 \cdot 10^{-3}$
5.0	$3.93 \cdot 10^{-3}$	$3.93 \cdot 10^{-3}$
10.0	$5.53 \cdot 10^{-4}$	$5.50 \cdot 10^{-4}$
15.0	$5.70 \cdot 10^{-5}$	$5.65 \cdot 10^{-5}$
20.0	$5.34 \cdot 10^{-6}$	$5.26 \cdot 10^{-6}$
25.0	$4.81 \cdot 10^{-7}$	$4.71 \cdot 10^{-7}$

Table 8.4 Calculated mean upcrossing rates for ten eigenvalues

$\eta = \zeta/\lambda_1$	$\hat{\nu}^+_{Z_2}(\zeta)$	$\check{\nu}^+_{Z_2}(\zeta)$	$\tilde{\nu}^+_{Z_2}(\zeta)$
2.0	$8.38 \cdot 10^{-3}$	$8.38 \cdot 10^{-3}$	$7.41 \cdot 10^{-3}$
5.0	$3.93 \cdot 10^{-3}$	$3.93 \cdot 10^{-3}$	$3.59 \cdot 10^{-3}$
10.0	$5.50 \cdot 10^{-4}$	$5.50 \cdot 10^{-4}$	$5.23 \cdot 10^{-4}$
15.0	$5.65 \cdot 10^{-5}$	$5.65 \cdot 10^{-5}$	$5.59 \cdot 10^{-5}$
20.0	$5.26 \cdot 10^{-6}$	$5.26 \cdot 10^{-6}$	$5.36 \cdot 10^{-6}$
25.0	$4.71 \cdot 10^{-7}$	$4.71 \cdot 10^{-7}$	$4.92 \cdot 10^{-7}$

Table 8.5 Calculated mean upcrossing rates for 50 eigenvalues

$\eta = \zeta/\lambda_1$	$\hat{\nu}^+_{Z_2}(\zeta)$	$\check{\nu}^+_{Z_2}(\zeta)$	$\tilde{\nu}^+_{Z_2}(\zeta)$
2.0	$6.55 \cdot 10^{-3}$	$6.55 \cdot 10^{-3}$	$5.93 \cdot 10^{-3}$
5.0	$3.25 \cdot 10^{-3}$	$3.25 \cdot 10^{-3}$	$2.98 \cdot 10^{-3}$
10.0	$5.04 \cdot 10^{-4}$	$5.04 \cdot 10^{-4}$	$4.70 \cdot 10^{-4}$
15.0	$5.44 \cdot 10^{-5}$	$5.44 \cdot 10^{-5}$	$5.28 \cdot 10^{-5}$
20.0	$5.19 \cdot 10^{-6}$	$5.19 \cdot 10^{-6}$	$5.20 \cdot 10^{-6}$
25.0	$4.71 \cdot 10^{-7}$	$4.71 \cdot 10^{-7}$	$4.86 \cdot 10^{-7}$

In this example, the focus is on the slow-drift response. Hence, only results for $Z_2(t)$ will be presented. In the tables, $\hat{\nu}^+_{Z_2}(\zeta)$, $\check{\nu}^+_{Z_2}(\zeta)$, $\tilde{\nu}^+_{Z_2}(\zeta)$, and $\overline{\nu}^+_{Z_2}(\zeta)$ denote the mean upcrossing rate of $Z_2(t)$ calculated by the full numerical method, the hybrid method, the Laplace approximation, and the simplified method of Eq. (8.41), respectively.

To highlight the effect of the increment parameter h, Table 8.3 compares the results obtained by the full numerical method for two values of h for ten eigenvalues, that is, for a response representation retaining the first ten terms. The CPU time differs by a factor of roughly 10 between the two choices of a value for h. Since the differences between the calculated crossing rates are fairly small, the larger value was chosen to save CPU time.

Tables 8.4 and 8.5 present the results obtained for 10 and 50 eigenvalues, respectively. It is apparent that there is some variability of the calculated mean upcrossing rates depending on the number of eigenvalues included in the analysis. Ideally, it would therefore be desirable to carry out the calculations with at least 50 eigenvalues.

To get a more detailed picture of how the crossing rate varies with the number of eigenvalues retained, the mean upcrossing rate was calculated for the level $\eta = 20$ as a function of the number of eigenvalues. The result is shown in Fig. 8.4. It was also

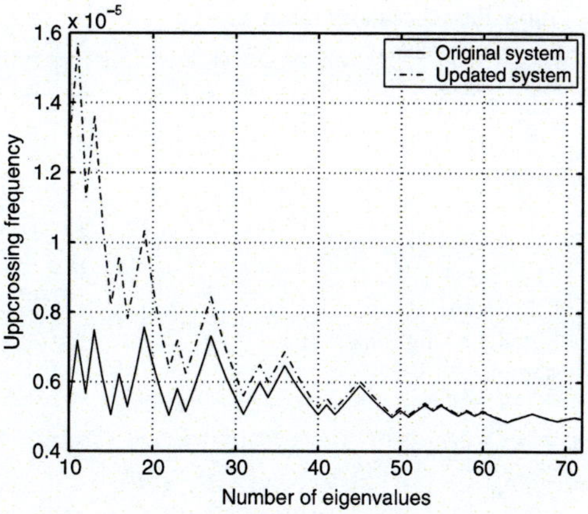

Fig. 8.4 The mean upcrossing rate of the level $\eta = 20$ as a function of the number of retained eigenvalues

decided to investigate the effect of updating the truncated response representation so that it had the correct variance. This was achieved by multiplying the retained eigenvalues by a suitable factor. The effect of this updating on the calculated upcrossing rate is also shown in Fig. 8.4. The figure indicates a couple of interesting conclusions. Updating for variance can lead to inaccurate results for the crossing rate for small to moderate number of eigenvalues retained. Comparing Figs. 8.3 and 8.4, it is seen that surprisingly accurate results are obtained for even a small number of retained eigenvalues when the truncation is done exactly where negative eigenvalues are followed by positive eigenvalues. This seems to provide the right balance between the terms in the response representation, and it indicates a useful criterion for truncating the response representation for crossing rate calculations.

It is also of great interest to observe that the simple Laplace approximation in fact provides quite accurate estimates of the mean upcrossing rates, and for this method the number of eigenvalues has practically no effect on the computational burden. Hence, from a practical point of view, this is an extremely appealing method. In Table 8.6, the results obtained by the hybrid method, the Laplace approximation, and also the simple approximation of Eq. (8.41) for 100 eigenvalues have been listed. It is seen that while there is excellent agreement between the hybrid method and the Laplace approximation, the simple approximation leads to significantly lower values. In terms of extreme value predictions, for the example structure at hand the Laplace approximation is within about 1% of the hybrid method, while the simple approximations would lead to an underestimation of typically 5–10% compared with the two more accurate methods.

8.5.3 Wind Excited Structure

The third example is a simple model of a wind excited structure. It could easily be adapted to the study of the response to wind loading of a high slender mast

Table 8.6 Calculated mean upcrossing rates for 100 eigenvalues

$\eta = \zeta/\lambda_1$	$\check{\nu}^+_{Z_2}(\zeta)$	$\tilde{\nu}^+_{Z_2}(\zeta)$	$\overline{\nu}^+_{Z_2}(\zeta)$
2.0	$6.17 \cdot 10^{-3}$	$5.59 \cdot 10^{-3}$	$6.03 \cdot 10^{-3}$
5.0	$3.03 \cdot 10^{-3}$	$2.78 \cdot 10^{-3}$	$2.74 \cdot 10^{-3}$
10.0	$4.71 \cdot 10^{-4}$	$4.40 \cdot 10^{-4}$	$3.65 \cdot 10^{-4}$
15.0	$5.11 \cdot 10^{-5}$	$4.96 \cdot 10^{-5}$	$3.44 \cdot 10^{-5}$
20.0	$4.88 \cdot 10^{-6}$	$4.90 \cdot 10^{-6}$	$2.94 \cdot 10^{-6}$
25.0	$4.44 \cdot 10^{-7}$	$4.63 \cdot 10^{-7}$	$2.44 \cdot 10^{-7}$

Fig. 8.5 A light-mast at a stadium

supporting a large light panel for illumination at a sports stadium. An example of such a structure is shown in Fig. 8.5.

The theory can be extended to cover the general case of an MDOF system as discussed by Benfratello et al. (1998). However, for the purpose of illustration, an SDOF model of structural response has been chosen.

Let the wind load on the structure be given as $F(t) = c\,U(t)^2$, where c is some constant, and $U(t)$ denotes a stationary Gaussian process representing the wind speed. Writing $U(t) = \bar{U} + V(t)$, where $\bar{U}$ is the mean wind speed, and $V(t)$ denotes a zero mean Gaussian process, the structural response to the loading process $\tilde{F}(t) = 2\,c\,\bar{U}\;V(t) + c\;V(t)^2$ will be considered. That is, the constant force term $c\,\bar{U}^2$ is neglected. For a linear dynamic model, its effect can be added at the end.

The structural response to $\tilde{F}(t)$ is assumed to be determined by the equation

$$\ddot{Z}(t) + 2\xi\,\omega_0\,\dot{Z}(t) + \omega_0^2\,Z(t) = M^{-1}\tilde{F}(t) = \frac{2c\bar{U}}{M}V(t) + \frac{c}{M}V(t)^2, \tag{8.47}$$

where ω_0 is the undamped natural frequency of the system, ξ is the relative damping, and M is the total mass.

The linear transfer function of this system is clearly given by $\hat{L}(\omega) = [-\omega^2 + 2i\,\xi\,\omega_0\,\omega + \omega_0^2]^{-1}$. For this particular force model, the Volterra series is of a degenerate kind. It can be shown that the QTF for the quadratic response component of Eq. (8.47) is given by the expression $c\,\hat{L}(\omega_i - \omega_j)/M$ (Naess 1987).

What is now needed to fully specify the Q-matrix of Eq. (8.12) is the wind velocity spectrum $S_V(\omega)$. Here, a Davenport spectrum is adopted

$$\frac{\omega\,S_V(\omega)}{\kappa\,\bar{U}^2} = \frac{4\theta^2}{(1+\theta^2)^{4/3}}, \tag{8.48}$$

where $\theta = \omega\,L/(2\pi\bar{U})$, κ is the roughness/drag coefficient, and L is a length scale. For the present example, a turbulence level of $\sigma/\bar{U} = 0.1$ and 0.3 is assumed. The corresponding drag coefficient κ is given by the relation $\sigma/\bar{U} = \sqrt{6\kappa}$. For the other parameters, $\bar{U} = 50$ [m/s] and $L = 1200$ [m] are selected. The damping ratio $\xi = 0.1$ and the natural frequency $\omega_0 = \pi$ [rad/s]. Also, $c/M = 0.5 \cdot 10^{-3}$. The frequency discretization of the problem ranges from −5 to 5 [rad/s] with an increment of 0.1 [rad/s].

In Table 8.7 are listed calculated values for the mean upcrossing rates $\nu_Z^+(\zeta)$ of the response process $Z(t)$ with no restrictions imposed on the covariance structure for the case $\sigma/\bar{U} = 0.3$. To get an idea about the influence of the correlation between W and $\dot{W}$, the upcrossing rate under the assumption that $\Sigma_{12} = 0$ has also been calculated. This crossing rate is denoted by $\tilde{\nu}_Z^+(\zeta)$. From Table 8.7, it is seen that $\nu_Z^+(\zeta)$ and $\tilde{\nu}_Z^+(\zeta)$ are almost equal, supporting the often assumed negligible influence on the crossing rate of the dependence between W and $\dot{W}$.

Both estimates $\tilde{\nu}_Z^+(\zeta)$ and $\nu_Z^+(\zeta)$ are then compared with the upcrossing rate $\nu_g^+(\zeta)$ of the linear part of the response, which is a Gaussian process. The spectrum of the linear part of the response is

$$S_Z(\omega) = |\hat{L}(\omega)|^2\left(\frac{2c\bar{U}}{M}\right)^2 S_V(\omega). \tag{8.49}$$

Comparing $\nu_g^+(\zeta)$ with $\nu_Z^+(\zeta)$, it is observed that the relative difference between them increases as the level ζ becomes higher, which is to be expected.

From a practical point of view, it is interesting to observe the effect on the predicted extreme responses of the full quadratic model for the wind loading as

Table 8.7 Mean upcrossing rates against different levels. $\tilde{\nu}_Z^+(\zeta)$: numerical method assuming $\boldsymbol{\Sigma}_{12} = 0$; $\nu_Z^+(\zeta)$: numerical method; and $\nu_g^+(\zeta)$: Gaussian approximation

ζ [meters]	$\tilde{\nu}_Z^+(\zeta)$	$\nu_Z^+(\zeta)$	$\nu_g^+(\zeta)$
0.1	0.29891	0.28714	0.22118
0.15	0.17232	0.16506	0.12021
0.2	0.08484	0.08092	0.05119
0.3	0.01454	0.01368	0.00446
0.4	0.00181	0.00167	$1.46{\cdot}10^{-4}$
0.5	$1.83{\cdot}10^{-4}$	$1.66{\cdot}10^{-4}$	$1.82{\cdot}10^{-6}$

opposed to a Gaussian approximation. For this purpose, the following extreme value distribution is adopted:

$$F_{ext}(\zeta) = \exp\Big(- \nu_Z^+(\zeta)\, T \Big), \tag{8.50}$$

where $F_{ext}(\zeta)$ is the cumulative distribution function of the extreme response during a time interval of length T, that is, $max\{Z(t); 0 \le t \le T\}$.

Taking into account also the contributing constant forcing term $c\,\bar{U}^2$ that was left out and which adds a displacement of 0.13 m, it can be verified that the extreme values at the exceedance probability level 10^{-2} are about 6–7% higher for the full quadratic model as compared to the Gaussian approximation when $\sigma/\bar{U} = 0.1$. The corresponding number is about 30% when $\sigma/\bar{U} = 0.3$. Clearly, these numbers depend on the specific examples considered, but they serve to indicate what the effect will be of neglecting the quadratic part of the wind loading.

Appendix 1: The Average Crossing Rate

A heuristic proof of Eqs. (8.19) and (8.20) will be given. For a rigorous derivation, the reader is referred to Naess (2000b). The derivation will be based on Parseval's formula, which relates pairs of Fourier transforms. First, equation (8.18) will be rewritten. For this purpose the Heaviside function $H(x)$ is introduced. It is defined as follows: $H(x) = 0$ for $x < 0$, $H(0) = 1/2$, and $H(x) = 1$ for $x > 0$. For simplicity, also write $p(x) = f_{Z\dot{Z}}(\zeta, x)$, where ζ will be fixed. Then,

$$\nu_Z^+(\zeta) = \int_{-\infty}^{\infty} x\, p(x)\, H(x)\, dx. \tag{8.51}$$

The Fourier transform $\mathscr{F}$ and its inverse $\hat{\mathscr{F}}$ are (formally) defined as follows:

$$\hat{f}(t) = \mathscr{F}[f(x)] = \int_{-\infty}^{\infty} f(x)\, e^{\mathrm{i}\,x\,t}\, dx, \tag{8.52}$$

and

$$f(x) = \hat{\mathscr{F}}[\hat{f}(t)] = \frac{1}{2\pi}\int_{-\infty}^{\infty} \hat{f}(t)\, e^{-\mathrm{i}xt}\, dt. \tag{8.53}$$

Let $G(x) = x\,p(x)$. Parseval's formula provides us with the following equation (* denotes complex conjugation):

$$\int_{-\infty}^{\infty} G(x)\, H(x)\, dx = \frac{1}{2\pi}\int_{-\infty}^{\infty} \hat{G}(t)\, \hat{H}(t)^*\, dt. \tag{8.54}$$

The Fourier transform $\hat{H}(t)$ of the Heaviside function can be shown to be given by the relation (Bracewell 1986)

$$\hat{H}(t) = \frac{\mathrm{i}}{t} + \pi\,\delta(t), \tag{8.55}$$

where $\delta(\cdot)$ denotes Dirac's delta function. The Fourier transform pair $H(x)$, $\hat{H}(t)$ can only be interpreted as such in a formal manner. A certain caution should therefore be exercised when they are used in calculations. That is, one should understand their limitations and proper use, which requires knowledge about how the relation given in Eq. (8.55) was established (Bracewell 1986).

If $\hat{p}(t)$ denotes the Fourier transform of $p(x)$, then $\hat{G}(t) = -\mathrm{i}\, d\hat{p}(t)/dt$. From the definition of the characteristic function, it is obtained that

$$\hat{p}(t) = \frac{1}{2\pi}\int_{-\infty}^{\infty} M_{Z\dot{Z}}(s,t)\, e^{-\mathrm{i}\zeta s}\, ds, \tag{8.56}$$

leading to

$$\frac{d\hat{p}(t)}{dt} = \frac{1}{2\pi}\int_{-\infty}^{\infty} \frac{\partial M_{Z\dot{Z}}(s,t)}{\partial t}\, e^{-\mathrm{i}\zeta s}\, ds. \tag{8.57}$$

Hence it is found that

$$\begin{aligned} \nu_Z^+(\zeta) &= \frac{-\mathrm{i}}{(2\pi)^2}\int_{-\infty}^{\infty}\int_{-\infty}^{\infty} \frac{\partial M_{Z\dot{Z}}(s,t)}{\partial t}\, e^{-\mathrm{i}\zeta s}\Big(-\frac{\mathrm{i}}{t} + \pi\,\delta(t)\Big)\, dt\, ds \\ &= -\frac{1}{(2\pi)^2}\int_{-\infty}^{\infty}\int_{-\infty}^{\infty} \frac{1}{t}\frac{\partial M_{Z\dot{Z}}(s,t)}{\partial t}\, e^{-\mathrm{i}\zeta s}\, dt\, ds \\ &\quad - \frac{\mathrm{i}}{4\pi}\int_{-\infty}^{\infty}\Big(\frac{\partial M_{Z\dot{Z}}(s,t)}{\partial t}\Big)_{t=0} e^{-\mathrm{i}\zeta s}\, ds, \end{aligned} \tag{8.58}$$

which is seen to agree with Eq. (8.19). A scrutiny of the derivation of the (formal) Fourier transform of the Heaviside function $H(\cdot)$ reveals that the integral with respect to t in the above expression should be understood as a principal value integral as described after Eq. (8.19).

In the case of a stationary process, $\nu_Z^+(\zeta) = \nu_Z(\zeta)/2$, where $\nu_Z(\zeta)$ denotes the average crossing rate of the level ζ. The corresponding Rice formula for the crossing rate is given as

$$\nu_Z(\zeta) = \int_{-\infty}^{\infty} x\, p(x)\, \mathrm{sign}(x)\, dx, \tag{8.59}$$

where $\mathrm{sign}(x) = -1$ for $x < 0$, $\mathrm{sign}(0) = 0$, and $\mathrm{sign}(x) = 1$ for $x > 0$. Since $\hat{\mathscr{F}}[\mathrm{sign}(x)] = 2\mathrm{i}/t$, it is seen from the derivation above that indeed

$$\begin{aligned} \nu_Z^+(\zeta) &= \frac{-\mathrm{i}}{(2\pi)^2} \int_{-\infty}^{\infty} \int_{-\infty}^{\infty} \frac{\partial M_{Z\dot{Z}}(s,t)}{\partial t}\, e^{-\mathrm{i}\zeta s} \Big(-\frac{\mathrm{i}}{t} \Big)\, dt\, ds \\ &= -\frac{1}{(2\pi)^2} \int_{-\infty}^{\infty} \int_{-\infty}^{\infty} \frac{1}{t} \frac{\partial M_{Z\dot{Z}}(s,t)}{\partial t}\, e^{-\mathrm{i}\zeta s}\, dt\, ds. \end{aligned} \tag{8.60}$$

Appendix 2: The Characteristic Function

The derivation of the formula for the characteristic function $M_{Z\dot{Z}}$ will be based on an integral equality which is cited, but not proved, in a less general form by Cramer (1946). The prime (') will be used to denote transposition of a matrix or vector, and $\det(A)$ denotes the determinant of a square matrix A.

Theorem 8.1 *Let A be a complex, symmetrical, and nonsingular $n \times n$ matrix, and assume that the real part of A is positive definite. Let d denote a complex $n \times 1$ vector. Then the following integral equality holds:*

$$\int_{-\infty}^{\infty} \cdots \int_{-\infty}^{\infty} \exp\Big(d'x - \frac{1}{2}\, x' A\, x \Big)\, dx_1 \ldots dx_n = \frac{(2\pi)^{n/2}}{\sqrt{\det(A)}} \exp\Big(\frac{1}{2}\, d' A^{-1} d \Big). \tag{8.61}$$

Proof Let $\mu = A^{-1} d$. It is then easily verified that $2 d' x - x' A x = -(x - \mu)' A (x - \mu) + d' A^{-1} d$. Hence, it follows that the desired result is obtained if the following equality is proved ($z = x - \mu$):

$$\int_{-\infty-\mathrm{i}\Im(\mu_1)}^{\infty-\mathrm{i}\Im(\mu_1)} \cdots \int_{-\infty-\mathrm{i}\Im(\mu_n)}^{\infty-\mathrm{i}\Im(\mu_n)} \exp\Big(-\frac{1}{2}\, z' A\, z \Big)\, dz = \frac{(2\pi)^{n/2}}{\sqrt{\det(A)}}, \tag{8.62}$$

where $\mu = (\mu_1, \ldots, \mu_n)'$, $\Im(\mu_j)$ denotes the imaginary part of μ_j, and $dz = dz_1 \ldots dz_n$.

By assumption, $A = P + \mathrm{i}Q$, where P and Q are real symmetric matrices with P positive definite. Then there exists a real unitary matrix U such that

$U'PU = D = \mathrm{diag}(\lambda_1, \ldots, \lambda_n)$, $\lambda_j > 0$ for $j = 1, \ldots, n$. $\mathrm{diag}(\cdot)$ denotes a diagonal matrix with the indicated arguments on the diagonal. The variable shift $z \to y$ is made, where $y = Uz$. Then, $z' A z = y' D y + \mathrm{i}\, y' \tilde{Q} y$, where $\tilde{Q} = U' Q U$ is a real symmetric matrix. A scaling of the coordinate variables is introduced by the variable shift $y \to u$, where $y = \sqrt{D} u$, and $\sqrt{D} = \mathrm{diag}(\sqrt{\lambda_1}, \ldots, \sqrt{\lambda_n})$. Then, $z' A z = u' u + \mathrm{i}\, u' \hat{Q} u$, where $\hat{Q} = \sqrt{D}^{-1} \tilde{Q} \sqrt{D}^{-1}$ is again a real symmetric matrix. Hence, a real unitary matrix V exists such that $V' \hat{Q} V = \Omega = \mathrm{diag}(\omega_1, \ldots, \omega_n)$. Let $E = I + \mathrm{i}\Omega = \mathrm{diag}(1 + \mathrm{i}\omega_1, \ldots, 1 + \mathrm{i}\omega_n)$. By the variable shift $u \to v$, where $v = V u$, it follows that $z' A z = v' E v = \sum_{j=1}^{n} (1 + \mathrm{i}\omega_j)\, v_j^2$, $v = (v_1, \ldots, v_n)'$. Since U and V are unitary matrices, it is observed that $dz = dy = (\sqrt{\det(D)})^{-1} du = (\sqrt{\det(D)})^{-1} dv$. Denoting the integral on the left hand side of Eq. (8.62) by I, it is obtained that

$$I = \frac{1}{\sqrt{\det(D)}} \int_{-\infty - \mathrm{i}c_1}^{\infty - \mathrm{i}c_1} \cdots \int_{-\infty - \mathrm{i}c_n}^{\infty - \mathrm{i}c_n} \exp\left(-\frac{1}{2} \sum_{j=1}^{n} (1 + \mathrm{i}\omega_j)\, v_j^2 \right) dv_1 \ldots dv_n, \tag{8.63}$$

where c_j, $j = 1, \ldots, n$, are suitable constants. Since the functions $f_j(z) = \exp\{-(1/2)\,(1 + \mathrm{i}\omega)\, z^2\}$ are entire functions in the complex variable $z = x + \mathrm{i}y$ and since $\lim_{|x| \to \infty} f(z) = 0$ uniformly in y for $|y| \leq$ const., it follows that the path of integration from $-\infty - \mathrm{i}c_j$ to $\infty - \mathrm{i}c_j$ in Eq. (8.63) can be replaced by the path from $-\infty$ to ∞ by Cauchy's theorem. From the standard result that $\int_{-\infty}^{\infty} \exp\{-a\, x^2\}\, dx = \sqrt{\pi / a}$ for a complex constant a provided that $\Re(a) > 0$, it now follows that

$$I = \frac{1}{\sqrt{\det(D)}} \prod_{j=1}^{n} \frac{\sqrt{2\pi}}{\sqrt{1 + \mathrm{i}\omega_j}} = \frac{1}{\sqrt{\det(D)}} \frac{(2\pi)^{n/2}}{\sqrt{\det(E)}}. \tag{8.64}$$

To prove Eq. (8.62), it only remains to show that $\det(D) \cdot \det(E) = \det(A)$. Invoking the fact that U and V are unitary, it is obtained that

$$\begin{aligned} \det(E) &= \det(I + \mathrm{i}\Omega) = \det(I + \mathrm{i}\hat{Q}) = \det(I + \mathrm{i}\sqrt{D}^{-1} \tilde{Q} \sqrt{D}^{-1}) \\ &= \det(\sqrt{D}^{-1} D \sqrt{D}^{-1} + \mathrm{i}\sqrt{D}^{-1} \tilde{Q} \sqrt{D}^{-1}) \\ &= (\det(D))^{-1} \det(D + \mathrm{i}\tilde{Q}) = (\det(D))^{-1} \det(P + \mathrm{i}Q) \\ &= \frac{\det(A)}{\det(D)}, \end{aligned} \tag{8.65}$$

which is what was needed to prove.

Our goal is to calculate the characteristic function

$$M_{Z\dot{Z}}(u, v) = E\left[\exp\left(\mathrm{i}u\, Z + \mathrm{i}v\, \dot{Z}\right)\right]. \tag{8.66}$$

Using conditional probabilities, one may write

$$M_{Z\dot{Z}}(u, v) = E\left[e^{\mathrm{i}u\,Z}\, E[e^{\mathrm{i}v\,\dot{Z}}|W]\right] = E\left[e^{\mathrm{i}u\,Z}\, E[e^{\mathrm{i}v\,(\dot{Z}|W)}]\right]. \tag{8.67}$$

From Eq. (8.17) it follows that $\dot{Z} = Y'\,\dot{W}$, where $Y = \beta + 2DW$, $\beta = (\beta_1, \ldots, \beta_n)'$, and $D = \mathrm{diag}(\lambda_1, \ldots, \lambda_n)$. Hence, $(\dot{Z}|W) = Y'\,(\dot{W}|W)$. $(\dot{W}|W = w)$, where $w = (w_1, \ldots, w_n)' \in \mathbf{R}^n$, is a Gaussian vector with mean value $\mu = \Sigma_{21}\,w$ and covariance matrix $C = \Sigma_{22} - \Sigma_{21}\,\Sigma_{12}$, cf. Anderson (1958). This means that $(\dot{Z}|W = w)$ is a scalar Gaussian variable with mean value $m = y'\,\mu = y'\,\Sigma_{21}\,w$ and variance $s^2 = y'\,C\,y$, where $y = \beta + 2Dw$. Invoking the expression for the characteristic function of a Gaussian variable, it follows that $E[\exp\{\mathrm{i}v\,(\dot{Z}|W)\}] = \exp\{\mathrm{i}v\,Y'\,\Sigma_{21}\,W - \frac{1}{2}\,v^2\,Y'\,C\,Y\}$. It is now obtained that

$$M_{Z\dot{Z}}(u, v) = E\left[\exp\left(\mathrm{i}u\,W'DW + \mathrm{i}u\,\beta'W + \mathrm{i}v\,Y'\,\Sigma_{21}\,W - \frac{1}{2}\,v^2\,Y'\,C\,Y\right)\right]. \tag{8.68}$$

The calculation of the expected value of this equation amounts to the calculation of the following integral:

$$M_{Z\dot{Z}}(u, v) = \frac{1}{(2\pi)^{n/2}} \int_{-\infty}^{\infty} \cdots \int_{-\infty}^{\infty} \exp\left(-\frac{1}{2}\,v^2\,\beta'\,C\,\beta + d'w - \frac{1}{2}\,w'\,B\,w\right) dw, \tag{8.69}$$

where

$$d = (\mathrm{i}\,u\,I + \mathrm{i}\,v\,\Sigma_{12} - 2\,v^2\,D\,C)\,\beta, \tag{8.70}$$

and

$$B = I - 2\mathrm{i}u\,D - 4\,\mathrm{i}\,v\,D\,\Sigma_{21} + 4\,v^2 D\,C\,D. \tag{8.71}$$

It is recognized by inspection that the matrix B is not in general symmetric. However, the expression $w'Bw$ is a scalar quantity, which implies that $w'Bw = (w'Bw)' = w'B'w$. Hence $w'Bw = w'Aw$, where A is the symmetrized version of B, that is, $A = (B + B')/2$. By invoking Theorem 8.1, it is therefore obtained:

Theorem 8.2 *Let the stochastic process $Z(t)$ be represented as given by Eq. 8.17. Then the characteristic function $M_{Z\dot{Z}}(u, v) = E\left[\exp\left(\mathrm{i}u\,Z + \mathrm{i}v\,\dot{Z}\right)\right]$ of the joint variable $(Z, \dot{Z})$ is given by the expression*

$$M_{Z\dot{Z}}(u, v) = \frac{1}{\sqrt{\det(A)}}\,\exp\left(-\frac{1}{2}\,v^2\,\beta'\,C\,\beta + \frac{1}{2}\,d'\,A^{-1}\,d\right), \tag{8.72}$$

where

$$d = (\mathrm{i}\, u\, I + \mathrm{i}\, v\, \Sigma_{12} - 2\, v^2\, D\, C)\, \beta, \tag{8.73}$$

and

$$A = I - 2\mathrm{i}u\, D\, - 2\,\mathrm{i}\, v\, \big(D\, \Sigma_{21} + \Sigma_{12}\, D\big) + 4\, v^2 D\, C\, D. \tag{8.74}$$

Note that in the absence of a linear component, that is, when $Z(t) = \sum_{j=1}^{n} \lambda_j\, W_j(t)^2$, then

$$M_{Z\dot{Z}}(u, v) = \frac{1}{\sqrt{\det(A)}}. \tag{8.75}$$

The marginal characteristic functions $M_Z(u) = E[\exp\{\mathrm{i}\, u\, Z\}]$ and $M_{\dot{Z}}(v) = E[\exp\{\mathrm{i}\, v\, \dot{Z}\}]$ are now easily obtained by the relations $M_Z(u) = M_{Z\dot{Z}}(u, 0)$ and $M_{\dot{Z}}(v) = M_{Z\dot{Z}}(0, v)$. In particular,

$$M_Z(u) = \frac{1}{\sqrt{\prod_{j=1}^{n}(1 - 2\,\mathrm{i}\, u\, \lambda_j)}} \exp\Big(- \frac{1}{2} \sum_{j=1}^{n} \frac{u^2\, \beta_j^2}{1 - 2\,\mathrm{i}\, u\, \lambda_j} \Big), \tag{8.76}$$

which is in agreement with previously derived results (Kac and Siegert 1947; Naess 1986). It is also obtained that

$$M_{\dot{Z}}(v) = \frac{1}{\sqrt{\det(\tilde{A})}} \exp\Big(- \frac{1}{2}\, v^2\, \beta'\, C\, \beta + \frac{1}{2}\, \tilde{d}'\, \tilde{A}^{-1}\, \tilde{d} \Big), \tag{8.77}$$

where

$$\tilde{d} = (\mathrm{i}\, v\, \Sigma_{12} - 2\, v^2\, D\, C)\, \beta, \tag{8.78}$$

and

$$\tilde{A} = I - 2\,\mathrm{i}\, v\, \big(D\, \Sigma_{21} + \Sigma_{12}\, D\big) + 4\, v^2 D\, C\, D. \tag{8.79}$$

Chapter 9
Monte Carlo Methods and Extreme Value Estimation

9.1 Introduction

The last decade has seen a dramatic increase in the use of Monte Carlo methods for solving stochastic engineering problems. There are primarily two reasons for this increase. First, the computational power available today, even for a laptop computer, is formidable and steadily increasing. Second, the versatility of Monte Carlo methods make them very attractive as a way of obtaining solutions to stochastic problems. The drawback of Monte Carlo methods for a range of problems has been that the required numerical calculations may take days, weeks, or even months to do. But this situation is changing, some numerical problems that required several days of computer time for their solution just a few years ago can now be solved in minutes or hours. This has really opened the door for the use of Monte Carlo-based methods for solving a wide array of stochastic engineering problems. In this chapter the focus is on adapting Monte Carlo methods for estimation of extreme values of stochastic processes encountered in various engineering disciplines.

9.2 Simulation of Stationary Stochastic Processes

The approach to the simulation of stationary stochastic processes favored in this book is the spectral representation method (Shinozuka and Deodatis 1991). The main reason for this is its simplicity and transparency for practical applications.

The procedure is described in Sect. 9.2.2, where Eq. (9.15) is a key result. Useful background information for this is contained in Sect. 9.2.1. Combining this with Example 9.2.4 provides a hands-on guide of how to simulate realizations of a stationary Gaussian process.

A. Naess, *Applied Extreme Value Statistics*,
https://doi.org/10.1007/978-3-031-60769-1_9

The use of the fast Fourier transform (FFT) technique can substantially speed up the production of realizations of stationary Gaussian processes, cf. Shinozuka (1974) and Newland (1991).

9.2.1 Realizations of Stochastic Processes

Monte Carlo methods can be used for any input-output system subjected to a stochastic process such that for every realization of the input stochastic process a corresponding realization of the output process can be calculated. The basic idea underlying the Monte Carlo method is that of producing a sample of output/response time histories from a sample of input/loading time histories. This makes it possible to estimate various statistics of the output/response process based on the available sample of realizations. Hence, a key element in the Monte Carlo method is therefore the realizations of a stochastic process. For their simplicity, the focus here is on a couple of widely used ways of representing stationary stochastic processes, and a discussion is given of how realizations of such processes are generated, see also Shinozuka and Jan (1972) and Shinozuka and Deodatis (1991).

A simple, but useful representation of a stationary stochastic process $X(t)$ can be obtained as follows,

$$X(t) = \sum_{j=1}^{N} \left\{A_j \cos \omega_j t + B_j \sin \omega_j t\right\}, \quad \infty < t < \infty, \tag{9.1}$$

where, for $j = 1, \ldots, N$, the ω_j are positive constants, the A_j and the B_j are random variables with $\mathbb{E}[A_j] = \mathbb{E}[B_j] = 0, \mathbb{E}[A_j^2] = \mathbb{E}[B_j^2] = \sigma_j^2$, and $\mathbb{E}[A_j B_j] = 0$ for $j = 1, \ldots, N$. Also, $\mathbb{E}[A_j A_k] = \mathbb{E}[B_j B_k] = \mathbb{E}[A_j B_k] = 0$ for $j \neq k$. When these conditions are satisfied, it can be shown that $X(t)$ is a (weakly) stationary process with,

$$m_X = \mathbb{E}[X(t)] = 0, \tag{9.2}$$

and

$$C_X(\tau) = \mathbb{E}[(X(t) - m_X)(X(t+\tau) - m_X)] = \sum_{j=1}^{N} \sigma_j^2 \cos \omega_j \tau. \tag{9.3}$$

The first step of the procedure is to specify what kind of random variables A_j and B_j are assumed to be. A common choice in many cases is to assume that these variables are independent and normally distributed with zero mean value and known standard deviations σ_j. A realization of $X(t)$ is then obtained when a set of outcomes of the random variables A_j and B_j has been generated. In this connection, it is useful to make the following observation: if $\tilde{A}_j$ is normally distributed with zero mean and standard deviation equal to 1.0, then $A_j = \sigma_j \tilde{A}_j$ is normally distributed with mean zero and standard deviation equal to σ_j. Similarly for B_j. Equation (9.1) may therefore be written in the form,

$$X(t) = \sum_{j=1}^{N} \sigma_j \left\{ \tilde{A}_j \cos(\omega_j t) + \tilde{B}_j \sin(\omega_j t) \right\}, \tag{9.4}$$

where $\tilde{A}_j$ and $\tilde{B}_j$, $j = 1, \ldots, N$, is now a set of independent, standard normally distributed variables. There are computer programs that may be used to generate independent outcomes of a standard, normally distributed variable. It is seen that $2N$ outcomes are needed for this example. Specifically, assume that $\tilde{a}_j$ and $\tilde{b}_j$, $j = 1, \ldots, N$ are the obtained outcomes from such a program. The corresponding realization $x(t)$ is then,

$$x(t) = \sum_{j=1}^{N} \sigma_j \left\{ \tilde{a}_j \cos(\omega_j t) + \tilde{b}_j \sin(\omega_j t) \right\}. \tag{9.5}$$

This procedure can be repeated as many times as needed to produce the requested sample size of realizations. It is, of course, understood here that each realization is generated independently of all others. This procedure is illustrated with an application to ocean waves in Example 9.2.4.

The following representation is also frequently used, especially when ergodic properties are desirable,

$$X(t) = \sum_{j=1}^{N} a_j \cos(\omega_j t + \Phi_j), \tag{9.6}$$

where a_j and ω_j, $j = 1, \ldots, N$ are positive constants, and $\{\Phi_j\}_{j=1}^{N}$ is a set of independent random variables that are uniformly distributed over $(0, 2\pi)$. It may be shown that $X(t)$ is ergodic with respect to the mean value and the autocorrelation.

Generating realizations of this process is now quite straightforward. It is seen that, in fact, only outcomes of the uniformly distributed phase angles Φ_j, $j = 1, \ldots, N$ are needed. If R_j, $j = 1, \ldots, N$, denote independent random variables uniformly distributed on $(0, 1)$, usually referred to as random numbers, then clearly one may put $\Phi_j = 2\pi R_j$. Hence, by invoking a random number generator, usually available on any computer, outcomes of Φ_j, $j = 1, \ldots, N$ can be easily generated. Let r_j, $j = 1, \ldots, N$, denote a set of independent random number outcomes. A realization $x(t)$ of $X(t)$ is then,

$$x(t) = \sum_{j=1}^{N} a_j \cos(\omega_j t + 2\pi r_j). \tag{9.7}$$

9.2.2 Variance Spectra

Fourier analysis is routinely used to decompose the time histories of load and response as sums or integrals of $\cos(\cdot)$ and $\sin(\cdot)$ terms over the frequency domain. It is known how the transfer function gives a direct connection between the amplitudes at each frequency for the load and response, provided the system is linear and time invariant. A time history that is periodic can be decomposed as a sum over a finite or countably infinite number of frequencies. If it is not periodic, the decomposition must be expressed as an integral. It can be shown that this is possible only if the time history dies out with time. Regarding the realizations of stationary stochastic processes, they will not generally be periodic. Nor will they decrease with time because the variance is constant, and this is a measure of the fluctuations around the mean value. A direct frequency decomposition of the realizations of a stationary process is therefore not directly feasible. This difficulty is circumvented by using the autocovariance function of a stationary process. This function will generally approach zero when its time argument increases.

Let $X(t)$ be a stationary process with autocovariance function $C_X(\tau)$. Assume that $C_X(\tau) \to 0$ when $\tau \to \infty$ sufficiently rapidly so that $\int_{-\infty}^{\infty} |C_X(\tau)|\, d\tau$ has a finite value, that is, $C_X(\tau)$ is assumed to be an integrable function.

The *variance spectrum* $S_X(\omega)$ of $X(t)$ is defined as the Fourier transform of $C_X(\tau)$ as follows,

$$S_X(\omega) = \frac{1}{2\pi} \int_{-\infty}^{\infty} C_X(\tau)\, e^{-i\omega\tau}\, d\tau. \tag{9.8}$$

It was shown already in the 1930s that $C_X(\tau)$ and $S_X(\omega)$ constituted a Fourier transform pair; that is, $C_X(\tau)$ is given by the inverse Fourier transform as,

$$C_X(\tau) = \int_{-\infty}^{\infty} S_X(\omega)\, e^{i\omega\tau}\, d\omega. \tag{9.9}$$

Equations (9.8) and (9.9) are often called the Wiener-Khintchine relations after the originators, cf. Chatfield (1989).

The quantity $S_X(\omega)$ is known by many names. A few names that seem appropriate to mention here are *energy spectrum, power spectral density, spectral density* or just *spectrum*. The name *variance spectrum*, ties in directly with the interpretation of $S_X(\omega)$. By putting $\tau = 0$ in Eq. (9.9), it is obtained that,

$$\sigma_X^2 = C_X(0) = \int_{-\infty}^{\infty} S_X(\omega)\, d\omega. \tag{9.10}$$

This equation shows that $S_X(\omega)$ can be interpreted as a distribution of variance along the frequency axis, provided that $S_X(\omega) \geq 0$, and in Sect. 9.2.5, it is shown that this is always the case. Negative frequencies are a mathematical convenience and have no real physical content. If $S_X(\omega)$ is to be interpreted as distribution of variance along the frequency axis, one would therefore expect that $S_X(-\omega) = S_X(\omega)$. This symmetry property of $S_X(\omega)$ can be shown directly from Eq. (9.8). It follows also from a rewriting of the Wiener-Khintchine relations to real form, which is of interest in itself. Because $C_X(\tau)$ is symmetric, it follows from Eq. (9.8) that,

$$\begin{aligned} S_X(\omega) &= \frac{1}{2\pi}\int_0^\infty C_X(\tau)\, e^{-\mathrm{i}\omega\tau}\, d\tau + \frac{1}{2\pi}\int_0^\infty C_X(-\tau)\, e^{\mathrm{i}\omega\tau}\, d\tau \\ &= \frac{1}{\pi}\int_0^\infty C_X(\tau)\, \cos\omega\tau\, d\tau\,, \end{aligned} \tag{9.11}$$

where Euler's relation $e^{\mathrm{i}x} = \cos x + \mathrm{i}\sin x$ is used. The last integral in the upper line is obtained by the change of variable $\tau \to -\tau$, while the lower line follows from the symmetry property $C_X(-\tau) = C_X(\tau)$.

From Eq. (9.11), it follows that,

$$S_X(-\omega) = S_X(\omega), \tag{9.12}$$

because $\cos(-\omega t) = \cos\omega t$. Equation (9.9) can then be rewritten as,

$$C_X(\tau) = 2\int_0^\infty S_X(\omega)\, \cos\omega\tau\, d\omega. \tag{9.13}$$

If $S_X(\omega)$ is a reasonably nice function, the integral on the rhs of Eq. (9.13) can be approximated by a finite sum, viz.

$$C_X(\tau) = 2\int_0^\infty S_X(\omega)\, \cos\omega\tau\, d\omega \approx \sum_{j=1}^{N} 2S_X(\omega_j)\, \Delta\omega\, \cos\omega_j\tau, \tag{9.14}$$

for a suitable choice of $\omega_1 < \ldots < \omega_N$, and sufficiently small $\Delta\omega = (\omega_N - \omega_1)/(N-1)$. Let us now recollect previous results and define a stationary process,

$$\tilde{X}(t) = \sum_{j=1}^{N} \left\{A_j\, \cos\omega_j t + B_j\, \sin\omega_j t\right\}, \tag{9.15}$$

where the random variables A_j and B_j satisfy the necessary conditions for stationarity, cf. Eq. (9.1). In addition, $\sigma_j^2 = \mathbb{E}[A_j^2] = \mathbb{E}[B_j^2] = 2S_X(\omega_j)\Delta\omega$. Then, according to Eq. (9.10), the autocovariance of $\tilde{X}(t)$ is given as,

$$C_{\tilde{X}}(\tau) = \sum_{j=1}^{N} 2S_X(\omega_j)\, \Delta\omega\, \cos\omega_j\tau. \tag{9.16}$$

A stationary process $\tilde{X}(t)$ is thus constructed with approximately the same variance distribution, that is, variance spectrum, as $X(t)$. In a certain sense, $\tilde{X}(t)$ can be said to represent $X(t)$. What has just been described, is a variant from a class of methods that is extensively used in practice to generate realizations of a given stationary process. To get a concrete realization of a process represented by Eq. (9.15), one has to generate outcomes of the random variables that enter the sum.

On the basis of Eq. (9.12) and the fact that negative frequencies do not really have any physical meaning, it is common practice in engineering to use the *one-sided* variance spectrum, which is denoted by $S_X^+(\omega)$ in this chapter, and defined as,

$$S_X^+(\omega) = \begin{cases} 2S_X(\omega)\,, & \omega \geq 0, \\ 0\,, & \omega < 0. \end{cases} \tag{9.17}$$

The distribution of variance is thereby concentrated to positive (physically realizable) frequencies. The variance expressed in terms of the one-sided variance spectrum is clearly,

$$\sigma_X^2 = \int_0^\infty S_X^+(\omega)\, d\omega. \tag{9.18}$$

A typical variance spectrum produced from measured data for the wave elevation at a point on the sea surface at a location in the North Sea is shown in Fig. 9.1. The somewhat jagged look is mostly due to low numerical resolution.

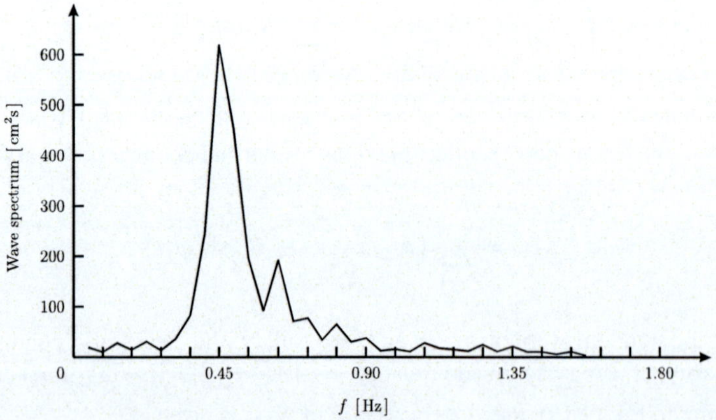

Fig. 9.1 Typical "wave spectrum" from the North Sea

9.2.3 Units of Variance Spectra

In Eq. (9.10), the units of $S_X(\omega)$ are the square of the units of $X(t)$ divided by radians per second. If $X(t)$ models the wave elevation at a location on the ocean surface and is measured in meters, then the units of $S_X(\omega)$ are m^{2}s/rad.

So far, the variance spectrum has only been considered as a function of circular frequency ω with units rad/s. It is also quite common to give the variance spectrum as a function of frequency f with units in Hertz (oscillations per second). The relation $\omega = 2\pi f$ implies a rescaling of the frequency axis when ω is replaced by f. It is therefore not correct to believe that $S_X(2\pi f)$ would represent the variance spectrum as a function of f; this would easily lead to the wrong variance. If $G_X^+(f)$ denotes the one-sided variance spectrum as a function of f in Hz, then to conserve the variance, it is needed to require that $G_X^+(f)df = S_X(\omega)d\omega$. This leads to the relation,

$$G_X^+(f) = 2\pi S_X^+(\omega). \tag{9.19}$$

Exactly the same relation applies to two-sided spectra. A factor of $2\pi = 6.28$ occurring erroneously in the variance can have a significant impact on the results in some cases. Thus, care should be exercised when adopting values of spectral moments by checking which kind of frequency is used in the variance spectrum.

9.2.4 Example: A Realization of a Wave Process

In this example, the procedure for producing a realization of a stationary stochastic process described in Sect. 9.2.1 is illustrated. In particular, it will be shown how a realization of a stochastic process with a given variance spectrum can be generated. To be even more specific, it will be assumed that the task is to produce an arbitrary realization of the wave elevation $X(t)$ at a given location in the North Sea with a recorded wave spectrum, as depicted in Fig. 9.1. Thus, it is assumed that this wave elevation can be represented as a stationary stochastic process $X(t)$ with a one-sided variance spectrum $G_X^+(f)$, as shown in Fig. 9.1. As mentioned in Sect. 9.6.5, there are several alternative methods that can be used to generate realizations of a stochastic process. Our choice here is to use the method described in Sect. 9.6.5, which amounts to approximating the process $X(t)$ with $\tilde{X}(t)$, and then generate realizations of $\tilde{X}(t)$ instead.

To proceed, it is necessary to specify what kind of random variables A_j and B_j to use in Eq. (9.15). The common choice in the case of wave processes on the open ocean is to assume that these variables are independent and normally distributed with zero mean value. To determine the standard deviation, it is necessary to decide on a suitable discretization of the frequency axis. For many practical purposes, it would be desirable to have a discretization that would give of the order of 10^3

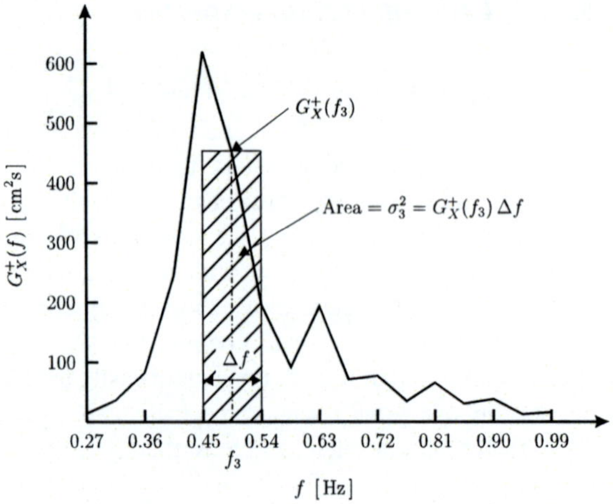

Fig. 9.2 Wave spectrum from the North Sea

frequencies in the frequency range where the waves have significant energy, e.g., from 0.27 to 0.99 Hz in this case, cf. the discussion at the end of the example. Because the objective here is to illustrate the procedure, a rather coarse frequency increment $\Delta f = 0.09$ Hz will do.. That gives 8 frequencies denoted by $f_1, \ldots, f_8$ ($f_1 = 0.315$ Hz, $f_{j+1} = f_j + 0.09$ Hz) in the specified frequency range. The process $\tilde{X}(t)$ can then be written as,

$$\tilde{X}(t) = \sum_{j=1}^{8} \left\{A_j \cos(2\pi f_j t) + B_j \sin(2\pi f_j t)\right\}. \tag{9.20}$$

Figure 9.2 shows the relevant part of the wave spectrum in Fig. 9.1 magnified along the frequency axis to clarify how the standard deviation of A_j and B_j is determined.

What remains to get an approximate realization of $X(t)$ is to generate outcomes of the random variables that enter $\tilde{X}(t)$. For this, an observation made in Sect. 9.2.1 is invoked: if $\tilde{A}_j$ is normally distributed with zero mean and standard deviation equal to 1.0, then $A_j = \sigma_j \tilde{A}_j$ is normally distributed with zero mean and standard deviation equal to σ_j. Similarly for B_j. Because $\sigma_j = \sqrt{G_X^+(f_j)\,\Delta f}$, Eq. (9.20) may be written in the form,

$$\tilde{X}(t) = \sum_{j=1}^{8} \sqrt{G_X^+(f_j)\,\Delta f}\,\left\{\tilde{A}_j \cos(2\pi f_j t) + \tilde{B}_j \sin(2\pi f_j t)\right\}, \tag{9.21}$$

where $\tilde{A}_j$ and $\tilde{B}_j$, $j = 1, \ldots, 8$, now constitute a set of independent, standard normally distributed variables. Computer programs are easily available for generating independent outcomes of a standard, normally distributed variable. It is seen that 16 is needed for the example.

Table 9.1 A table of the numerical information needed to produce a realization of the stochastic process given by Eq. (9.21)

j	f_j[Hz]	$\sqrt{G_X^+(f_j)\,\Delta f}$ [cm]	r_{2j-1}	r_{2j}	$\tilde{a}_j$	$\tilde{b}_j$
1	0.315	1.90	0.10097	0.32533	−1.2760	−0.4529
2	0.405	5.45	0.76520	0.13586	0.7232	−1.0991
3	0.495	6.57	0.34673	0.54876	−0.3942	0.1225
4	0.585	3.15	0.80959	0.09117	0.8764	−1.3335
5	0.675	2.23	0.39292	0.74945	−0.2717	0.6728
6	0.765	1.56	0.37542	0.04805	−0.3175	−1.6641
7	0.855	1.44	0.64894	0.74296	0.3825	0.6525
8	0.945	1.41	0.24805	0.24037	−0.6807	−0.7051

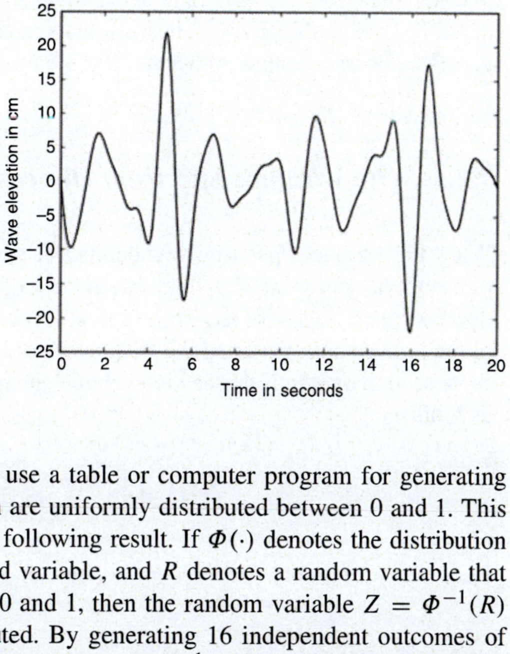

Fig. 9.3 A realization of the wave elevation

An alternative procedure is to use a table or computer program for generating (pseudo-)random numbers, which are uniformly distributed between 0 and 1. This can also be used by invoking the following result. If $\Phi(\cdot)$ denotes the distribution of a standard, normally distributed variable, and R denotes a random variable that is uniformly distributed between 0 and 1, then the random variable $Z = \Phi^{-1}(R)$ is standard and normally distributed. By generating 16 independent outcomes of R: $r_1, \ldots, r_{16}$, then $z_1 = \Phi^{-1}(r_1), \ldots, z_{16} = \Phi^{-1}(r_{16})$ will be 16 independent outcomes of a standard, normally distributed variable. This procedure is used here, and the results are shown in Table 9.1, where $\tilde{a}_j = \Phi^{-1}(r_{2j-1})$ is an outcome of $\tilde{A}_j$ and $\tilde{b}_j = \Phi^{-1}(r_{2j})$ is an outcome of $\tilde{B}_j$.

A piece of the corresponding realization is shown in Fig. 9.3, and one may get a similar impression as when observing irregular seas out on the oceans. In practice, there is often a need to generate many realizations to perform statistical analyses. It is then necessary to repeat the procedure just described the requisite number of times, and for each realization, a new set of outcomes independent of the previous ones are chosen.

If a piece of the realization that has been generated was shown with a duration which was twice as long, it would have become clear that the wave pattern is repeating itself. This is due to the way it was constructed, which indeed makes it periodic. The period is determined by the greatest common divisor of the frequency increment and the initial frequency of the discretization of the frequency range that is chosen. In this case, the period becomes 1/0.045 = 22.2 s. The practical consequence of this is that one must choose a discretization that is in correspondence with the required length of a realization. It may be worth mentioning that for most practically relevant discretizations, one may say that a corresponding realization will have a period that can be assumed to be given as $1/\Delta f$. One does not fully avoid the problem related to periodicity by choosing an (almost) irrational ratio between the starting frequency and the frequency increment, or other "smart" tricks such as choosing the frequencies randomly within each subinterval of the discretization.

Another point worth noting is that the process $\tilde{X}(t)$ is not ergodic. If it is desirable to ensure this property, one may use the method described in Sect. 9.2.1. However, the difference in the practical results obtained by using this method versus the one described here, is usually rather small if the discretization is properly done.

9.2.5 *The Variance Spectrum Directly from the Realizations*

When the variance spectrum was defined, it was mentioned that a realization $x(t)$ of a stationary process $X(t)$ does not have a Fourier transform because it does not decrease toward zero for large t. Intuitively, one would nevertheless expect that most of the information regarding the frequency content of $x(t)$ should be contained in a finite section of $x(t)$ if this section is large enough. A section of $x(t)$ can be defined as follows,

$$x_T(t) = \begin{cases} x(t), & 0 \le t \le T, \\ 0, & \text{elsewhere.} \end{cases} \tag{9.22}$$

Because $x_T(t)$ is zero outside a finite interval, it has a Fourier transform, viz.

$$X_T(\omega) = \frac{1}{2\pi}\int_{-\infty}^{\infty} x_T(t)\, e^{-\mathrm{i}\omega t}\, dt = \frac{1}{2\pi}\int_{0}^{T} x(t)\, e^{-\mathrm{i}\omega t}\, dt. \tag{9.23}$$

It would seem natural to expect that there is a connection between $X_T(\omega)$ and the variance spectrum $S_X(\omega)$. And it turns out that the connection is, in fact, quite simple. It is written in the following way: assume that $x_j(t)$, $j = 1, 2, \ldots$ are realizations of a stationary process. Then the following equation applies,

$$S_X(\omega) = \lim_{T\to\infty} \lim_{N\to\infty} \frac{2\pi}{TN} \sum_{j=1}^{N} |X_{j,T}(\omega)|^2, \tag{9.24}$$

where $X_{j,T}(\omega)$ denotes the Fourier transform of $x_{j,T}(t)$, which equals $x_j(t)$ for $0 \le t \le T$ and zero elsewhere. From Eq. (9.24), it is also immediately seen that $S_X(\omega) \ge 0$.

Equation (9.24) is based on the availability of an ensemble of realizations. As previously discussed, there are many situations where only one realization is available. By assuming that the process is ergodic, it can be shown that the variance spectrum can be determined in the following way. Let $x(t)$ denote a realization of the assumed ergodic process $X(t)$. Define a set of truncated Fourier transforms $X_{(j),T}(\omega)$ over the intervals $\big((j-1)T, jT\big)$, $j = 1, 2, \ldots$, as follows,

$$X_{(j),T}(\omega) = \frac{1}{2\pi}\int_{(j-1)T}^{jT} x(t)\, e^{-i\omega t}\, dt. \tag{9.25}$$

It is then obtained that,

$$S_X(\omega) = \lim_{T\to\infty}\lim_{N\to\infty}\frac{2\pi}{TN}\sum_{j=1}^{N}|X_{(j),T}(\omega)|^2. \tag{9.26}$$

Equation (9.24) or (9.26) can be regarded as the basis for the *fast Fourier transform* (FFT) algorithm for calculating spectra. This method, developed around 1965, has assumed a dominating position among numerical methods for calculating Fourier transforms. This is primarily due to the fact that the method is much faster than traditional methods. An extensive discussion of the FFT method for calculating Fourier transforms is given by Newland (1991).

9.3 Monte Carlo Simulation of Load and Response

The linear or linearized equations of motion for marine structures considered in this book and their solutions are discussed in Naess and Moan (2013). For such equations, the comments of the previous section apply. An example of a nonlinear dynamic model that is often adopted for an offshore structure, can be written in the following form,

$$\boldsymbol{M}\,\ddot{\boldsymbol{X}}(t) + \boldsymbol{C}\,\dot{\boldsymbol{X}}(t) + \boldsymbol{f}\big(\boldsymbol{X}(t), \dot{\boldsymbol{X}}(t), t\big) = \boldsymbol{P}(t)\,, \tag{9.27}$$

where $\boldsymbol{M}$ and $\boldsymbol{C}$ are suitable mass and damping matrices, respectively; $\boldsymbol{f}(\cdot,\cdot) = \big(f^{kl}(\cdot,\cdot)\big)$, $k, l = 1, \ldots, n$, is a nonlinear matrix function; $\boldsymbol{P}(t) = (P_1(t), \ldots, P_n(t))'$ denotes a stochastic loading process; and $\boldsymbol{X}(t) = (X_1(t), \ldots, X_n(t))'$ is the corresponding response process.

The Monte Carlo method applied to such a system would consist of generating a statistical sample of specified size N, say, of response time histories by first generating a sample of time histories of the same size of the loading process, and then solve Eq. (9.27) for each of the load time histories in the sample using methods of numerical integration of dynamical system equations, cf. Argyris and Mlejnek (1991) and Naess and Moan (2013). When the desired sample of response time histories is produced, the statistical analysis of the response may then proceed as discussed in the remaining part of this chapter.

9.4 Sample Statistics of Simulated Response

An important element for graphic representation of sampled data from a statistical population for which the underlying distribution function is unknown is the so-called plotting position formula. It is based on the notion of order statistics.

Let us start by assuming that X is a continuous random variable with a distribution $F(x)$ and a probability density $f(x)$. The given sample of independent observations $x_1, x_2, \ldots, x_n$ is now ordered in an increasing sequence $x_{(1)} \le x_{(2)} \le \ldots \le x_{(n)}$. The random variable $X_{(m)}$ corresponding to $x_{(m)}$ is called the mth order statistic, $m = 1, \ldots, n$. The probability density $f_m(x)$ of $X_{(m)}$ follows from the observation that $X_{(m)} = x$ implies the event that there are $m-1$ outcomes of X with values less than (or equal to) x, and $n-m$ outcomes with values greater than x. According to the binomial distribution, the probability of this event equals $n!/\big((m-1)!(n-m)!\big)\, F(x)^{m-1}\big(1-F(x)\big)^{n-m}$. Hence, it is obtained that (Casella and Berger 2002),

$$f_m(x) = \frac{n!}{(m-1)!(n-m)!}\, F(x)^{m-1}\big(1-F(x)\big)^{n-m}\, f(x). \tag{9.28}$$

It is now required to calculate $\mathbb{E}[F(X_{(m)})]$. This is given by,

$$\begin{aligned}\mathbb{E}[F(X_{(m)})] &= \int_x F(x) f_m(x)\,dx = m\binom{n}{m}\int_x F(x)^m\big(1-F(x)\big)^{n-m} f(x)\,dx \\ &= m\binom{n}{m}\int_0^1 F^m(1-F)^{n-m}\,dF = m\binom{n}{m}\frac{m!(n-m)!}{(n+1)!} = \frac{m}{n+1}.\end{aligned} \tag{9.29}$$

Similarly, the variance of $F(X_{(m)})$ is calculated to be,

$$\begin{aligned}\mathrm{Var}[F(X_{(m)})] &= \mathbb{E}[F(X_{(m)})^2] - \mathbb{E}[F(X_{(m)})]^2 \\ &= \frac{m(m+1)}{(n+1)(n+2)} - \frac{m^2}{(n+1)^2} = \frac{m(n+1-m)}{(n+1)^2(n+2)}.\end{aligned} \tag{9.30}$$

The results of Eqs. (9.29) and (9.30) are useful because they provide a means of plotting the sample of observations $x_1, x_2, \ldots, x_n$ in order to estimate the distribution $F(x)$ empirically. Equation (9.29) states that the expected value of the distribution function evaluated at the observation of order m is equal to $m/(n+1)$. This result suggests that an optimal plotting strategy is obtained by plotting the points $\big(x_{(m)}, m/(n+1)\big)$, $m = 1, \ldots, n$. Equation (9.30) provides information on the variance of $F(X_{(m)})$, that is, the ordinate of the plotting point. Due to the symmetry of the expression, it attains its maximum $1/[4(n+2)]$ at the median and decreases symmetrically to $n/[(n+1)^2(n+2)]$ toward the ends of the interval [0, 1]. Because the distribution function itself is not known, these results are called distribution-free results.

The preceding results are complemented by calculating the probability that the mth observation is not exceeded by a future observation. Now, the conditional probability that a single observation will not exceed $X_{(m)}$ given that $X_{(m)} = x$ is equal to $F(x)$. The corresponding unconditional probability p_m, say, is then obtained by using the law of total probability, which gives exactly the same result as given by Eq. (9.29), that is,

$$p_m = \int_x F(x) f_m(x)\, dx = \frac{m}{n+1}. \tag{9.31}$$

This result shows that a new observation of the continuous random variable X has equal probability of assuming a value in any of the $n+1$ intervals defined by the previous n observations. This lends further support to the optimality of the plotting position formula expressed by Eq. (9.29).

It may be pointed out that several alternative plotting position formulas for specific classes of distributions have been suggested over the years, which primarily aim at correcting for sample bias. A discussion of this topic is not pursued here, but the reader is rather referred to the literature. A discussion of particular interest related to the estimation of return periods can be found in Makkonen (2006, 2008), where it is argued for the use of Eq. (9.29).

A useful diagnostic tool to check the accuracy of an assumed statistical distribution F for the observed data is obtained by comparing the fitted distribution $\hat{F}$ with the data on a quantile plot (QQ-plot) or a probability plot (PP-plot). Assuming that $\hat{F}$ is strictly increasing and continuous, the QQ-plot is obtained by comparing the ordered data with the corresponding quantiles of the fitted distribution by plotting

$$\Big(\hat{F}^{-1}\big(\frac{m}{n+1}\big), x_{(m)}\Big). \tag{9.32}$$

The name *QQ-plot* derives from the fact that both $\hat{F}^{-1}\big(\frac{m}{n+1}\big)$ and $x_{(m)}$ are estimates of the $m/(n+1)$th quantile of F. If F is a good choice for the distribution of the data, the QQ-plot should be close to the straight line of slope 1 passing through the origin. Alternatively, the fit of $\hat{F}$ to the data can be checked by the PP-plot, which is obtained by plotting the points

$$\Big(\hat{F}\big(x_{(m)}\big), \frac{m}{n+1}\Big). \tag{9.33}$$

A good fit is again demonstrated if the plotted graph is approximately a straight line. The main difference between the two plots is that the QQ-plot gives a more clear impression of the fit of the tail data, which may be of particular significance for extreme value statistics.

It is worth noting that for a range of distribution functions QQ-plots can be constructed without having to estimate distribution parameters. This typically applies to distributions characterized by a scale and a location parameter, e.g., the normal distribution. In such cases the intercept of the line fitted to the QQ-plot would represent location, while the slope represents scale. For example, a QQ-plot for a normal distribution can be achieved by plotting $\big(\Phi^{-1}(m/(n+1)), x_{(m)}\big)$.

9.5 Latin Hypercube Sampling

Latin hypercube sampling (LHS) (McKay et al. 1979) is a method for effectively reducing the sample size for Monte Carlo simulations of stochastic response processes that depend on many random parameters. In cases where the parameters of a dynamic model, such as mass, damping, and stiffness are modeled as random variables, the response process will then also depend on these random variables. Specifically, let us assume that the model depends on the random parameters $Y_1, \ldots, Y_m$. To highlight the dependence on these parameters, a random response process $X(t)$ of this model may then be written as $X(t) = X(t; Y_1, \ldots, Y_m)$. Because the external loading is often modeled as a stochastic process, the response $X(t; y_1, \ldots, y_m)$ becomes a stochastic process for each sample $y_1, \ldots, y_m$ of the random parameters. If m is not small, then the number of samples needed to provide good sample statistics for the response process may become huge if no consideration is made on how to effectively represent the statistical variability of the random parameters. LHS is a very good and simple method for this purpose.

LHS starts by selecting k different values from each of the m random variables in the following manner. The interval (0, 1) is divided into k equally long intervals $I_j = ((j-1)/k, j/k)$, $j = 1, \ldots, k$. Let $F_i(y)$ denote the distribution function of Y_i, $i = 1, \ldots, m$. For each i, k independent outcomes $u_1, \ldots, u_k$ of the random number U, which is uniformly distributed on (0,1), are produced. The resulting ordered sample for Y_i is $y_{i,1} < \ldots < y_{i,k}$ where $y_{i,j} = F_i^{-1}((j-1)/k + u_j/k)$. Note that $(j-1)/k+u_j/k$ is nothing but an outcome of a random number uniformly distributed on the interval $((j-1)/k, j/k)$. A Latin hypercube (LH) presample for Y_i is then obtained as $y_{i,r_1}, \ldots, y_{i,r_k}$, where $r_1, \ldots, r_k$ is a random reordering of $1, \ldots, k$. Finally, an LH sample for $Y_1, \ldots, Y_m$, which will also be of size k, is now represented by the $m \times k$ array or matrix (y_{i,r_j}), where each column is an element in the LH sample.

It is tacitly assumed that the random parameters are independent. If this is not the case, LHS can also deal with correlated parameters. Standard statistical software packages usually offer LHS as an optional sampling technique.

9.6 Estimation of Extreme Response

The view that estimation of extreme values by Monte Carlo methods is generally prohibitive in terms of computer time, is a truth in contention. In fact, with present-day computational power, it is possible to perform simulations on a scale that allows estimation of extreme structural response for a range of problems. Admittedly, it is not difficult to describe a dynamic model for which the required simulation time for direct estimation of extreme response would be beyond any acceptable bounds, but the systems for which this is the case are steadily diminishing in step with the development of increasingly powerful computers. This being the case, it is good reason to believe that Monte Carlo simulation-based methods will become much more frequently used even for extreme value estimation than what is the case today. It is therefore considered appropriate to discuss such methods here. Note that the methods described in this section are equally applicable to measured response time histories, obtained, for instance, from real life experiments. In this section, only the Gumbel and the point process methods are applied. For simplicity, only the case of positive extreme values will be discussed. The necessary modifications to deal with other situations are usually obvious.

9.6.1 The Gumbel Method

It was pointed out in Chap. 5 that for response processes relevant for many engineering structures, the appropriate extreme value distribution would almost always be the Gumbel distribution. Therefore, let us assume that this is indeed the case for the response process $X(t)$, which can be simulated by a suitable procedure. Now, assume that N independent response time histories, each of duration T, have been simulated for a given environmental condition. For the Gumbel method, the extreme response is then identified for each time series or block of data. These extreme value data are assumed to be Gumbel distributed, and plotting the obtained data set of extreme values using a Gumbel probability plot should then ideally result in a straight line. In practice, one cannot expect this to happen, but on the premise that the data follow a Gumbel distribution, a straight line can be fitted to the data. Due to its simplicity, a popular method for fitting this straight line is the method of moments, cf. Sect. 2.6. That is, writing the Gumbel distribution of the extreme value $M(T)$ as,

$$\text{Prob}(M(T) \leq \xi) = \exp\left\{- \exp\left(- a(\xi - b)\right)\right\}, \tag{9.34}$$

it was shown that the parameters $a > 0$ and b are related to the mean value m_M and standard deviation σ_M of $M(T)$ as follows: $b = m_M - 0.5772\, a^{-1}$ and $a = 1.2826/\sigma_M$ (Bury 1975). The estimates of m_M and σ_M obtained from the available sample therefore provide estimates of a and b, which leads to the fitted Gumbel distribution by the method of moments.

Typically, a specified fractile value of the fitted Gumbel distribution is then extracted and used in a design consideration. To be specific, let us assume that the requested fractile value is the $100(1-\alpha)\%$ fractile, where α is usually a small number, for example $\alpha = 0.1$. To quantify the uncertainty associated with the obtained $100(1-\alpha)\%$ fractile value based on a sample of size N, the 95% confidence interval of this value is often used. A good estimate of this confidence interval can be obtained by using a parametric bootstrapping method (Efron and Tibshirani 1993; Davison and Hinkley 1997), cf. Sect. 2.8. In our context, this simply means that the initial sample of N extreme values is assumed to have been generated from an underlying Gumbel distribution, whose parameters are, of course, unknown. If this Gumbel distribution had been known, it could have been used to generate a large number of (independent) samples of size N. For each sample, a new Gumbel distribution would be fitted and the corresponding $100(1-\alpha)\%$ fractile value identified. If the number of samples had been large enough, an accurate estimate of the 95% confidence interval on the $100(1-\alpha)\%$ fractile value based on a sample of size N could be found. Because the true parameter values of the underlying Gumbel distribution are unknown, they are replaced by the estimated values obtained from the initial sample. This fitted Gumbel distribution is then used as previously described to provide an approximate 95% confidence interval. Note that the assumption that the initial N extreme values are actually generated with good approximation from a Gumbel distribution cannot be easily verified, which is a drawback of this method. As has been pointed out, compared with the POT method, the Gumbel method would also seem to use much less of the information available in the data. This may explain why the POT method has become increasingly popular over the past years, but the Gumbel method is still widely used in practice.

9.6.2 The Point Process Method

It is known from Eq. (4.31) that an approximation of the distribution of the extreme value $M(T)$ is obtained from the formula,

$$\text{Prob}(M(T) \le \xi) = \exp\{-\nu^+(\xi)\,T\}, \tag{9.35}$$

where $\nu^+(\xi)$ denotes the mean upcrossing rate of a stationary process $X(t)$.

The method to be discussed here, relies on this particular approximation. This implies that the mean upcrossing rate needs to be estimated from the simulated time series. Assuming the so-called ergodic mean value property, it is obtained that

$$\nu^+(\xi) = \lim_{T\to\infty} \frac{1}{T}\, n^+(\xi;T)\,, \tag{9.36}$$

where $n^+(\xi; T)$ denotes a realization of $N^+(\xi; T)$; that is, $n^+(\xi; T)$ denotes the counted number of upcrossings during time T from a particular simulated time history. In practice, k time histories of a suitable length T_0, say, are provided by simulation. The appropriate ergodic mean value estimate of $\nu^+(\xi)$ is then,

$$\hat{\nu}^+(\xi) = \frac{1}{k\,T_0} \sum_{j=1}^{k} n_j^+(\xi; T_0)\,, \tag{9.37}$$

where $n_j^+(\xi; T_0)$ denotes the counted number of upcrossings of the level ξ from time history no. j. This will often be the chosen approach to the estimation of the mean upcrossing rate.

For a suitable number k, e.g., $k \geq 20 - 30$, and provided that T_0 is sufficiently large, a fair approximation of the 95% confidence interval ($\mathrm{CI}_{0.95}$) for the value $\nu^+(\xi)$ can be obtained as,

$$\mathrm{CI}_{0.95}(\xi) = \Big(\hat{\nu}^+(\xi) - 1.96\,\frac{\hat{s}(\xi)}{\sqrt{k}}\,,\ \hat{\nu}^+(\xi) + 1.96\,\frac{\hat{s}(\xi)}{\sqrt{k}}\Big)\,, \tag{9.38}$$

where the empirical standard deviation $\hat{s}(\xi)$ is given as,

$$\hat{s}(\xi)^2 = \frac{1}{k-1} \sum_{j=1}^{k} \Big(\frac{n_j^+(\xi; T_0)}{T_0} - \hat{\nu}^+(\xi)\Big)^2. \tag{9.39}$$

Note that k and T_0 may not necessarily be the number and length of the actually simulated response time series. Rather, they were chosen to optimize the estimate of Eq. (9.39). If, initially, $\tilde{k}$ time series of length $\tilde{T}$ are simulated, then $k = \tilde{k}\,k_0$ and $\tilde{T} = k_0\,T_0$. That is, each initial time series of length $\tilde{T}$ was divided into k_0 time series of length T_0, assuming, of course, that $\tilde{T}$ is large enough to allow for this in an acceptable way. The consistency of the estimates obtained by Eq. (9.39) can be checked by the observation that $\mathrm{Var}[N^+(\xi; t)] = \nu^+(\xi)\,t$ when $N^+(\xi; t)$ becomes a Poisson random variable, which by assumption occurs for large values of ξ. This leads to the equation,

$$s(\xi)^2 = \frac{1}{k}\,\mathrm{Var}\left[\sum_{j=1}^{k} \frac{N_j^+(\xi; T_0)}{T_0}\right] = \frac{\nu^+(\xi)}{T_0}, \tag{9.40}$$

where $\{N_1^+(\xi; T_0), \ldots, N_k^+(\xi; T_0)\}$ denotes a random sample with a possible outcome $\{n_1^+(\xi; T_0), \ldots, n_k^+(\xi; T_0)\}$. Hence, $\hat{s}(\xi)^2/k \approx \hat{\nu}^+(\xi)/(k\,T_0)$. Because this last relation is consistent with the adopted assumptions (for large ξ), it could have been used as the empirical estimate of the variance in the first place. It is also insensitive to the blocking of data discussed previously because $k\,T_0 = \tilde{k}\tilde{T}$. However, the accuracy of this approach may be poor for small to moderate values of

ξ, where the Poisson assumption about the upcrossing events may fail. In contrast, the advantage of Eq. (9.39) is that it does not rely on any specific assumptions about the statistical distributions involved.

The idea underlying the development of the approach described here is based on the observation that for dynamic models relevant for most engineering structures, the mean ξ-upcrossing rate as a function of the level ξ is highly regular in a particular way (Naess and Gaidai 2008). As is shown later, the mean upcrossing rate tail, say, for $\xi \geq \xi_0$, behaves similarly to $\exp\{-a(\xi - b)^c\}$ ($\xi \geq \xi_0$), where $a > 0$, $b \leq \xi_0$, and $c > 0$ are suitable constants. Hence, as discussed in detail by Naess and Gaidai (2008), it may be assumed that,

$$\nu^+(\xi) \approx q(\xi)\, \exp\{-a(\xi - b)^c\}\,,\ \xi \geq \xi_0, \tag{9.41}$$

where the function $q(\xi)$ is slowly varying compared with the exponential function $\exp\{-a(\xi - b)^c\}$. Equation (9.41) can be rewritten as,

$$\ln\big|\ln\big(\nu^+(\xi)/q(\xi)\big)\big| \approx c\, \ln(\xi - b) + \ln a\,,\ \xi \geq \xi_0\,. \tag{9.42}$$

It follows that by plotting $\ln\big|\ln\big(\nu^+(\xi)/q(\xi)\big)\big|$ versus $\ln(\xi - b)$, it is expected that an almost perfectly linear tail behavior will be obtained. Now, as it turns out, the function $q(\xi)$ can be largely considered as a constant q, say, for tail values of ξ. This suggests using a method for identifying the parameters q and b by optimizing the linear fit in the tail. When this is achieved, the corresponding values of a and c can then be extracted from the plot. This is discussed at some length in Naess and Gaidai (2008). A plot of $\ln\big|\ln\big(\nu^+(\xi)/q\big)\big|$ versus $\ln(\xi - b)$ for optimal parameters b and q will be referred to as an optimal transformed plot. Examples of such plots will be given for some of the examples to follow, mostly for the purpose of demonstrating the validity of assumptions. An alternative, more extensive method for optimizing the fit to the data is described in Chap. 5.

In engineering applications, it is quite common to assume that the observed extreme value response data do follow a Gumbel distribution, cf. Sect. 9.6.1. The problem with this approach is that classical extreme value theory cannot be used to decide to what extent the asymptotic distribution is actually valid for a given set of extreme value data. Note that the asymptotic Gumbel distribution given by Eq. (9.34) corresponds to an asymptotic upcrossing rate that is purely exponential, that is, with $c = 1$ in Eq. (9.41). Hence, by adopting a much more general class of functions, with the purely exponential functions as a subclass, to represent the upcrossing rate, as done here, the ability to capture subasymptotic behavior is greatly enhanced. By this, the necessity to adopt a strictly asymptotic extreme value distribution of questionable validity is avoided. However, note that for any $c > 0$, the corresponding extreme value model will be asymptotically Gumbel.

Note also that the so-called Weibull method for extreme value prediction, which is based on the assumption that the local peak values follow a three parameter Weibull distribution is, in fact, basically a prejudiced version of the point process method in the sense that the parameter q is a priori given the value 1.

In cases where the approximation implied by Eq. (9.35) may be questioned, the ACER method should be applied.

9.6.3 A Comparison of Methods

In this subsection, the performance of the point process method is compared with that of the Gumbel method. This is done for the particular case of the horizontal deck response of a jacket structure installed on the Kvitebjørn field in the North Sea. For this kind of response process, the point process approach is very accurate because a plot of the ACER functions shows that beyond the second ACER function, which corresponds to the upcrossing rate for suitably sampled time series, there are no dependence effects that need to be accounted for.

Figure 9.4 depicts the Kvitebjørn jacket platform with the superstructure removed together with the corresponding three-dimensional computer model used for the Monte Carlo simulations, see Naess et al. (2007) for details.

For the simulations discussed here, two long-crested sea states described by a JONSWAP wave spectrum as listed in Table 9.2, were used. Twenty independent response time histories, each of 3 hours' duration, were simulated for each sea state. For the Gumbel method, the extreme horizontal deck response in the wave direction is identified for each time series. These extreme value data are then assumed to be Gumbel distributed, and plotting each data set as a Gumbel probability plot results

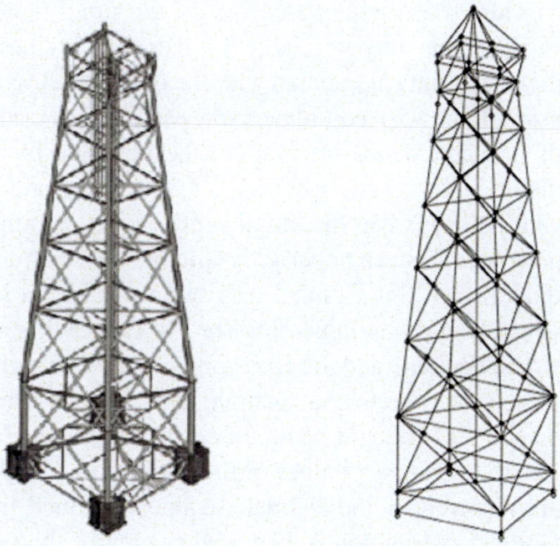

Fig. 9.4 Left: Sketch of the Kvitebjørn platform with the superstructure removed. Right: Computer model of the Kvitebjørn platform (Karunakaran et al. 2001)

Table 9.2 Representative sea states

H_s (m)	T_p (s)
12.0	12.0
14.7	16.5

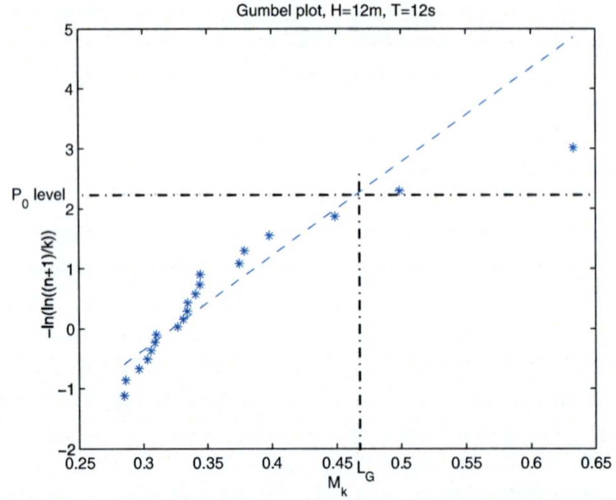

Fig. 9.5 Empirical Gumbel plot of the 20 simulated 3-hour extremes of the horizontal deck displacement for the sea state H_s = 12 m and T_p = 12 s together with the fitted Gumbel distribution (– – –)

in Figs. 9.5 and 9.6. Specifically, the observed 3-hour extremes M_k are plotted versus $-\ln\big(\ln(21/k)\big)$, for $k = 1, \ldots, 20$. The fitted straight line in each figure, which represents the fitted Gumbel distribution, is based on the moment estimation method, cf. Sect. 9.6.1.

The 90% fractile value L_G of the fitted Gumbel distribution is identified and shown in each figure. Table 9.3 lists the obtained 90% fractile values. To quantify the uncertainty associated with the obtained 90% fractile value based on a sample of size 20, the 95% confidence interval ($\mathrm{CI}_{0.95}$) of this value is used. A good estimate of this confidence interval can be obtained by using a parametric bootstrapping method (Efron and Tibshirani 1993; Davison and Hinkley 1997). In our context, this simply means that the initial sample of 20 extreme values is assumed to have been generated from an underlying Gumbel distribution, whose parameters are, of course, unknown. If this Gumbel distribution had been known, it could have been used to generate many samples of size 20. For each sample, a new Gumbel distribution would be fitted, and the corresponding 90% fractile value identified. If the number of samples had been large enough, an accurate estimate of the 95% confidence interval on the 90% fractile value based on a sample of size 20 could be found. Because the true parameter values of the underlying Gumbel distribution are unknown, they are replaced by the estimated values obtained from the initial sample. This fitted Gumbel distribution is then used as previously described to provide an approximate 95% confidence interval. Note that the assumption that the initial 20 extreme values are actually generated from a Gumbel distribution is quite accurate in this case, as discussed later.

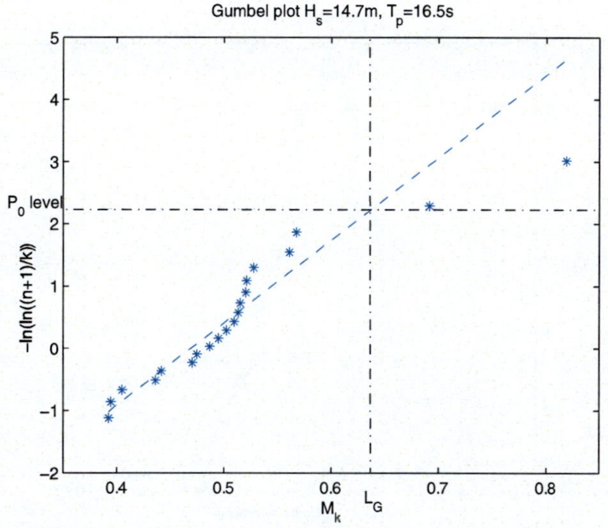

Fig. 9.6 Empirical Gumbel plot of the 20 simulated 3-hour extremes of the horizontal deck displacement for the sea state H_s = 14.7 m and T_p = 16.5 s together with the fitted Gumbel distribution (− − −)

Table 9.3 90% fractile values L_G of the fitted Gumbel distributions

H_s (m)	T_p (s)	L_G (m)	$CI_{0.95}$
12.0	12.0	0.47	(0.40, 0.54)
14.7	16.5	0.63	(0.56, 0.73)

Invoking the parametric bootstrap, the 95% confidence interval is estimated for each case based on 100,000 bootstrap samples. The obtained results are listed in Table 9.3. This way of estimating the 90% fractile value of the 3-hour extreme value distribution is referred to as the Gumbel method. The empirical densities obtained for the predicted 90% fractile values with the $CI_{0.95}$ indicated are shown in Figs. 9.7 and 9.8.

Let us now compare the results provided by the Gumbel method previously discussed, with the results obtained by the point process method. Using a Levenberg–Marquardt least squares optimization method, cf. Chap. 5, leads to the results shown in Figs. 9.9 and 9.10.

As shown in Figs. 9.11 and 9.12, when the mean upcrossing rate is plotted on a logarithmic scale, the tails are closely linear. This means that the associated extreme value distribution can be expected to be similar to a Gumbel distribution, which would correspond to exactly linear tails, cf. Eq. (4.31). Although it may not be obvious that the data plotted in Figs. 9.5 and 9.6 come from a distribution very close to a Gumbel distribution, the approximately linear exponential decay of the crossing rate strongly supports this assumption. This is yet another indication of the usefulness of the mean upcrossing rate function.

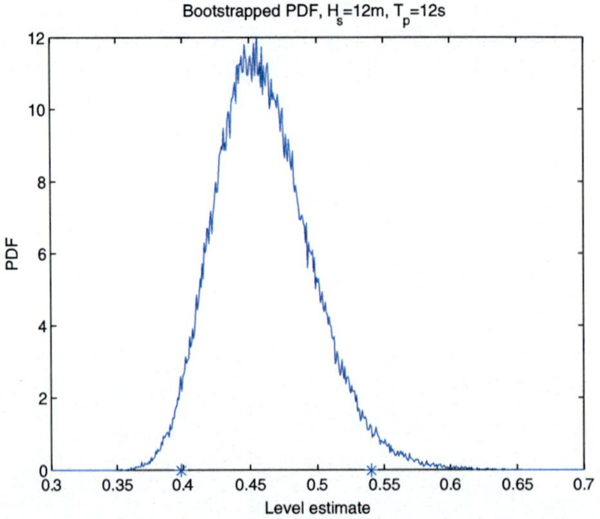

Fig. 9.7 Empirical density of the predicted 90% fractile value based on sample of size 20 for the sea state with $H_s = 12\,\mathrm{m}$, $T_p = 12\,\mathrm{s}$. The $*$ indicates the limits of $\mathrm{CI}_{0.95}$

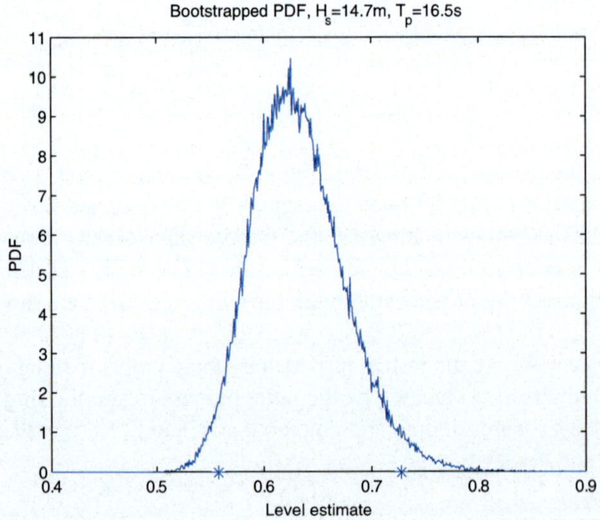

Fig. 9.8 Empirical density of the predicted 90% fractile value based on sample of size 20 for the sea state with $H_s = 14.7\,\mathrm{m}$, $T_p = 16.5\,\mathrm{s}$. The $*$ indicates the limits of $\mathrm{CI}_{0.95}$

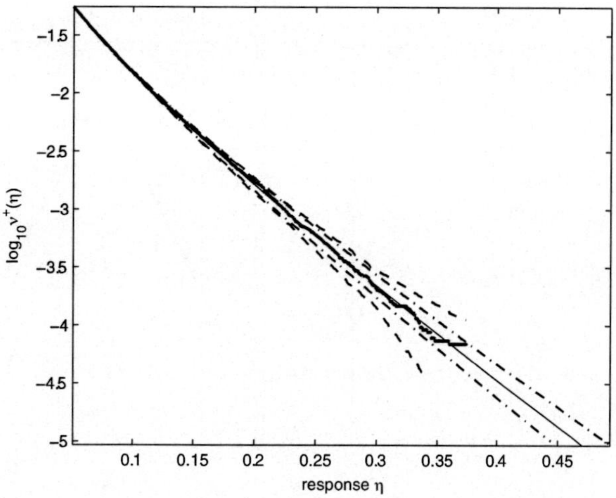

Fig. 9.9 Log plot of the empirical and fitted upcrossing rate with the reanchored 95% empirical confidence band (− −) and fitted confidence band (− ·) for the sea state with $H_s = 12$ m, $T_p = 12$ s. 90% fractile estimate = 0.47 m, with $\mathrm{CI}_{0.95}$ = (0.45, 0.50)

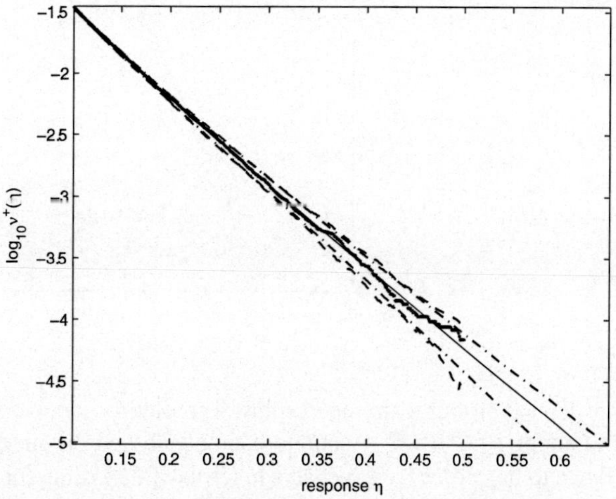

Fig. 9.10 Log plot of the empirical and fitted upcrossing rate with the reanchored 95% empirical confidence band (− −) and fitted confidence band (− ·) for the sea state with $H_s = 14.7$ m, $T_p = 16.5$ s. 90% fractile estimate = 0.62 m, with 95% $\mathrm{CI}_{0.95}$ = (0.58, 0.65)

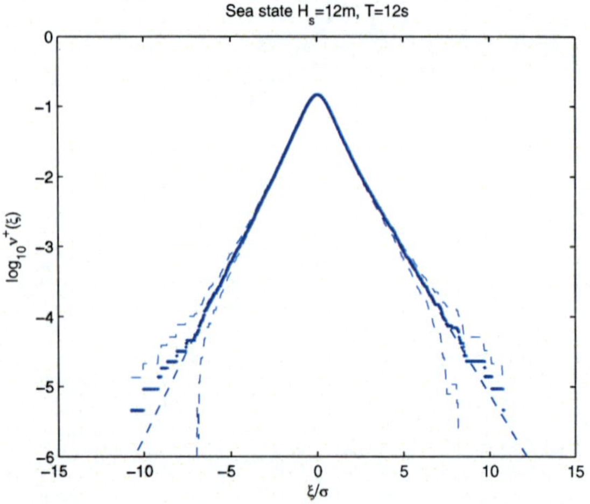

Fig. 9.11 Mean upcrossing rate statistics along with 95% confidence bands (−−) for the sea state with $H_s = 12$ m, $T_p = 12$ s, and response standard deviation $\sigma = 0.047$ m. $*$: Monte Carlo; − − − : linear fit

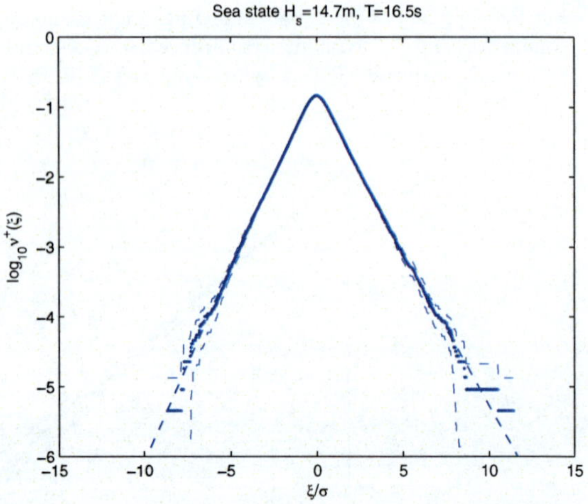

Fig. 9.12 Mean upcrossing rate statistics along with 95% confidence bands (−−) for the sea state with $H_s = 14.7$ m, $T_p = 16.5$ s, and response standard deviation $\sigma = 0.068$ m. $*$: Monte Carlo; − − − : linear fit

Aiming at $T = $ 3-hour extreme response prediction, one needs upcrossing rates down to about 10^{-4}–10^{-6}. Accurate estimates based on direct Monte Carlo simulation down to this order are expensive in terms of CPU time for the considered structure. It is therefore convenient when accurately estimated upcrossing rates down to about 10^{-3} can be used as a basis for extrapolation down to the appropriate response level ξ (with $\nu^+(\xi) \approx 10^{-5}$), as illustrated in Figs. 9.9 and 9.10.

Returning now to the specific prediction of the 90% fractile of the 3-hour extreme value distribution, L_{CR}, Figs. 9.9 and 9.10 lead to the estimates listed in Table 9.4. The estimated 95% confidence intervals are given in Table 9.4, and indicated in the figures. They are significantly smaller than those obtained by the Gumbel method. The prediction accuracy is thus significantly higher for the proposed method. However, it is also observed that there is good agreement between the L_{CR}-

Table 9.4 90% fractile values L_{CR} of the fitted Gumbel distributions

H_s (m)	T_p (s)	L_{CR} (m)	$CI_{0.95}$
12.0	12.0	0.47	(0.45, 0.50)
14.7	16.5	0.62	(0.58, 0.65)

values and the L_G-values, which is to be expected because the exact extreme value distribution is very close to a Gumbel distribution.

9.6.4 Combination of Multiple Stochastic Load Effects

A prominent problem in the design of structures subjected to random loads is to find methods for the combination of resulting load effects at high and extreme response levels. In codified design, this is usually implemented as linear combination rules of specified characteristic values of the individual load effects (Madsen et al. 1986; Melchers 1999). For nonlinear dynamic structures, the precision level of such procedures would seem highly questionable. One of the reasons for adopting such simplified procedures is the complexity of the task to accurately predict the extreme value statistics of the combined load effects, even in the case of linear combinations. Over the years, several simplified procedures have been suggested for the linear combination of load effects, most notably the Ferry Borges-Castanheta method (Ditlevsen and Madsen 1996), Turkstra's rule (Turkstra 1970; Madsen et al. 1986), the load coincidence method (Wen 1990; Melchers 1999), the SRSS method (Wen and Pearce 1981; Wen 1990), and the point crossing approximation method (Larrabee and Cornell 1981; Madsen et al. 1986). A main shortcoming of these combination procedures is that they apply only to the case of independent load effect components. A method for lifting this restriction from the point process approximation is proposed by Toro (1984). An effort to extend Turkstra's rule to dependent processes is described by Naess and Royset (2000). In this section, the use of the point process method for stochastic load effect combination problems is illustrated. The illustrations in this section are from Naess and Gaidai (2009).

The general formulation of the load effect combination problem to be studied here is the following:

$$H(t) = h[X_1(t), \ldots, X_N(t)], \tag{9.43}$$

where the stochastic load effect component processes $X_1(t), \ldots, X_N(t)$ are combined according to a specified deterministic function h to produce the load effect combination process $H(t)$. The component processes may derive from a vector solution process of a dynamic model for the structural response of an offshore platform to random waves. They may often be modeled as stationary stochastic processes, but that is not a requirement for the application of the point process method.

The typical problem to answer concerning the load effect combination process $H(t)$ is to determine the probability of exceeding a critical threshold h_c during a specified time interval T. Let us call this the failure probability and denote it by $p_f = p_f(T)$. Defining $M(T) = \max\{H(t) : 0 \leq t \leq T\}$, it is realized that the goal is to calculate,

$$p_f = 1 - \text{Prob}(M(T) \leq h_f)\,. \tag{9.44}$$

In many practical applications, the structure of the process $H(t)$ is quite involved and the dimension N can be high. This makes a direct analytical approach virtually impossible. In such cases, Monte Carlo simulations of some sort would seem to be the most attractive way to provide estimates of the failure probability.

Two different load effect combination examples are used for illustration purposes: von Mises yielding stress and linear combination of non-Gaussian load effects. In both cases, the load components are correlated stochastic processes. A Newmark integration method was used to produce accurate response time series, cf. Naess and Moan (2013).

The load effect components $X_i(t)$ are modeled as stationary processes, being the response of Duffing-type systems to the same stationary Gaussian white noise excitation $W(t)$ with $\mathbb{E}[W(t)W(t+\tau)] = \delta(\tau)$, where $\delta(\cdot)$ denotes the Dirac delta function. That is,

$$\ddot{X}_i + 2\zeta_i\omega_i\dot{X}_i + \omega_i^2(X_i + \varepsilon X_i^3) = W(t)/m_i\,, \tag{9.45}$$

with specific damping constants ζ_i and resonance frequencies $\omega_i = 2\pi/T_i$, and m_i represent masses, $i = 1, \ldots, N$. In this subsection N is taken to be 3.

First, the linear case, i.e., $\varepsilon = 0$ in Eq. (9.45), is considered. In this case, the (two-sided) PSD of the process $X_i(t)$ is given as $S_i(\omega) = |A_i(\omega)|^2$, where,

$$A_i(\omega) = \frac{1}{\sqrt{2\pi}\, m_i(-\omega^2 + 2\mathrm{i}\zeta_i\omega_i\omega + \omega_i^2)}\,, \tag{9.46}$$

with $i = 1, 2, 3$. The correlation coefficient $\rho_{ij} = E[X_i(t)X_j(t)]/\sigma_i\sigma_j$ between $X_i(t)$ and $X_j(t)$ is given by,

$$\rho_{ij} = \int_{-\infty}^{\infty} A_i(\omega)\, A_j^*(\omega)\, d\omega/\sigma_i\sigma_j\,. \tag{9.47}$$

Here, $\sigma_i^2 = \text{Var}[X_i(t)] = \int_{-\infty}^{\infty} S_i(\omega)d\omega$ is the variance of $X_i(t)$, $i = 1, 2, 3$.

von Mises Stress Combination

Let $\sigma_x, \sigma_y, \sigma_z$ be axial stresses and $\tau_{xy}, \tau_{xz}, \tau_{yz}$ be shear stresses in a structural element. According to the von Mises yield criterion (Madsen et al. 1986), yielding occurs if,

$$(\sigma_x - \sigma_y)^2 + (\sigma_x - \sigma_z)^2 + (\sigma_y - \sigma_z)^2 + 6(\tau_{xy}^2 + \tau_{xz}^2 + \tau_{yz}^2) \geq 2\sigma_Y^2 , \tag{9.48}$$

where σ_Y is the yield stress. In many cases, in practice, several stress components are zero. In the analysis carried out here, only xy-plane stresses are encountered, meaning that $\sigma_z = \tau_{xz} = \tau_{yz} = 0$. According to the von Mises criterion, yielding then occurs if,

$$(\sigma_x - \sigma_y)^2 + \sigma_x^2 + \sigma_y^2 + 6\tau_{xy}^2 \geq 2\sigma_Y^2 . \tag{9.49}$$

The load effect vector in three-dimensional space is introduced as,

$$(X_1(t), X_2(t), X_3(t)) \propto (\sigma_x(t), \sigma_y(t) , \tau_{xy}(t)) , \tag{9.50}$$

where the components $X_k(t)$ are determined by Eq. (9.45).

The von Mises criterion (Eq. (9.49)) states that yielding (failure) occurs when,

$$H_{vM}(t) = \left((X_1 - X_2)^2 + X_1^2 + X_2^2 + 6X_3^2\right)^{1/2} \geq h_f , \tag{9.51}$$

where $h_f \propto \sqrt{2}\sigma_Y$.

In the linear case, ($\varepsilon = 0$ in Eq. (9.45)), the process $H_{vM}(t)^2$ becomes a quadratic expression in correlated stationary Gaussian processes. Hence, the saddle point method described by Naess and Karlsen (2004) is applicable. Because this method is very accurate, it provides an opportunity to check the accuracy and efficiency of the point process method. Table 9.5 lists the damping values and resonance periods used in Eq. (9.45) for the load effect processes $X_i(t)$.

For the linear case ($\varepsilon = 0$), it is straightforward to calculate the standard deviations and correlation coefficients for the X_k, cf. Eq. (9.47): $\sigma_1 = 0.38$, $\sigma_2 = 0.45$, $\sigma_3 = 0.52$, $\rho_{12} = 0.36$, $\rho_{13} = 0.14$, $\rho_{23} = 0.41$. There is significant correlation between some of the load effect components. For each example, 20 time series of length approximately 0.92 hours each were simulated. The total computation time for each example was less than a minute, including simulation time and optimization.

Table 9.5 Model parameters

i	ζ_i	T_i (s)	m
1	0.04	1.8	1
2	0.04	2.0	1
3	0.04	2.2	1

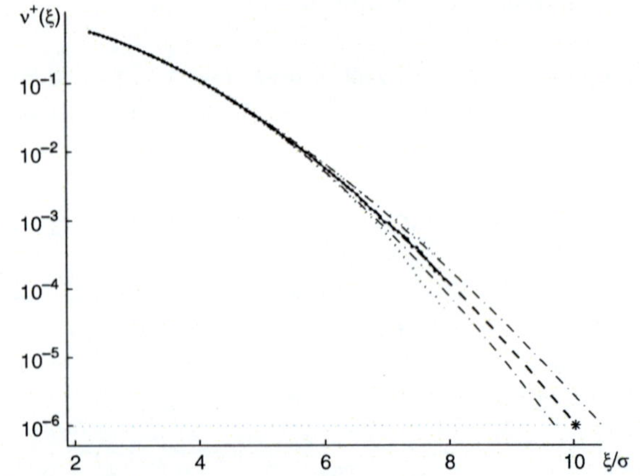

Fig. 9.13 Log plot of the mean upcrossing rate of $H_{vM}(t)$: Monte Carlo (•), reanchored empirical 95% confidence band (- -), fitted curve (– –), fitted confidence band (– ·). Linear case: $\varepsilon = 0$. $a = 0.795$, $b = 0.996$, $c = 1.607$, $\ln q = -0.249$, $\sigma = 0.682$

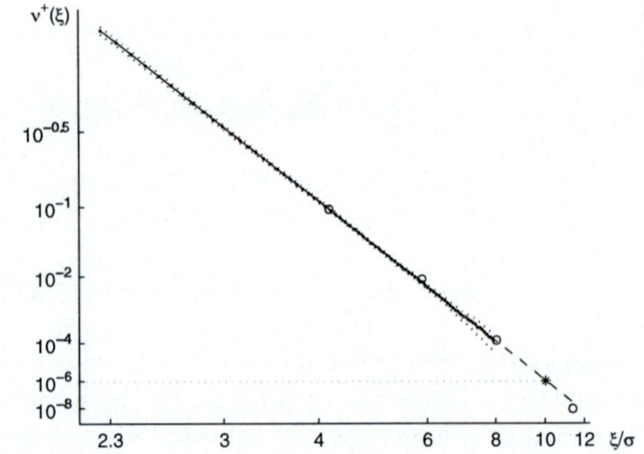

Fig. 9.14 Optimal transformed plot of the mean upcrossing rate of $H_{vM}(t)$: Monte Carlo (•), empirical confidence band (- -), saddle point results (o), linear extrapolation (– –). Linear case: $\varepsilon = 0$. $b = 0.996$, $\ln q = -0.249$, $\sigma = 0.682$

The log plot presented in Fig. 9.13 shows the optimized fitted parametric curve for the case of linear dynamics, i.e., $\varepsilon = 0$, together with the confidence band generated by the fitted parametric curves to the borders of the reanchored empirical 95% confidence band. For illustration purposes, the predicted value given by $\nu^+(\xi) = 10^{-6}$, which corresponds to the 99% fractile value of a 3-hour extreme value distribution is indicated. Predicted value = 6.84 $=$ 10.03 σ, with 95% CI = (6.60, 7.12).

In Fig. 9.14, it is demonstrated that when the fitted parametric curve shown in Fig. 9.13 is replotted as an optimal transformed plot, which yields a straight line, the empirical curve is also largely indistinguishable from a straight line, verifying the validity of our assumption about the representation of the upcrossing rate function. The results from saddle point calculations, which are practically exact, are also plotted in Fig. 9.14, and they are seen to agree very well with the extrapolated straight line results.

Figure 9.15 shows the optimized fitted parametric curve for the case of nonlinear dynamics, i.e., $\varepsilon = 1$, together with the confidence band as generated previously.

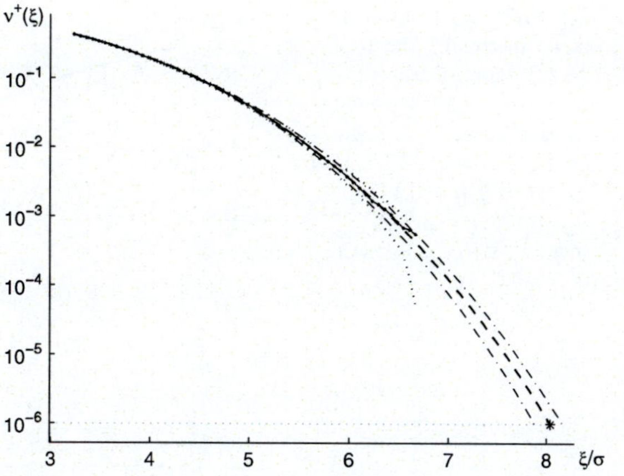

Fig. 9.15 Log plot of the mean upcrossing rate of $H_{vM}(t)$: Monte Carlo (•), reanchored empirical 95% confidence band (- -), fitted curve (– –), fitted confidence band (– ·). Nonlinear case: $\varepsilon = 1.0$. $a = 0.082$, $b = -0.382$, $c = 3.334$, $\ln q = 0.124$, $\sigma = 0.534$

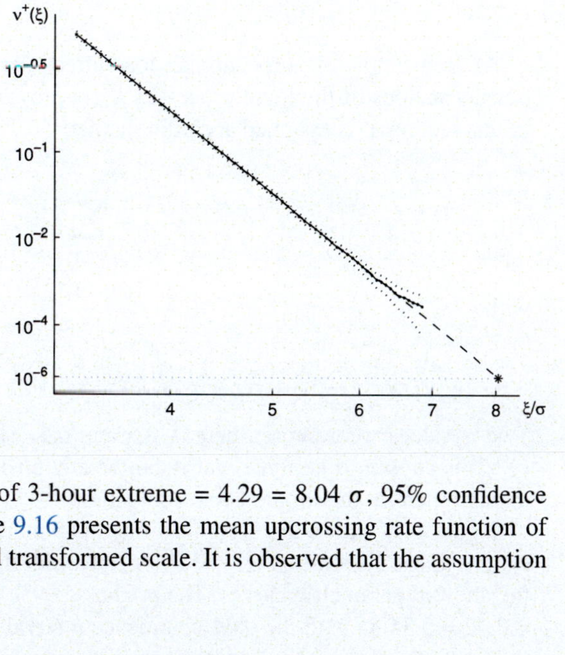

Fig. 9.16 Optimal transformed plot of the mean upcrossing rate of $H_{vM}(t)$: Monte Carlo (•), empirical confidence band (- -), linear extrapolation (–). Nonlinear case: $\varepsilon = 1.0$. $b = -0.382$, $\ln q = 0.124$, $\sigma = 0.534$

Predicted 99% fractile value of 3-hour extreme = 4.29 = 8.04 σ, 95% confidence interval = (4.21, 4.36). Figure 9.16 presents the mean upcrossing rate function of $H_{vM}(t)$ plotted on the optimal transformed scale. It is observed that the assumption is again fully verified.

Linear Combination of Non-Gaussian Load Effects

To get a flexible model that also provides a convenient way of investigating the effect of statistical dependence between load components, an example from Naess and Royset (2000) is used. The input load components are again assumed to be given by Eq. (9.45), with the same parameters as in the previous case study, cf. Table 9.5.

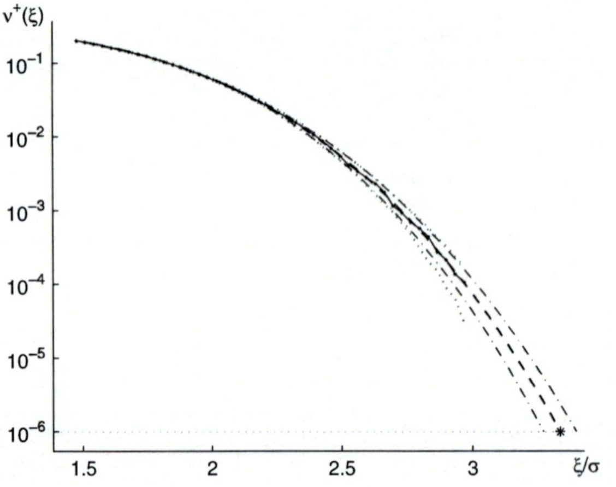

Fig. 9.17 Log plot of the mean upcrossing rate of $H_{lc}(t)$: Monte Carlo (•), empirical confidence band (- -), fitted curve (– –), fitted confidence band (– ·). Linear case: $\varepsilon = 0$. $a = 0.147$, $b = 0.129$, $c = 3.742$, $\ln q = -1.103$, $\sigma = 1.026$, $\alpha = -0.5$

As an example of non-Gaussian load effect component processes, memoryless transformations of the input processes $X_i(t)$ provided by Eq. (9.45) is considered. In this section, it is assumed accordingly that,

$$H_{lc}(t) = \sum_{i=1}^{3} Z_i(t), \tag{9.52}$$

with

$$Z_i(t) = X_i(t)|X_i(t)|^{\alpha}, \quad -1 < \alpha < 1. \tag{9.53}$$

Two α-values were chosen here, -0.5 and 0.5. The number of terms in the sum (9.52) is chosen to be three, but it can be any positive integer because it does not matter much for the Monte Carlo simulation.

Figure 9.17 shows the optimized fitted parametric curve for $\alpha = -0.5$ and the case of linear dynamics, i.e., $\varepsilon = 0$, together with the confidence band generated by the allowed parametric curves. The predicted 99% fractile value of 3 hour extreme = 3.42 = 3.34 σ, with the 95% confidence interval = (3.36, 3.49). In Fig. 9.18 it is demonstrated that when the fitted parametric curve shown in Fig. 9.17 is replotted as an optimal transformed plot, the empirical curve is largely indistinguishable from a straight line, supporting the assumption about the representation of the upcrossing rate function.

Figure 9.19 shows the optimized fitted parametric curve for $\alpha = 0.5$ and the case of linear dynamics, together with the confidence band generated by the allowed parametric curves. The predicted 99% fractile value of 3 hour extreme = 5.15 = 7.69 σ, with the 95% confidence interval = (4.95, 5.37). In Fig. 9.20 it is demonstrated that when the fitted parametric curve shown in Fig. 9.19 is replotted as an optimal transformed plot, the empirical curve is largely indistinguishable from a straight

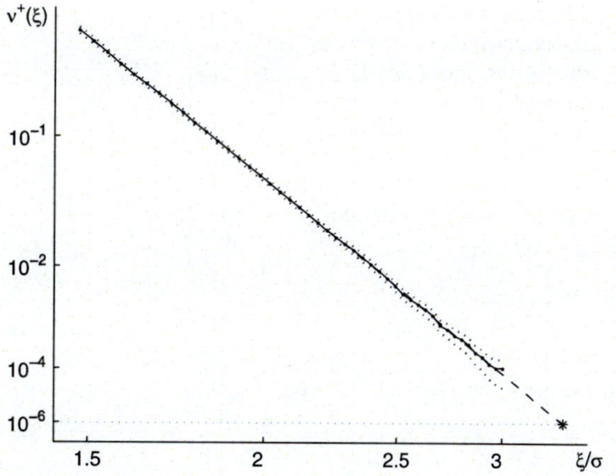

Fig. 9.18 Optimal transformed plot of the mean upcrossing rate of $H_{lc}(t)$: Monte Carlo (•), confidence band (- -), linear extrapolation (–). Linear case: $\varepsilon = 0$. $b = 0.129$, $\ln q = -1.103$, $\sigma = 1.026$, $\alpha = -0.5$

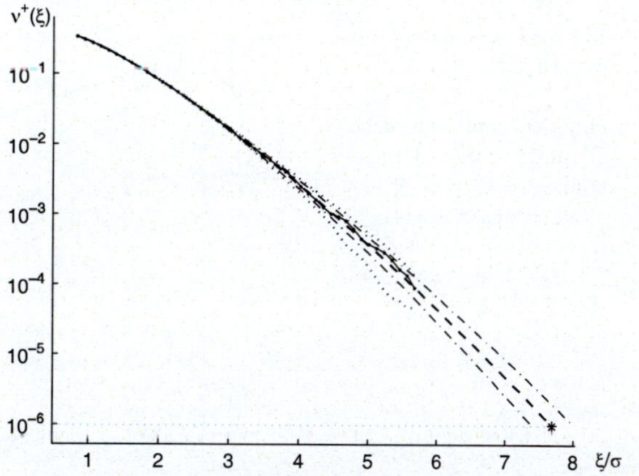

Fig. 9.19 Log plot of the mean upcrossing rate of $H_{lc}(t)$: Monte Carlo (•), empirical confidence band (- -), fitted curve (– –), fitted confidence band (– ·). Linear case: $\varepsilon = 0$. $a = 1.771$, $b = 0.449$, $c = 1.280$, $\ln q = -0.980$, $\sigma = 0.670$, $\alpha = 0.5$

line, again supporting the assumption about the representation of the upcrossing rate function.

Figures 9.21 and 9.22 show the log plot and the optimal transformed plots for the case $\alpha = -0.5$, under nonlinear dynamics with $\varepsilon = 1$. The predicted 99% fractile value of the 3-hour extreme = 3.04 = 3.18 σ, with the 95% confidence interval = (3.00, 3.09). By way of comment to the rather high optimal value for c that was found in this particular case, it may be of interest to observe that the function $F(q^*(c), a^*(c), b^*(c), c)$ is almost constant for a range of values from about $c = 4$ to 7, where the calculations stopped. From the transformed optimal plot, the assumption is also still verified for this case.

Figures 9.23 and 9.24 show the log plot and the optimal transformed plots for the case $\alpha = 0.5$, under nonlinear dynamics with $\varepsilon = 1$. The predicted 99% fractile value of 3-hour extreme = 3.06 = 6.147 σ, with the 95% confidence interval = (2.97, 3.12). From the transformed optimal plot, it is again seen that the assumption is fully verified also for this case.

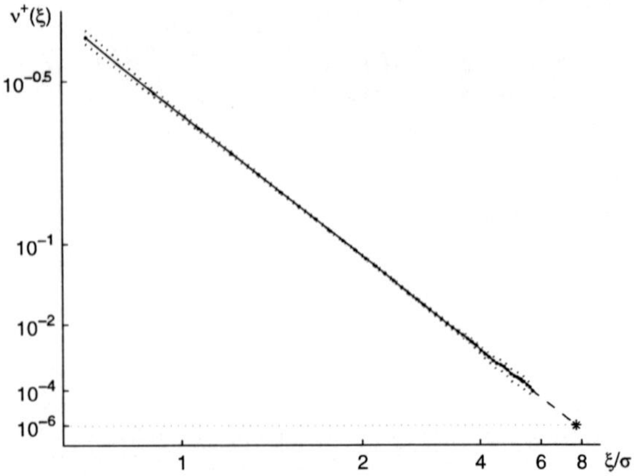

Fig. 9.20 Optimal transformed plot of the mean upcrossing rate of $H_{lc}(t)$: Monte Carlo (•), confidence band (- -), linear extrapolation (–). Linear case: $\varepsilon = 0$. $b = 0.449$, $c = 1.280$, $\ln q = -0.980$, $\sigma = 0.670$, $\alpha = 0.5$

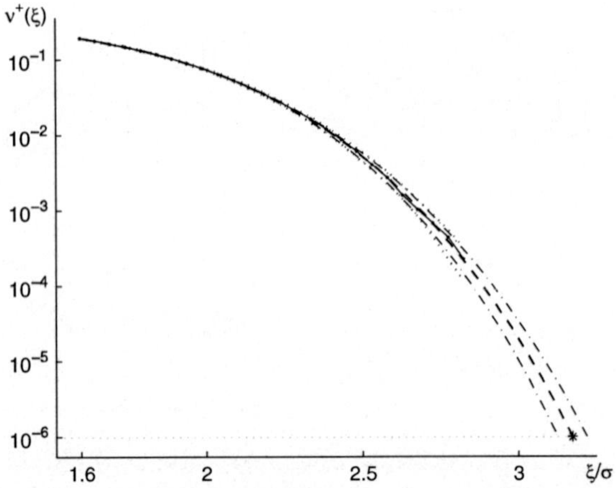

Fig. 9.21 Log plot of the mean upcrossing rate of $H_{lc}(t)$: Monte Carlo (•), empirical confidence band (- -), fitted curve (– –), fitted confidence band (– ·). Nonlinear case: $\varepsilon = 1$. $a = 0.014$, $b = -0.456$, $c = 5.423$, $\ln q = -1.048$, $\sigma = 0.958$, $\alpha = -0.5$

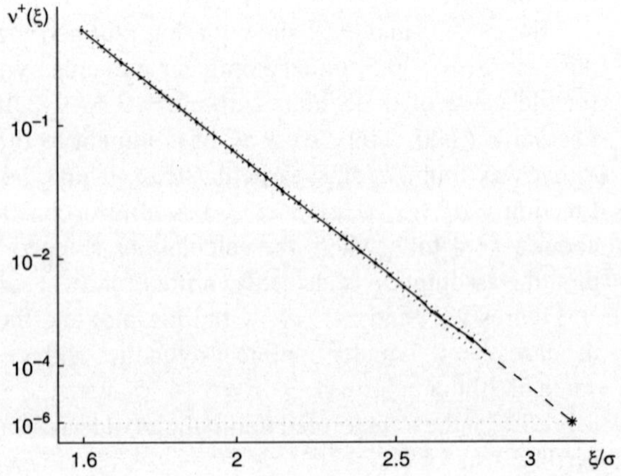

Fig. 9.22 Optimal transformed plot of the mean upcrossing rate of $H_{lc}(t)$: Monte Carlo (•), confidence band (- -), linear extrapolation (–). Nonlinear case: $\varepsilon = 1$. $b = -0.456$, $\ln q = -1.048$, $\sigma = 0.958$, $\alpha = -0.5$

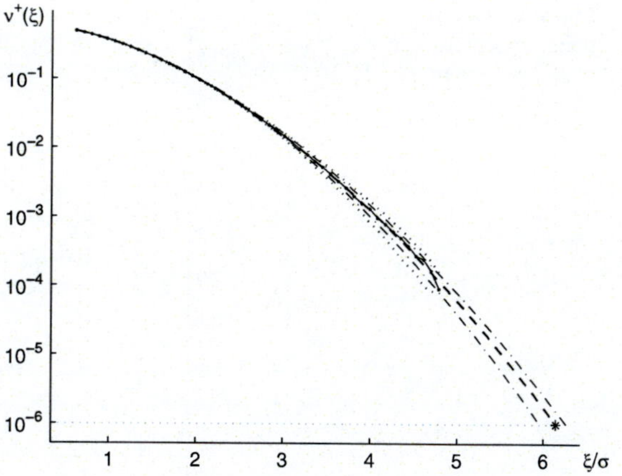

Fig. 9.23 Log plot of the mean upcrossing rate of $H_{lc}(t)$: Monte Carlo (•); empirical confidence band (- -); fitted curve (– –); fitted confidence band (– ·). Nonlinear case: $\varepsilon = 1$. $a = 1.872$, $b = 0.030$, $c = 1.767$, $\ln q = -0.527$. $\sigma = 0.498$, $\alpha = 0.5$

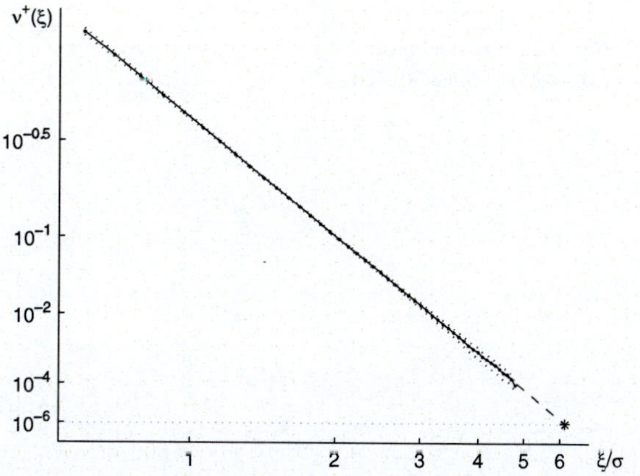

Fig. 9.24 Optimal transformed plot of the mean upcrossing rate of $H_{lc}(t)$: Monte Carlo (•), confidence band (- -), linear extrapolation (–). Nonlinear case: $\varepsilon = 1$. $b = 0.030$, $\ln q = -0.527$. $\sigma = 0.498$, $\alpha = 0.5$

20 maxima M_j, $j = 1, \ldots, 20$ are extracted, one from each realization, in order to view it on a Gumbel plot. The latter is a plot of M_j versus $-\ln(\ln \frac{20+1}{j})$, $j = 1, \ldots, 20$.

A 95% confidence interval for the response level L_{90} of the Gumbel distribution not being exceeded during time $T = 500 \max(T_1, T_2, T_3)$ (cf. Table 9.5), with probability 90% based on a sample of size 20 can be obtained by the MC technique and with parametric bootstrapping from the fitted Gumbel distribution (Davison and Hinkley 1997). One hundred thousand bootstrap samples where used to estimate the density of the 90% fractile, and from this density, the desired confidence interval was extracted. Figures 9.25 and 9.26 present the L_{90} estimates by the point process and Gumbel methods for the nonlinear system $\varepsilon = 1$, $\alpha = 0.5$. Figure 9.27 presents the parametrically bootstrapped density of L_{90} from the Gumbel distribution.

The estimate of L_{90} from the point process method is 3.70 with 95% confidence interval (3.65, 3.75). For the Gumbel method, the estimate of L_{90} is 3.55 with 95% confidence interval (3.23, 3.85).

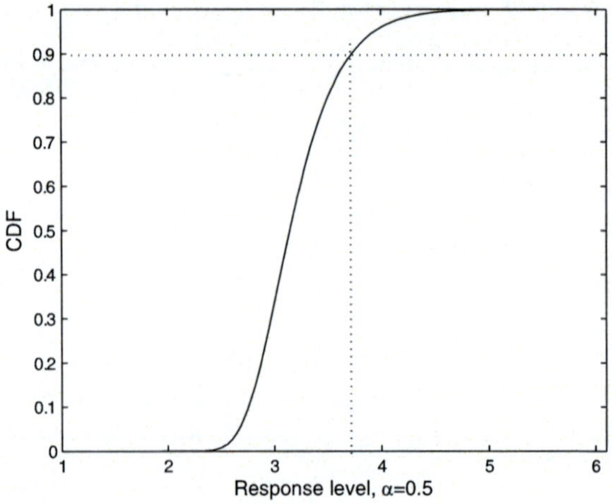

Fig. 9.25 Distribution by the point process method; nonlinear system, $\varepsilon = 1$, $\alpha = 0.5$

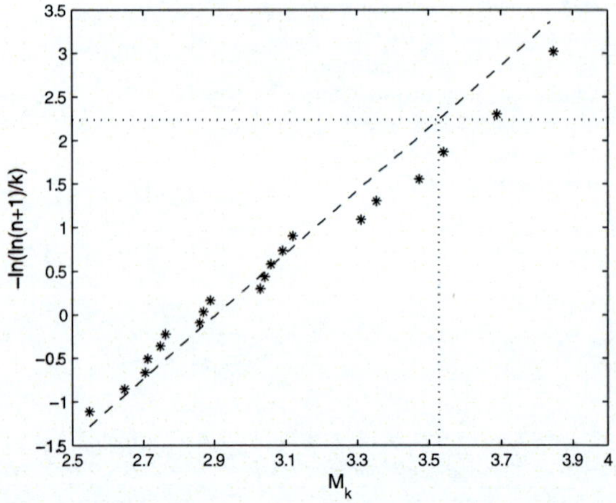

Fig. 9.26 Gumbel plot of 20 maxima; nonlinear system, $\varepsilon = 1, \alpha = 0.5$

9.6.5 *Total Surge Response of a TLP*

The next example illustrates the problem of predicting the total surge response of a tension leg platform (TLP) in random waves. The TLP concept was developed for production of oil at offshore fields. A simple rendition of a TLP structure is presented in Fig. 9.28. With the tools developed in this chapter, the problem of response prediction may be solved in a rather satisfactory manner. The presented material is largely taken from Naess et al. (2007). A Monte Carlo-based approach to the investigation of the response statistics of an offshore structure is also presented in Sagrilo et al. (2011).

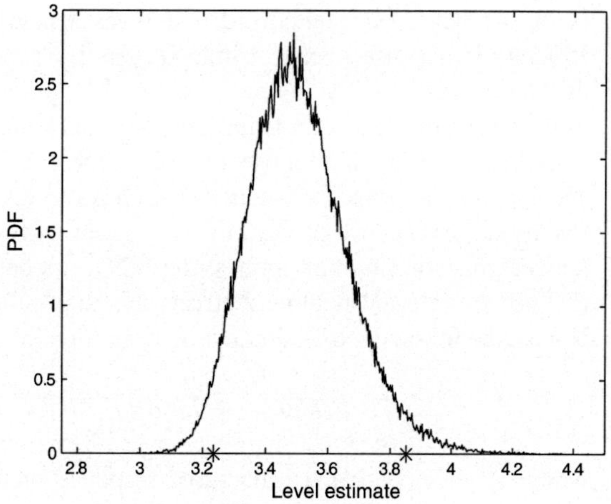

Fig. 9.27 Bootstrapped density; nonlinear system, $\varepsilon = 1, \alpha = 0.5$

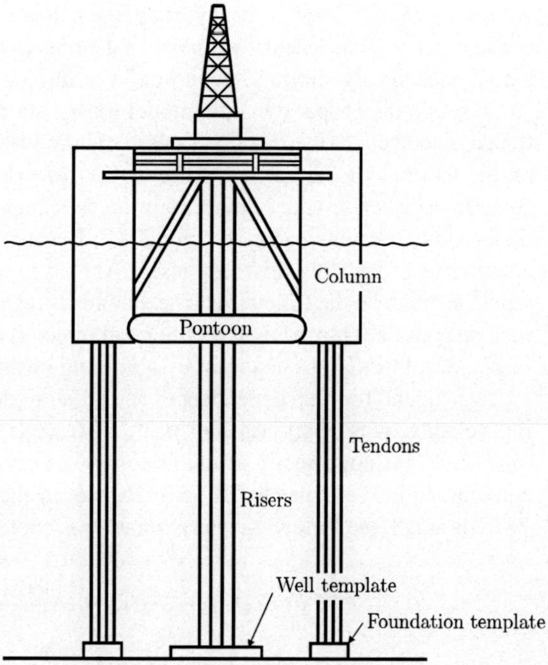

Fig. 9.28 A sketch of a TLP structure

The equations of motion for a floating, rigid-body TLP structure subjected to environmental forces such as wind, waves, and current would generally be written as,

$$\boldsymbol{M}\,\ddot{\boldsymbol{Z}}(t) + \boldsymbol{H}(\boldsymbol{Z}(t), \dot{\boldsymbol{Z}}(t), t) = \boldsymbol{F}(t)\,. \tag{9.54}$$

Here, $\boldsymbol{M}$ denotes a generalized 6×6 mass matrix, $\boldsymbol{Z} = (Z_1, \dots, Z_6)^T$ = the structure's response vector, while $\boldsymbol{H}$ denotes a nonlinear vector function. $\boldsymbol{F}(t)$ denotes a stochastic loading process, which in general also depends on the response of the structure. This point is further discussed later.

Because the main purpose of this section is to illustrate the versatility and accuracy of the proposed method, we chose to discuss a simplified SDOF model for the surge response of the TLP in random waves. Except for interaction effects between different motion modes, the SDOF model allows for the introduction of most of the relevant nonlinear effects that may influence the response of the TLP. Hence, the following SDOF equation of motion is studied,

$$M\ddot{Z}(t) + H(Z(t), \dot{Z}(t), t) = F(t), \tag{9.55}$$

where $Z = Z(t)$ denotes the surge response of the TLP; M is the mass of the platform, including added mass; and H is a nonlinear function to be specified. As discussed in Chap. 8, the hydrodynamic loading process $F(t)$ is assumed to consist of two components: a linear, first-order (wave frequency) term $F_1(t)$, and a nonlinear, slowly varying, second-order term $F_2(t)$.

To set up the proper dynamic model for the surge response $Z(t)$, it is necessary to take into account the fact that hydrodynamic loading on a floating body depends on the motions of the body. In case of the slow-drift motions of the TLP, it is of some importance to take into account the dependence of the slow-drift force $F_2(t)$ on the slowly varying surge velocity $\dot{Z}_2(t)$. Because of a nonlinear dynamic model, a definition of the slow-drift response $Z_2(t)$ has to introduced. A suitable definition would seem to be the following: The slow-drift response $Z_2(t)$ is obtained from the total response $Z(t)$ by a low-pass filter that removes all wave frequency components. In practice, this may be achieved by a running mean operator used iteratively.

To account for the dependence of the slow-drift force on the slowly varying velocity, it is appropriate to write $F_2(t, \dot{Z}_2(t))$ rather than $F_2(t)$. However, $F_2(t, \dot{Z}_2(t))$ is not directly available to us, but only $F_2(t) \equiv F_2(t, 0)$ given by an equation entirely similar to Eq. (8.7). Since, in the context of slow-drift motions, $\dot{Z}_2(t)$ is small, the following approximation is adopted,

$$F_2(t, \dot{Z}_2(t)) \approx F_2(t, 0) + \frac{\partial F_2(t, 0)}{\partial \dot{Z}_2} \dot{Z}_2(t). \tag{9.56}$$

It is shown by Naess and Johnsen (1993) that for a TLP structure $\frac{\partial F_2(t,0)}{\partial \dot{Z}_2} \approx -c\,F_2(t, 0) \equiv -c\,F_2(t)$ for a suitable constant $c > 0$ may to some extent serve as a useful approximation to capture qualitatively the time-variant damping effect, which is the result of the expansion in Eq. (9.56).

The first dynamic model adopted for $Z(t)$ is now the following,

$$\tilde{M}\ddot{Z} + C\dot{Z} + K(Z + \varepsilon Z^3) = F_1(t) + F_2(t, \dot{Z}_2) \approx F(t) - cF_2(t)\,\dot{Z}_2. \tag{9.57}$$

Here, $\tilde{M} = M + \tilde{m}$, where $\tilde{m}$ is an appropriately chosen (constant) added mass. C, K, and ε are suitably chosen positive constants. This equation is rewritten in the equivalent form,

$$\ddot{Z} + 2\omega_e \zeta \dot{Z} + 2\omega_e \tilde{c} F_2(t)\, \dot{Z}_2 + \omega_e^2 (Z + \varepsilon Z^3) = \frac{F(t)}{\tilde{M}}, \tag{9.58}$$

where $\omega_e^2 = K/\tilde{M}$, $\zeta = C/(2\omega_e \tilde{M})$, and $\tilde{c} = c/(2\omega_e \tilde{M})$.

Thus, the dynamic system is nonlinear with a Duffing-type hardening stiffness nonlinearity and time-varying damping. For the TLP the relative damping coefficient ζ is usually small. As a consequence, the contribution from the time-varying term is non-negligible, especially for severe seas for which the slow drift response is significant. The third-order term in the restoring force, generally referred to as the *set down* effect, is caused by the fact that the tethers will induce the TLP to act like an inverted pendulum. Note that the set down effect will also have an influence on the hydrodynamic loading process, which depends not only on $\dot{Z}$, but also on Z. Even if this dependence could have been taken into account, it was neglected here because it is of minor importance.

For the numerical simulations, a particular model of a TLP is considered, and the corresponding LTF and QTF are computed using the second-order diffraction program (WAMIT 2008). For simplicity, unidirectional seas are used, meaning that the directional argument β is skipped. This simplification should have no effect on the conclusions based on the comparison of accuracy. The combined first-order and second-order slowly varying surge deck motion is studied applying the single-degree-of-freedom model. The TLP particulars are detailed in Table 9.6, and the subsurface part of the structure is shown in Fig. 9.29, but without the vertical tethers.

The values in Table 9.6 are used to obtain the second-order response. This means that for the second-order response, a simplified version of Eq. (9.57) was used, where mass $\tilde{M}$, stiffness K, and damping coefficient C are frequency independent, which is a good approximation for the slow drift motion. The time-invariant damping part ζ is considered to be 5%.

Two versions of Eq. (9.58) are used. The first version is a linear, time-invariant model obtained by putting $\tilde{c} = \varepsilon = 0$. The second version is the fully nonlinear model where the parameter $\tilde{c}$ in Eq. (9.58) is chosen such that $\mathrm{Var}[2\omega_e \tilde{c} F_2(t)\, \dot{Z}_2(t)]$ is about 10% of $\mathrm{Var}[2\omega_e \zeta \dot{Z}(t)]$. The parameter ε is estimated from the condition that $0.2Z(t) \geq \varepsilon Z^3(t)$ when $Z(t) \leq 6\sigma_Z$, i.e., even in the extreme response region, stiffness hardening contributes not more than 20% relative to the linear part for severe sea, which lead to $\varepsilon = 1.36 \cdot 10^{-4}$. Finally, the following approximate values

Table 9.6 Particulars of the TLP

Column diameter D (m)	10.0
Eigenperiod surge T_e (s)	128.8
Relative damping ζ	0.05
Total mass (incl. added mass) $\tilde{M}$ (kg)	$1.5 \cdot 10^7$

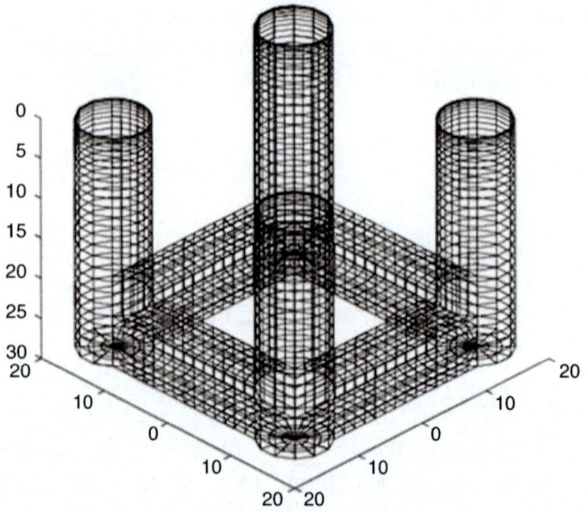

Fig. 9.29 Sketch of the submerged part of the TLP. Units in meters

were found: $\tilde{c} = 30/(\tilde{M}g)$ for moderate seas, and $\tilde{c} = 90/(\tilde{M}g)$ for severe seas, where $g = 9.81\,\mathrm{m/s^2}$, cf. Eq. (9.60). The adopted parameter values are largely arbitrary, but the choices made seem to provide a reasonable model for the chosen TLP structure.

To get an accurate representation of the response process, there is a specific requirement that must be observed. Because the damping ratio is only 5%, the frequency resolution $\Delta\omega$ must secure a sufficient number of frequency values over the resonance peak. This will ensure that the second-order, difference-frequency response component captures the TLP surge dynamics with sufficient accuracy. It is commonly required that there are at least 5 discrete frequencies over the frequency range where $|\hat{L}(\omega)|^2$ is equal to or higher than half of the resonance peak height $\max(|\hat{L}(\omega)|^2)$, where,

$$\hat{L}(\omega) = \big(-\omega^2 + 2\mathrm{i}\zeta\,\omega_e\,\omega + \omega_e^2\big)^{-1}. \tag{9.59}$$

For the surge force QTF $\hat{K}_2(\omega, \omega')$, a suitable initial frequency grid must be chosen for which the values of the force QTF are calculated. The calculation of the force QTF is generally the most time-consuming part of the numerical analysis. Therefore, the initial grid is usually rather coarse to avoid excessive computer time. For the calculations at hand, the discrete frequency range was the following: $\omega_1 = 2\pi/30.0, \ldots, \omega_n = 2\pi/4.0$ (rad/s), $n = 30$. This necessitates the use of an interpolation procedure to be able to provide values of the QTF on a much finer grid than the initial one to comply with the requirement of sufficient frequency resolution to capture the dynamics of slow-drift motion. In this chapter cubic spline interpolation is used. In the particular case considered here, the resolution requirement led to the choice $\Delta\omega = 0.0018$ rad/s and $L = 760$ interpolated discrete frequencies.

The random stationary sea state is specified by a JONSWAP spectrum, which is given as follows,

$$S_\eta(\omega) = \frac{\alpha g^2}{\omega^5} \exp\left\{ -\frac{5}{4}\left(\frac{\omega_p}{\omega}\right)^4 + \ln\kappa \, \exp\left[-\frac{1}{2\chi^2}\left(\frac{\omega}{\omega_p} - 1\right)^2\right]\right\}, \tag{9.60}$$

where $g = 9.81\,\mathrm{ms}^{-2}$, ω_p denotes the peak frequency in rad/s, and κ and χ are parameters affecting the spectral shape. $\chi = 0.07$ when $\omega \leq \omega_p$, and $\chi = 0.09$ when $\omega > \omega_p$. The parameter κ is chosen to be equal to 3.3, which is a rather typical value. The parameter α is determined from the following empirical relationship,

$$\alpha = 5.06\left(\frac{H_s}{T_p^2}\right)^2 (1 - 0.287 \ln\kappa), \tag{9.61}$$

where H_s denotes the significant wave height and $T_p = 2\pi/\omega_p$ is the spectral peak wave period. Table 9.7 presents the sea state parameters, along with the corresponding response standard deviations. For more results, cf. Naess et al. (2007).

Figure 9.30 shows the LTF for the wave exciting force amplitude, while Fig. 9.31 depicts the spline interpolated QTF. Because the QTF is complex valued, only its absolute value is plotted.

For the chosen sea state, Figs. 9.32 and 9.33 present the corresponding response tail crossing rates obtained by Monte Carlo simulation for the linear system given

Table 9.7 Representative sea state, along with response standard deviations for the linear and nonlinear TLP model

H_s (m)	T_p (s)	σ_Z (m), lin.	σ_Z (m), nonlin.
10.0	11	9.3	8.2

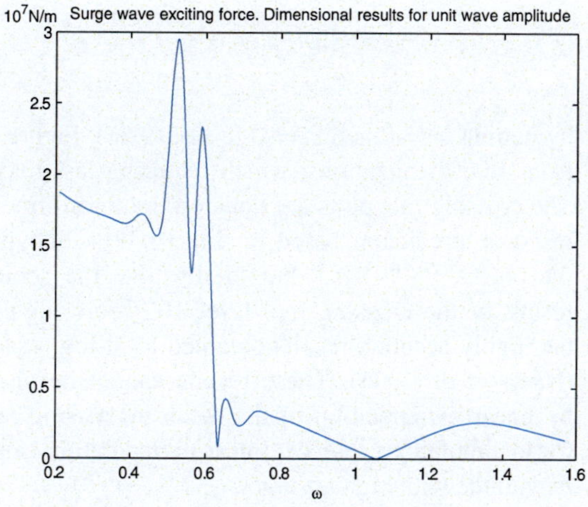

Fig. 9.30 Wave exciting force amplitude, surge LTF

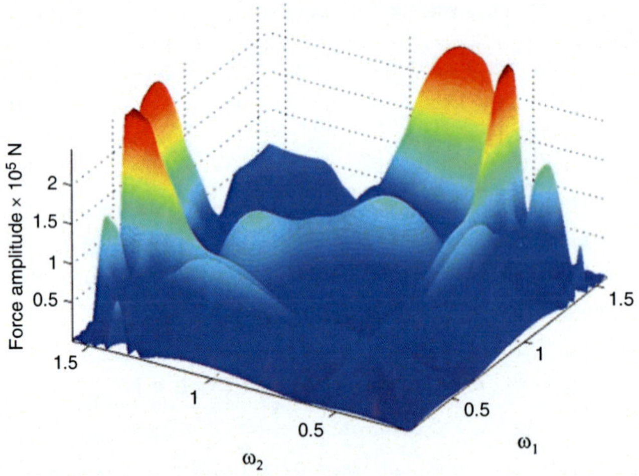

Fig. 9.31 Wave exciting force amplitude, surge QTF

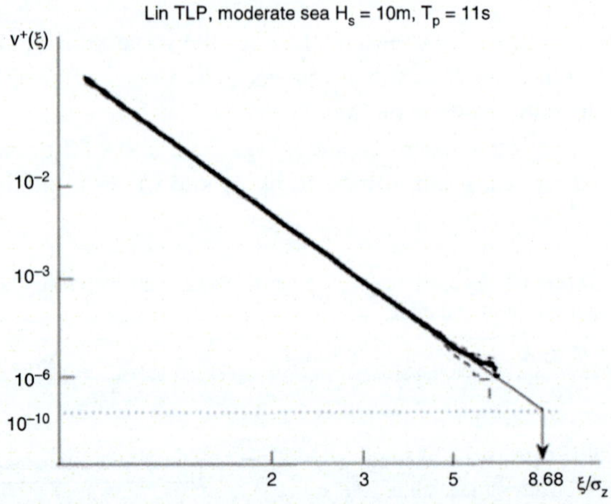

Fig. 9.32 Optimal transformed plot of the empirical crossing rates by Monte Carlo simulation (∗) with 95% confidence bands (−−) based on 1,000 hours of response time histories for the case of linear dynamics ($\tilde{c} = \varepsilon = 0$). Saddle point integration results (——) coincide with the optimized linear fit with $b = 0.75\,\sigma_Z$, $q = 0.0205$, where $\sigma_Z = 9.3$ m (see Table 9.7)

by putting $\varepsilon = 0$ and $\tilde{c} = 0$ in Eq. (9.58). Figure 9.32 shows the results obtained from 1,000 realizations, which requiring less than 1 hour on a laptop computer. The crossing rate plots are done on the transformed scale, see Eq. (9.42). Extreme response prediction based on Eq. (9.35) will typically involve crossing rates of the orders 10^{-6}–10^{-7}, but to illustrate the achieved accuracy the extrapolated results at the crossing rate level 10^{-10} are highlighted. Figure 9.32 also shows the highly accurate results obtained by using a saddle point integration technique (Naess et al. 2006). These results cannot be distinguished from those obtained by linear extrapolation of the mean upcrossing rate function provided by Monte Carlo simulations. Hence, linear extrapolation can be done over several orders of magnitude with high accuracy.

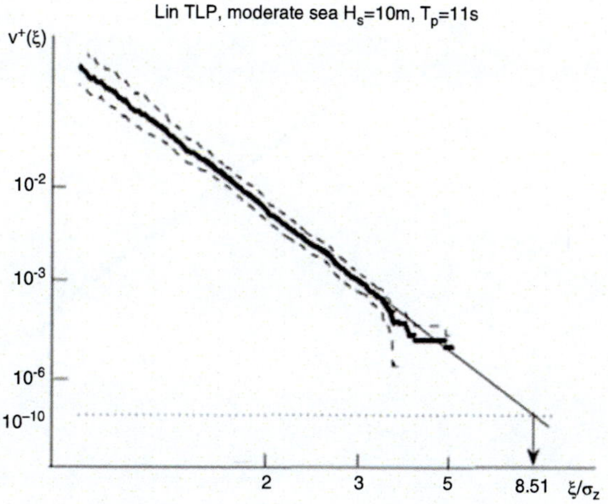

Fig. 9.33 Optimal transformed plot of the empirical crossing rates by Monte Carlo simulation ($*$) with 95% confidence bands (– –) based on 25 hours of response time histories for the case of linear dynamics ($\tilde{c} = \varepsilon = 0$); optimized linear fit (——) with $b = 0.75\,\sigma_Z$, $q = 0.020$, where $\sigma_Z = 9.3$ m (see Table 9.7), for the case of linear dynamics ($\tilde{c} = \varepsilon = 0$)

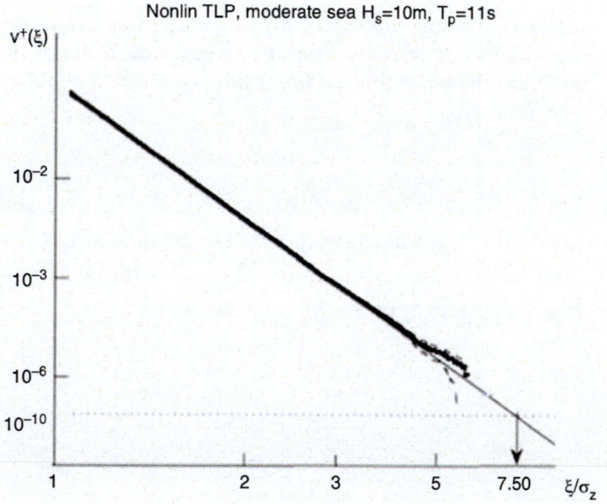

Fig. 9.34 Optimal transformed plot of the empirical crossing rates by Monte Carlo simulation ($*$) with 95% confidence bands (– –) based on 1,000 hours of response time histories for the fully nonlinear model; optimized linear fit (——) with $b = 0.54\,\sigma_Z$, $q = 0.0248$, where $\sigma_Z = 8.2$ m (see Table 9.7)

To illustrate the fact that good accuracy can be obtained with much shorter time simulation records, Fig. 9.33 shows the results of using only 25 hours of simulated response time histories, which required less than 10 minutes on a standard laptop PC.

Figures 9.34 and 9.35 present response tail crossing rates for the chosen sea state obtained by Monte Carlo simulation for the fully nonlinear model given by Eq. (9.58). In this case the Monte Carlo simulation results are the only results available for verification of the extrapolation method for the nonlinear model, but the experience from the linear case indicates that the results obtained from Fig. 9.34 are very accurate. Again, the crossing rate plots are done on the transformed scale, see

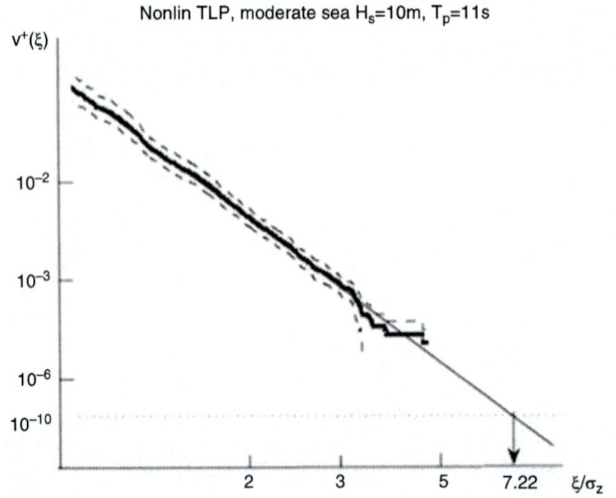

Fig. 9.35 Optimal transformed plot of the empirical crossing rates by Monte Carlo simulation (∗) with 95% confidence bands (−−) based on 25 hours of response time histories for the fully nonlinear model; optimized linear fit (——) with $b = 0.54\,\sigma_Z$, $q = 0.0234$, where $\sigma_Z = 8.2\,\text{m}$ (see Table 9.7)

Eq. (9.42), and for illustration purposes the extrapolated results at the crossing rate level 10^{-10} are highlighted. These results can be compared with the corresponding results obtained from only 25 h of simulated response time histories shown in Fig. 9.35. The agreement is again very good.

Chapter 10
Bivariate Extreme Value Distributions

10.1 Introduction

The title of this chapter is deliberately chosen to focus on the bivariate case instead of the general multivariate. The reason is mainly one of expediency, because the general multivariate case would easily embroil us in the necessity to roll out a heavy machinery of notation without contributing to a deeper understanding of the issues involved. For a discussion of the general multivariate case, the reader may consult the book by Beirlant et al. (2004). The extension of extreme value statistics from the univariate to the multivariate case meets with several challenges. First of all, there is no direct simple generalization of the univariate extreme value types theorem to the multivariate case, and in particular, this also applies to the bivariate case.

Developed on the basis of Gumbel's logistic and mixed models (Gumbel 1960a,b, 1961; Gumbel and Mustafi 1967), the later results on possible asymptotic bivariate extreme value distributions became, in a sense, too general, which poses severe problems for practical applications. Significant efforts have been made to model and estimate a function which describes the dependence structure between extreme components, cf. e.g., Coles and Tawn (1991, 1994). However, there are no precise estimation tools that allow us to decide on the joint distribution of the bivariate extremes from a given set of bivariate data. Of course, the marginal data sets can be used to derive estimates of the marginal extreme value distributions, as in Zachary et al. (1998) and de Haan and de Ronde (1998), but the joint distribution is still a long way off.

A popular method of trying to cope with the problem of bivariate extremes is to adopt a copula to represent the joint distribution structure. This copula is then usually combined with asymptotic extreme value distributions to represent the marginal distributions, typically of the GEV type (Coles 2001). For this purpose, a range of different copulas have been proposed (Tawn 1988; Waal and van Gelder 2005), see also Castillo et al. (2005) for a compendious treatment of bivariate copulas. Even in the case of the bivariate extreme value copula (Pickands 1981;

A. Naess, *Applied Extreme Value Statistics*,
https://doi.org/10.1007/978-3-031-60769-1_10

Balakrishnan and Lai 2009), due to the properties of the dependence function, generally speaking, there are an infinite number of models. Therefore, the main problem with this approach is that it is rather ad hoc. That is, there seems to be no appropriate theoretical justification for choosing one particular copula over the other.

It is therefore of considerable interest to note that the ACER method can easily be extended to several dimensions, in particular, to two (Naess 2011). By this fact, a vehicle is obtained for providing a nonparametric statistical estimate of the bivariate extreme value distribution inherent in a bivariate time series. It will be seen that the bivariate ACER function is able to cover both spatial and temporal dependence characteristics of the given time series. Thus, it covers all simultaneous and non-simultaneous extreme events. From a practical point of view, this makes it possible to investigate the true behavior of the bivariate extreme value distribution for a particular case, and at the same time check the validity of the proposed copula models for bivariate extremes.

As a first effort in investigating the functional representation of the empirically estimated bivariate ACER surface, the bivariate extreme value copula approach will be adopted. Specifically, the Asymmetric logistic and Gumbel logistic models are used, combined with asymptotically consistent marginal extreme value distributions based on the univariate ACER functions. Since the univariate ACER functions have proved to portray accurately the marginal tail behavior, this will offer an opportunity to verify to some extent the viability of copula models to capture the dependence structure of bivariate extreme value distributions.

The performance of the bivariate ACER method will be illustrated by application to two measured bivariate time series. The first consists of simultaneous wind speed data measured at two stations off the coast of Norway. The second time series consists of simultaneous wind speed and wave height data measured at an offshore location in the North Sea. For some other examples on the application of the bivariate ACER method, see, e.g., Gaidai et al. (2019) and Xu et al. (2022).

10.2 Componentwise Extremes

Assume that $(X_1, Y_1), (X_2, Y_2), \ldots$ is a sequence of iid bivariate random variables having distribution function $F(x, y)$. Let $M_{x,N} = \max_{1\le i\le N} X_i$ and $M_{y,N} = \max_{1\le i\le N} Y_i$. Then $M_N = (M_{x,N}, M_{y,N})$ is a vector of componentwise maxima, where the index i for which the component time series $X_1, X_2, \ldots$ and $Y_1, Y_2, \ldots$ assume their extreme may be different between the two series. Hence, M_N does not necessarily correspond to an observed vector of the original time series.

The search for the limiting forms of the bivariate extreme value distributions follows more or less the pattern of the univariate case by studying M_N as $N \to \infty$. By way of a first observation, the marginal distributions of the asymptotic bivariate extreme value distribution should by necessity be given by the asymptotic univariate extreme value distributions. Since the marginals are known, the representations may, in fact, be simplified a little by assuming that both X_i and Y_i have the standard Fréchet distribution $F(z) = \exp(-1/z)$, $z > 0$. By transformation of variables, any other marginal distribution can be obtained. With this specific choice of marginal, a special case of the GEV distribution with parameters $\mu = 0$, $\sigma = 1$ and $\gamma = 1$ is at hand. Now, it follows that $\text{Prob}(M_{x,N} \le z) = \exp(-N/z)$, or, equivalently, $\text{Prob}(M_{x,N}/N \le z) = \exp(-1/z)$, $z > 0$. The same result applies to $M_{y,N}$. Hence, to obtain standard univariate results for each margin, the re-scaled vector,

$$M_N^* = (M_{x,N}/N, M_{y,N}/N), \tag{10.1}$$

should be considered.

The main result of this section is the following:

Let $M_N^* = (M_{x,N}^*, M_{y,N}^*)$ be defined by Eq. (10.1) where the (X_i, Y_i) are iid random vectors with standard Fréchet marginal distributions. Then if,

$$\text{Prob}(M_{x,N}^* \le x, M_{y,N}^* \le y) \to G(x, y), \text{ as } N \to \infty, \tag{10.2}$$

where G is a non-degenerate distribution function, G has the form,

$$G(x, y) = \exp\{-V(x, y)\}, \; x > 0, \; y > 0, \tag{10.3}$$

where

$$V(x, y) = 2 \int_0^1 \max\left(\frac{w}{x}, \frac{1-w}{y}\right) dH(w), \tag{10.4}$$

where H is a distribution function on $[0, 1]$ satisfying the mean value constraint

$$\int_0^1 w \, dH(w) = 1/2. \tag{10.5}$$

The distributions obtained as limits in Eq. (10.2) are called the class of bivariate asymptotic extreme value distributions. According to this result, this class is in a 1-1 correspondence with the set of distributions H on $[0, 1]$ satisfying Eq. (10.5). So, this rather remarkable result tells us that for any such distribution function on $[0, 1]$, a valid bivariate asymptotic extreme value distribution is obtained.

Two simple examples may serve as illustration. Denote by W a random variable with distribution function H. First, let W have the two possible outcomes 0 and 1, and $\text{Prob}(W = 0) = \text{Prob}(W = 1) = 0.5$. Hence the distribution function H has jumps at 0 and 1, so that $dH(0) = 0.5$ and $dH(1) = 0.5$. The condition given by Eq. (10.5) is satisfied, and $V(x, y) = x^{-1} + y^{-1}$ by Eq. (10.4), so that the corresponding bivariate extreme value distribution becomes,

$$G(x, y) = \exp\{-(x^{-1} + y^{-1})\} = \exp\{-x^{-1}\}\exp\{-y^{-1})\}, \; x > 0, \; y > 0, \tag{10.6}$$

which clearly illustrates the case of two independent variables with standard Fréchet marginals. The opposite case of two fully dependent variables is obtained by considering a degenerate random variable with unit mass at $W = 0.5$. The corresponding distribution function then has one jump at $w = 0.5$, i.e., $dH(0.5) = 1$. Equation (10.5) is satisfied, and the corresponding bivariate extreme value distribution is,

$$G(x, y) = \exp\{-\max(x^{-1}, y^{-1})\}, \quad x > 0, \; y > 0, \tag{10.7}$$

where the marginals are still standard Fréchet.

To obtain the general situation for any GEV marginal, it is now only necessary to transform the marginals from standard Fréchet to the required members of the GEV family. Specifically, by defining,

$$\tilde{x} = \left\{1 + \gamma_x\left(\frac{x - \mu_x}{\sigma_x}\right)\right\}^{1/\gamma_x} \quad \text{and} \quad \tilde{y} = \left\{1 + \gamma_y\left(\frac{y - \mu_y}{\sigma_y}\right)\right\}^{1/\gamma_y}, \tag{10.8}$$

it follows that the complete set of bivariate asymptotic extreme value distributions is determined by distribution functions of the form,

$$G(x, y) = \exp\{-V(\tilde{x}, \tilde{y})\}, \tag{10.9}$$

provided $[1 + \gamma_x(x - \mu_x)/\sigma_x] > 0$ and $[1 + \gamma_y(y - \mu_y)/\sigma_y] > 0$, and where the function V satisfies Eq. (10.4) for some distribution function H, satisfying Eq. (10.5). The marginal distributions are GEV with parameters $(\mu_x, \sigma_x, \gamma_x)$ and $(\mu_y, \sigma_y, \gamma_y)$, respectively.

10.3 Bivariate ACER Functions

Consider a bivariate stochastic process $Z(t) = \big(X(t), Y(t)\big)$ with dependent component processes, which has been observed over a time interval, $(0, T)$ say. Assume that the sampled values $(X_1, Y_1), \ldots, (X_N, Y_N)$ are allocated to the (usually equidistant) discrete times $t_1, \ldots, t_N$ in $(0, T)$. Our goal in this section is to construct a methodology that allows us to accurately determine empirically the joint distribution function of the extreme value vector $M_N = \big(M_{x,N}, M_{y,N}\big)$, where $M_{x,N} = \max\big\{X_j \,;\, j = 1, \ldots, N\big\}$, and with a similar definition of $M_{y,N}$. Specifically, the goal is to estimate $P(\xi, \eta) = \text{Prob}\big(M_{x,N} \le \xi, M_{y,N} \le \eta\big)$ accurately for large values of ξ and η.

For notational convenience, it is expedient to introduce the non-exceedance event $\mathscr{C}_{kj}(\xi, \eta) = \{X_{j-1} \le \xi, Y_{j-1} \le \eta, \ldots, X_{j-k+1} \le \xi, Y_{j-k+1} \le \eta\}$ for $2 \le k \le j \le N+1$. Then, from the definition of $P(\xi, \eta)$ it follows that,

$$\begin{aligned} P(\xi, \eta) &= \text{Prob}\big(\mathscr{C}_{N+1,N+1}(\xi, \eta)\big) \\ &= \text{Prob}\big(X_N \le \xi, Y_N \le \eta \,|\, \mathscr{C}_{NN}(\xi, \eta)\big) \cdot \text{Prob}\big(\mathscr{C}_{NN}(\xi, \eta)\big) \\ &\;\vdots \\ &= \prod_{j=2}^{N} \text{Prob}\big(X_j \le \xi, Y_j \le \eta \,|\, \mathscr{C}_{jj}(\xi, \eta)\big) \cdot \text{Prob}\big(\mathscr{C}_{22}(\xi, \eta)\big). \end{aligned} \tag{10.10}$$

Following the pattern of the derivations of the univariate ACER functions in Chap. 5, it will be shown in Sect. 10.5.2, based on Eq. (10.10) and the properties of conditional probability, that a sequence of approximations may be introduced which converges to the target distribution $P(\xi, \eta)$. In practice, following the derivations in Sect. 10.5.2, it is therefore assumed that the following representation applies for a suitably chosen k, and for large values of ξ and η, cf. Eq. (10.34),

$$P(\xi, \eta) \approx \exp\Big\{ -\sum_{j=k}^{N} \big(\alpha_{kj}(\xi; \eta) + \beta_{kj}(\eta; \xi) - \gamma_{kj}(\xi, \eta)\big)\Big\}, \tag{10.11}$$

where $\alpha_{kj}(\xi; \eta) = \text{Prob}\big(X_j > \xi \,|\, \mathscr{C}_{kj}(\xi, \eta)\big)$, $\beta_{kj}(\eta; \xi) = \text{Prob}\big(Y_j > \eta \,|\, \mathscr{C}_{kj}(\xi, \eta)\big)$ and $\gamma_{kj}(\xi, \eta) = \text{Prob}\big(X_j > \xi, Y_j > \eta \,|\, \mathscr{C}_{kj}(\xi, \eta)\big)$. Note that Eq. (10.11) applies equally well to stationary and non-stationary time series. This is due to the fact that the possible time dependence of the conditional exceedance probabilities $\alpha_{kj}(\xi; \eta)$, $\beta_{kj}(\eta; \xi)$ and $\gamma_{kj}(\xi, \eta)$ has been retained, which is reflected in the presence of the time parameter j.

From Eq. (10.11) it emerges that for the estimation of the bivariate extreme value distribution, it is necessary and sufficient to estimate the sequence of functions $\{\alpha_{kj}(\xi; \eta) + \beta_{kj}(\eta; \xi) - \gamma_{kj}(\xi, \eta)\}_{j=k}^{N}$. To get a more compact representation, it is expedient to introduce the concept of a k'th order bivariate average conditional exceedance rate (ACER) function as follows, cf. Eq. (10.35),

$$\mathscr{E}_k(\xi, \eta) = \frac{1}{N-k+1} \sum_{j=k}^{N} \big(\alpha_{kj}(\xi; \eta) + \beta_{kj}(\eta; \xi) - \gamma_{kj}(\xi, \eta)\big); \;\; k = 1, 2, \ldots \tag{10.12}$$

Hence, when $N \gg k$, and for large values of ξ and η,

$$P(\xi, \eta) \approx \exp\{-(N-k+1)\, \mathscr{E}_k(\xi, \eta)\}\,. \tag{10.13}$$

The numerical estimation of the bivariate ACER function for the observed stationary or non-stationary time series consists in counting of the appropriate exceedance events. The estimation procedure is derived in more details in Sect. 10.5.2.

10.4 Functional Representation of the Empirically Estimated Bivariate ACER Functions

From the definition of $\mathscr{E}_k(\xi, \eta)$ follows that the product $\mathscr{E}_k(\xi, \eta) \cdot (N - k + 1)$ represents the expected number of the bivariate observations $Z_j = (X_j, Y_j)$ such that their components exceed corresponding levels ξ and η (both simultaneous and non-simultaneous) and follow after at least $k - 1$ previous simultaneous non-exceedances. Therefore the bivariate ACER function $\mathscr{E}_k(\xi, \eta)$ is able to capture the temporal and spatial dependence structure of the considered bivariate time series. Moreover, as is discussed in Sect. 10.5.2, in practice the existence of an effective $k_e \ll N$ such that $P(\xi, \eta) = P_{k_e}(\xi, \eta) = \exp\{-(N - k_e + 1)\,\mathscr{E}_{k_e}(\xi, \eta)\}$ can be assumed.

This provides a means to obtain high quantiles of the bivariate extreme value distribution. The joint T-year return period contour associated with the event that either $M_{x,N}$ or $M_{y,N}$ or both is exceeded, that is, $\{(M_{x,N} > \xi^T) \cup (M_{y,N} > \eta^T) \cup (M_{x,N} > \xi^T \cap M_{y,N} > \eta^T)\}$, is represented by,

$$1 - F^{1yr}(\xi^T, \eta^T) = \frac{1}{T}, \tag{10.14}$$

where $F^{1yr}(\xi, \eta)$ is the joint distribution function of the annual maxima. Assuming that the duration of the observation period of the bivariate process $Z(t)$ is n_y years, then, with $k = k_e$,

$$F^{1yr}(\xi, \eta) = \exp\left\{-\frac{N - k + 1}{n_y}\,\mathscr{E}_k(\xi, \eta)\right\}. \tag{10.15}$$

From Eqs. (10.14) and (10.15) follows that the joint T-year return levels (ξ^T, η^T) are obtained as solution of the implicit equation:

$$\mathscr{E}_k(\xi^T, \eta^T) = -\log\left(1 - \frac{1}{T}\right)\frac{n_y}{N - k + 1}. \tag{10.16}$$

It is evident that the empirically estimated k-th order bivariate ACER rarely provides enough information for estimation of high quantiles of the joint extreme value distribution. In addition, the exact behavior of the bivariate ACER as a continuous function of two variables cannot be decided using available statistical data. Therefore, the subasymptotic functional form of the ACER surface $\mathscr{E}_k(\xi, \eta)$

can possibly be obtained approximately by the copula representation of a bivariate extreme value distribution.

From the result by Sklar (1959), for any pair of random variables (X, Y) with marginal distribution functions $F_x(\xi)$ and $G_y(\eta)$, the joint distribution function $H_{xy}(\xi, \eta) = \text{Prob}(X \le \xi, Y \le \eta)$ can be presented by a bivariate copula $C(u, v)$ as follows: $H_{xy}(\xi, \eta) = C\big(F_x(\xi), G_y(\eta)\big)$, cf. e.g., Nelsen (2006) or Balakrishnan and Lai (2009). This result applies to any bivariate extreme value distribution as well.

Considering the above and using the result of Pickands (1981), any bivariate extreme value distribution $H_{xy}(\xi, \eta)$ with marginal univariate extreme value distributions $F_x(\xi)$ and $G_y(\eta)$ is given by the formula,

$$H_{xy}(\xi, \eta) = \exp\left\{ \log\big(F_x(\xi)G_y(\eta)\big) \cdot \mathscr{D}\left(\frac{\log\big(F_x(\xi)\big)}{\log\big(F_x(\xi)G_y(\eta)\big)} \right) \right\}, \tag{10.17}$$

where the Pickands dependence function $\mathscr{D}(\cdot)$ is a convex function and satisfies $\mathscr{D}(x) : [0, 1] \longmapsto [\max(x, 1-x), 1]$, cf. Gudendorf and Segers (2010).

We assume that asymptotically consistent marginal extreme value distributions $F_x(\xi)$ and $G_y(\eta)$ are represented by the corresponding univariate ACER functions, that is,

$$\begin{aligned} F_x(\xi) &\approx \exp\left\{ -(N-k+1)\varepsilon_k^x(\xi) \right\}, \quad \xi \ge \xi_1, \\ G_y(\eta) &\approx \exp\left\{ -(N-k+1)\varepsilon_k^y(\eta) \right\}, \quad \eta \ge \eta_1, \end{aligned} \tag{10.18}$$

where the subasymptotic functional form of the univariate ACER function is represented as $\varepsilon_k^x(\xi) = q_k^x \exp\{-a_k^x(\xi - b_k^x)^{c_k^x}\}$ with a similar definition of $\varepsilon_k^y(\eta)$, cf. Chap. 5.

Now, substituting Eq. (10.18) into Eq. (10.17) the following representation of the bivariate extreme value distribution applies:

$$H_{xy}(\xi, \eta) = \exp\left\{ -(N-k+1)\left(\varepsilon_k^x(\xi) + \varepsilon_k^y(\eta)\right) \cdot \mathscr{D}\left(\frac{\varepsilon_k^x(\xi)}{\varepsilon_k^x(\xi) + \varepsilon_k^y(\eta)} \right) \right\}. \tag{10.19}$$

On the other hand, as it has been discovered before in Eq. (10.13), the bivariate extreme value distribution can be expressed through the bivariate ACER function: $H_{xy}(\xi, \eta) = \exp\{-(N-k+1)\,\mathscr{E}_k(\xi, \eta)\}$. Thereby, the functional form of the bivariate ACER surface can possibly be obtained by:

$$\mathscr{E}_k(\xi, \eta) = \left(\varepsilon_k^x(\xi) + \varepsilon_k^y(\eta)\right) \mathscr{D}\left(\frac{\varepsilon_k^x(\xi)}{\varepsilon_k^x(\xi) + \varepsilon_k^y(\eta)} \right). \tag{10.20}$$

Consequently, our aim now is to find the dependence function $\mathscr{D}(\cdot)$ that would provide optimal fit of the parametrical surface defined by Eq. (10.20) to the empirical bivariate ACER $\hat{\mathscr{E}}_k(\xi, \eta)$.

Subject to the form of the dependence function $\mathscr{D}(\cdot)$, different parametric differentiable and non-differentiable models can be considered. By setting $\mathscr{D}(x) = \theta x^2 - \theta x + 1$, where $0 \leq \theta \leq 1$, the Type A bivariate extreme value distribution, or Gumbel mixed model, is obtained, cf. e.g., Gumbel (1960a,b), Gumbel and Mustafi (1967). Another differentiable model is acquired by setting $\mathscr{D}(x) = [x^m + (1 - x)^m]^{1/m}$ for $m \geq 1$. This is the Type B distribution or Gumbel logistic model (Gumbel 1961; Hougaard 1986). The functional form of the bivariate ACER surface in the Gumbel logistic case becomes,

$$\mathscr{G}_k(\xi, \eta) = \left[\left(\varepsilon_k^x(\xi)\right)^m + \left(\varepsilon_k^y(\eta)\right)^m\right]^{\frac{1}{m}}. \tag{10.21}$$

The Type C distribution, also known as the biextremal model (Tiago de Oliveira 1984), can be considered as an example of a non-differentiable model. The dependence function in this case is $\mathscr{D}(x) = \max(x,\ 1 - \theta x)$ for $0 \leq \theta \leq 1$.

In the literature, differentiable models have usually been of most interest, and they have been used to analyze bivariate environmental events. Yue et al. (1999) and Yue (2000), apply the Gumbel mixed model to rainfall data in order to provide storm frequency analysis. Yue (2001a,b) has also studied the Gumbel logistic model with application to flood peak—flood volume pair of bivariate data.

In the work by Yue and Wang (2004), a comparison between the Gumbel mixed and the Gumbel logistic models has been made. The authors argued that both models are appropriate and give similar estimates of the joint distribution of two Gumbel-distributed random variables whose Pearson product-moment correlation coefficient is: $0 \leq \rho \leq 2/3$. When $\rho > 2/3$ the Gumbel mixed model cannot be applied, see also Tiago de Oliveira (1982). For this reason, it was decided to consider the Gumbel logistic model as one that fits better the objectives of the present work.

In the dependence function $\mathscr{D}(\cdot)$ for the Gumbel logistic model, Tawn (1988) added extra parameters ϕ and θ to get further flexibility. This leads to the Asymmetric logistic model, which sets $\mathscr{D}(x) = [\phi^m x^m + \theta^m (1-x)^m]^{1/m} + (\theta - \phi)x + 1 - \theta$ with $0 \leq \theta \leq 1,\ 0 \leq \phi \leq 1,\ m \geq 1$. The functional form of the bivariate ACER surface in the Asymmetric logistic case is obtained as,

$$\mathscr{A}_k(\xi, \eta) = \left[\left(\phi\varepsilon_k^x(\xi)\right)^m + \left(\theta\varepsilon_k^y(\eta)\right)^m\right]^{\frac{1}{m}} + (1 - \phi)\varepsilon_k^x(\xi) + (1 - \theta)\varepsilon_k^y(\eta). \tag{10.22}$$

The optimal parameters m^* for $\mathscr{G}_k(\xi, \eta)$ and θ^*, ϕ^* and m^* for $\mathscr{A}_k(\xi, \eta)$, can be found by minimizing a mean square error function. This is discussed only for the case of $\mathscr{A}_k(\xi, \eta)$, since the case of $\mathscr{G}_k(\xi, \eta)$ then follows easily. Specifically, the mean square error function for $\mathscr{A}_k(\xi, \eta)$ is defined as,

$$F(m, \theta, \phi) = \sum_{j=1}^{N_\eta} \sum_{i=1}^{N_\xi} w'_{ij} \Big(\log \hat{\mathscr{E}}_k(\xi_i, \eta_j) - \log \mathscr{A}_k(\xi_i, \eta_j) \Big)^2 , \tag{10.23}$$

where N_ξ, N_η are numbers of levels ξ and η, respectively, at which the ACER function have been empirically estimated, and $w'_{ij} = w_{ij} / \sum \sum w_{ij}$ with,

$$w_{ij} = \big(\log CI^+(\xi_i, \eta_j) - \log CI^-(\xi_i, \eta_j) \big)^{-2} , \tag{10.24}$$

denoting normalized weight factors that put more emphasis on the more reliable estimates.

The constrained optimization problem with the objective function F defined in Eq. (10.23) is written as,

$$\begin{cases} F(m, \theta, \phi) \to \min , \\ \{m, \theta, \phi\} \in \mathscr{S} , \end{cases} \tag{10.25}$$

with a constraints domain,

$$\mathscr{S} = \Big\{ \{m, \theta, \phi\} \in \mathbb{R}^3 \;\Big|\; \theta,\ \phi \in [0,\ 1];\ m \in [1,\ +\infty) \Big\}. \tag{10.26}$$

10.5 Numerical Examples

10.5.1 Wind Speed Measured at Two Adjacent Weather Stations

In this section, the results obtained by Naess and Karpa (2015a) will be discussed. The simultaneous wind speed data measured along the Norwegian coast at Sula and Nordøyan Fyr weather stations, were analyzed to obtain numerical estimates of bivariate extreme wind speeds. Figure 10.1 shows the geographical locations of the measurement sites. The hourly maximum of the three seconds wind gust (10 m above the ground) were recorded during 13 years (1999–2012).

Figure 10.2 demonstrates the plot of the observed data. This plot reveals a rather strong dependence between the two time series.

The Pearson product-moment correlation coefficient is found to be $\rho = 0.73$. The Kendall's rank correlation coefficient is $\tau = 0.5$, while Spearman's ρ equals 0.68, which also indicates a nonlinear relationship between Sula and Nordøyan wind speeds.

Fig. 10.1 Map of a part of Norway with the marked weather stations: A—Sula station, B—Nordøyan Fyr station

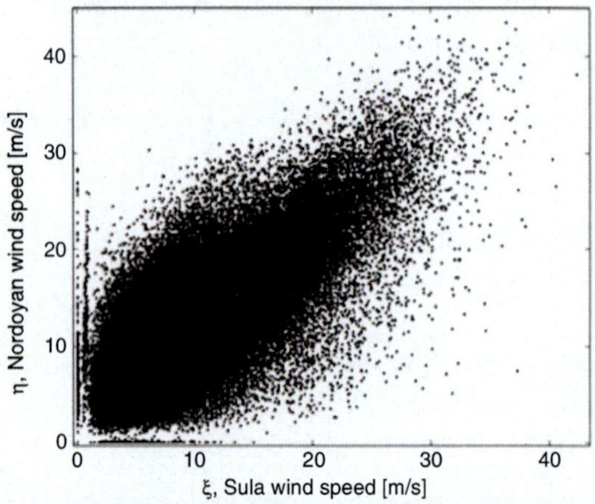

Fig. 10.2 Coupled observations of wind speed data observed at the Sula station (ξ axis) and at the Nordøyan Fyr station (η axis)

It was decided to divide the data series into 13 one-year records for the analysis. By this, also the standard deviation of the ACER function estimates can be calculated fairly accurately.

The univariate ACER functions were estimated first, using the Matlab-based standalone downloadable application (Karpa 2012). In Figs. 10.3 and 10.4 the sequence of $\hat{\varepsilon}_1 \ldots \hat{\varepsilon}_{96}$ are plotted versus different wind speed levels. Both figures reveal that there is significant temporal dependence between consecutive data. It is also seen that this dependence effect is largely accounted for by $k = 24$ since there is a marked degree of convergence in the tail of $\hat{\varepsilon}_k$ for $k \geq 24$ in both cases. Here, for $k = 96$, which corresponds to conditioning on data recorded up to 4 days earlier, $\hat{\varepsilon}_{96}$ is considered to represent the final converged results, since $\hat{\varepsilon}_{96} \approx \hat{\varepsilon}_k$ for $k > 96$ in the tail. Therefore, there is no need to consider conditioning of an even higher order than 96. So, effectively, $k_e = 96$ for the recorded data. Also note that 4 days is a typical separation of wind speed data adopted in the declustering process to achieve independence between the data used in, e.g., a peaks-over-threshold

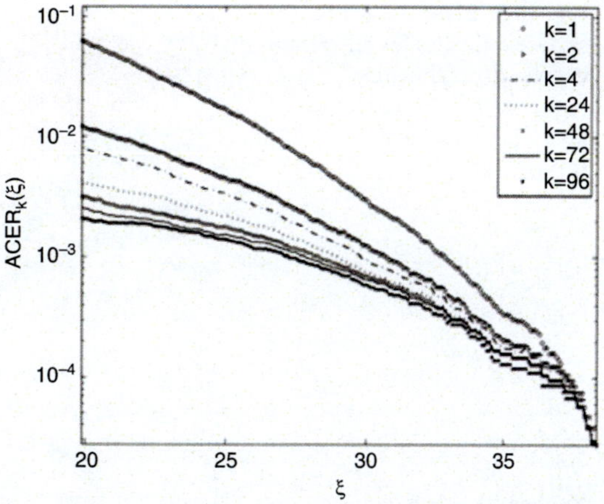

Fig. 10.3 ACER estimates for different degrees of conditioning. Wind speed data from the Sula station, cf. Fig. 6.9

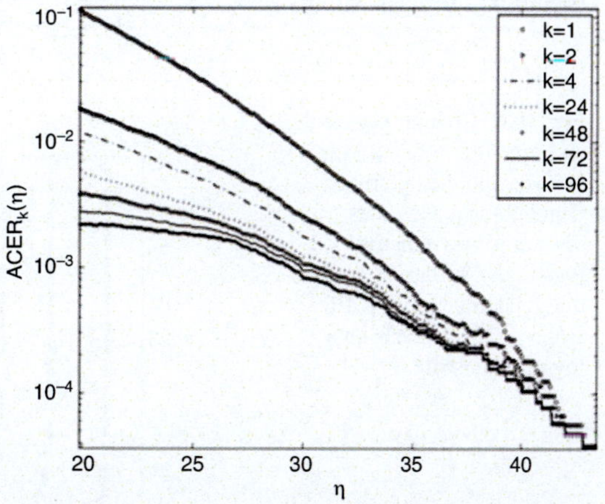

Fig. 10.4 ACER estimates for different degrees of conditioning. Wind speed data from the Nordøyan Fyr station

analysis. Figures 10.3 and 10.4 also demonstrate that for extreme value estimation, $\hat{\varepsilon}_1$ can be used since the ACER functions all coalesce in the far tail. This makes it possible to choose $k = 1$, which makes much more data available for estimation, with a possible reduction of uncertainty in estimation as a result.

The sequence of estimated bivariate ACER surfaces $\hat{\mathscr{E}}_k(\xi, \eta)$ is shown in Fig. 10.5. Matlab programs for ACER 2D analyses are available, cf. Karpa (2014). $\hat{\mathscr{E}}_k(\xi, \eta)$ with $k = 1$ is the uppermost. As it is seen from the figure, the cross-section of the surfaces at a high value of the wind speed level η gives the univariate ACER functions of the wind speed data from the Sula station, while the cross-section at a high level of ξ represents the univariate ACER of the time series from the Nordøyan Fyr.

Due to the observed convergence, the ACER surface for $k = 96$ is very close to the surface obtained by taking the logarithm of the exact bivariate extreme value distribution.

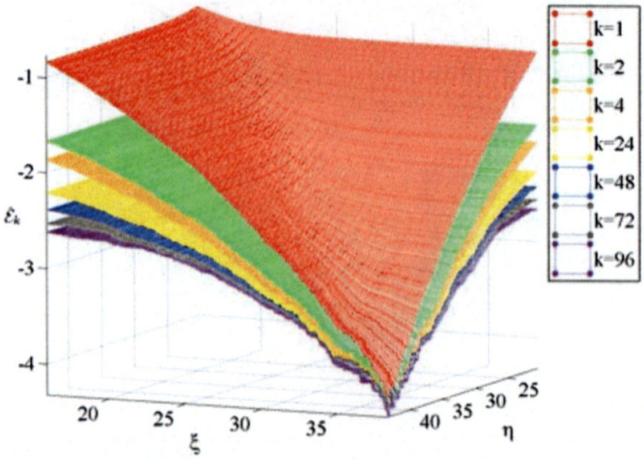

Fig. 10.5 Bivariate ACER surface estimates for different degrees of conditioning. $\hat{\mathcal{E}}_k(\xi, \eta)$ surfaces are plotted on a logarithmic scale

Table 10.1 Optimal parameters of AL and GL fits

k	AL	GL
1	$m_A = 2.44,\ \theta = 0.86,\ \phi = 0.92$	$m_G = 2.01$
96	$m_A = 4.53,\ \theta = 0.97,\ \phi = 0.97$	$m_G = 3.87$

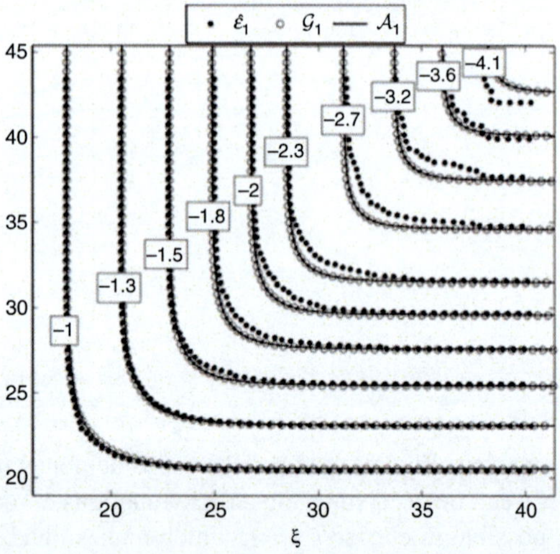

Fig. 10.6 Contour plot of the empirically estimated $\hat{\mathcal{E}}_1$ surface, and the optimized Gumbel logistic $\mathcal{G}_1$ and optimized Asymmetric logistic $\mathcal{A}_1$ surfaces based on marginal univariate ACER. Boxes indicate levels on a logarithmic scale

Parameters of the optimal Asymmetric logistic (AL) and Gumbel logistic (GL) surfaces are presented in Table 10.1.

Figures 10.6 and 10.7 show the contour plots of the optimized Asymmetric logistic $\mathcal{A}_k(\xi, \eta)$ and the optimized Gumbel logistic $\mathcal{G}_k(\xi, \eta)$ fits to the data for $\hat{\mathcal{E}}_k(\xi, \eta)$ surface for $k = 1$ and $k = 96$, respectively. The contour lines of three surfaces are plotted for those levels of ξ and η, where the bivariate ACER surface $\hat{\mathcal{E}}_k(\xi, \eta)$ have been empirically estimated.

The figures reveal that the empirical bivariate ACER surface $\hat{\mathcal{E}}_k$ captures high correlation between the data, and so do the optimally fitted $\mathcal{G}_k$ and $\mathcal{A}_k$ surfaces. It

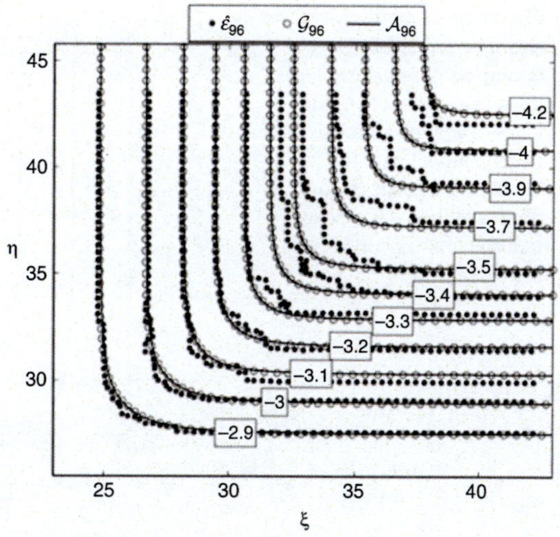

Fig. 10.7 Contour plot of the empirically estimated $\hat{\mathcal{E}}_{96}$ surface, and the optimized Gumbel logistic $\mathcal{G}_{96}$ and optimized Asymmetric logistic $\mathcal{A}_{96}$ surfaces based on marginal univariate ACER. Boxes indicate levels on a logarithmic scale

is also seen that the behavior of the estimated ACER surface in the case $k = 96$ in Fig. 10.7 affirms high uncertainty due to deficiency of data. However, the optimal surfaces $\mathcal{G}_{96}$ and $\mathcal{A}_{96}$ capture the statistical properties of the bivariate observations.

It is noticeable that the level of agreement between the estimated bivariate ACER and both the optimized Asymmetric logistic and the Gumbel logistic surfaces is equally significant. Yet, it is also important to keep in mind that the empirical bivariate ACER $\hat{\mathcal{E}}_k$ is the only discrete surface. The MATLAB (2009) built-in routine contourc that has been used to obtain the figures calculates the contour lines by producing a regularly spaced grid determined by the dimensions of a surface. Therefore, it evidently generates a certain spacing inaccuracy of the ACER $\hat{\mathcal{E}}_k$ surface level lines plot. In addition, the figures ascertain that the optimized Asymmetric logistic and Gumbel logistic surfaces conform at a level sufficient to affirm that they actually coincide.

Thereby, the optimized Gumbel logistic model with asymptotically consistent marginals obtained from the optimized univariate ACER, can be used as the parametric representative of the bivariate ACER surface estimated from the given data set.

Also, comparison of the contour lines of $\mathcal{G}_1$ and $\mathcal{G}_{96}$ that correspond to the same return period levels shows fairly good agreement in the tail considering the high uncertainty for the case $k = 96$. To highlight the results that would be obtained by adopting the common approach of assuming Gumbel marginal extreme value distributions combined with a suitable copula model, in Fig. 10.8 are plotted the 50 and 100-year return period levels obtained by using the Asymmetric logistic model with asymptotically consistent marginals obtained from the optimized univariate ACER marginals together with the corresponding return levels obtained by using the Gumbel logistic model with Gumbel marginals fitted by the method of moments. It is clear that the discrepancy is significant, which is primarily caused by the

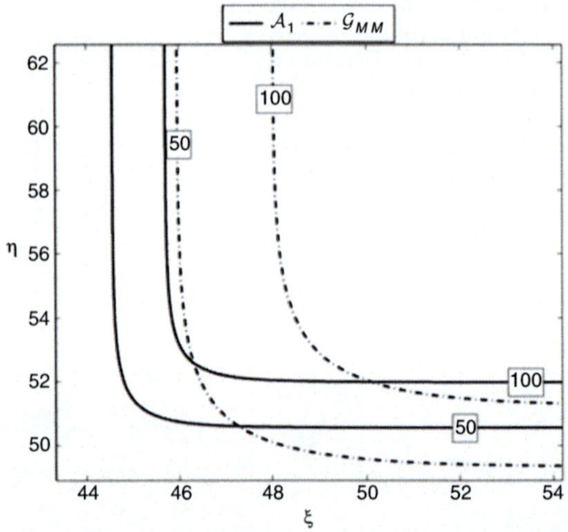

Fig. 10.8 Contour plot of the return period levels for the optimized Asymmetric logistic $\mathscr{A}_1$ surfaces (solid line —) and the Gumbel logistic model with the Gumbel marginals $\mathscr{G}_{MM}$ (dash-dot line –·–). Boxes indicate return period levels in years

use of asymptotic Gumbel marginals. Finally, it is noted that the bivariate ACER methodology has been studied in more detail in an initial study on synthetic data with known extreme value distribution, and therefore a known 100-year return period level $\left(\xi^{100yr}, \eta^{100yr}\right)$. To get an idea about the performance of the ACER method and the existing Gumbel logistic model with Gumbel marginals fitted by the method of moments, Monte Carlo simulations were carried out to produce 100 bivariate data samples. It was observed that the predicted 100-year return period levels were consistently better for the ACER method.

10.5.2 Wind Speed and Wave Height Measured at a North Sea Weather Station

This example presents results obtained by Naess and Karpa (2015b). Wind speed (WS—3 hours mean [m/s]) and significant wave height (Hs—total sea [m]) data measured in the Norwegian sea at location N 65.29, E 7.32 were analyzed to obtain numerical estimates of the bivariate extreme value distribution. Figure 10.9 shows the geographical position of the measurement site. The data were recorded during 54 years (1957–2011), 8 times per day (every three hours).

Figure 10.10 demonstrates the plot of the observed data. As it is seen from the plot, there is a rather strong dependence between the two time series.

The Pearson product-moment correlation coefficient is found to be $\rho = 0.79$. The Kendall's rank correlation coefficient is $\tau = 0.56$ and Spearman's ρ is equal to 0.7, which also indicates a nonlinear connection between WS and Hs.

It should be noted that the available bivariate observations have low accuracy. This especially concerns the significant wave height data, where the graduating

Fig. 10.9 Map of a part of Norway with the marked location

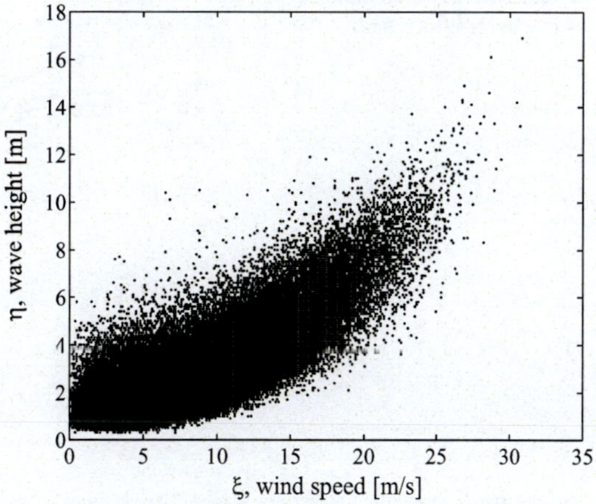

Fig. 10.10 Coupled observations of wind speed data (ξ axis) and significant wave height data (total sea, η axis)

mark is 0.1 m, and on average, there are 98 unique numerical values of the Hs data per year. Obviously, for fairly accurate estimation of the bivariate ACER functions, more data would be required. In order to get a good sample size of records (realizations), it was decided to divide the data series into 18 three-years records for the analysis. By this procedure, also the standard deviation of the ACER function estimates can be calculated fairly accurately.

As in the previous example, the univariate ACER functions $\hat{\varepsilon}_k$ were estimated first, using the Matlab-based standalone downloadable application (Karpa 2012). In Figs. 10.11 and 10.12, $\hat{\varepsilon}_k$ is plotted versus different levels of wind speeds and wave heights, respectively, for different values of k. From both figures, it is clearly seen that there is significant time dependence between WS observations, as well as between Hs data. It is also understood that this dependence effect is largely

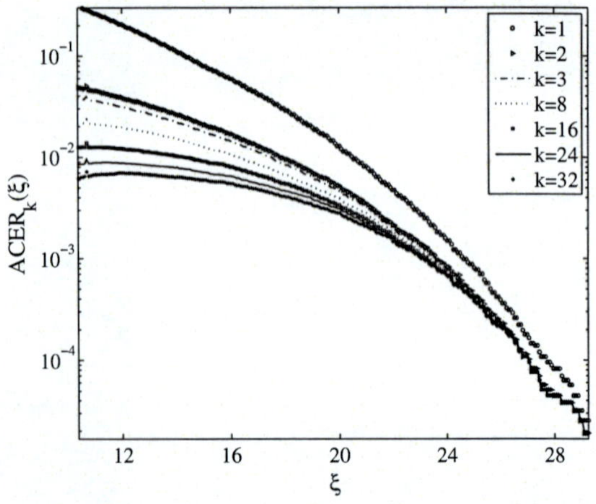

Fig. 10.11 ACER estimates for different degrees of conditioning. Wind speed data

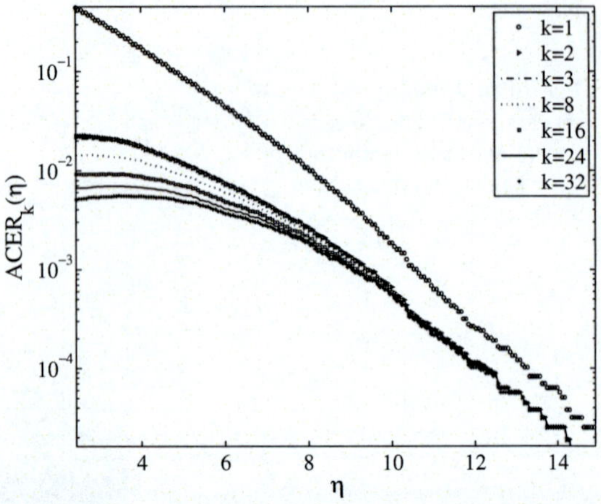

Fig. 10.12 ACER estimates for different degrees of conditioning. Significant wave height (total sea)

accounted for by $k = 16$ since there is a marked degree of convergence in the tail of $\hat{\varepsilon}_k$ for $k \geq 16$ in both cases. Obviously, $k = 16$ corresponds to exceedances separated by at least two days of non-exceedances for three hours observations. For $k \geq 32$, which corresponds to four days declustered data, full convergence has been achieved. Figures 10.11 and 10.12 also demonstrate that for extreme value estimation, $\hat{\varepsilon}_2$ can be used since the ACER functions for $k \geq 2$ all converge in the far tail. Again, this clearly demonstrates the power of an ACER function plot as a diagnostic tool to decide on the value of k needed for extreme value estimation in a particular case. In spite of significant dependence effects for the WS and Hs data with lower magnitudes, for the extreme values this is largely absent. This makes it possible to choose $k = 2$, which makes much more data available for estimation, with a possible reduction of uncertainty in estimation as a result.

Figures 10.13 and 10.14 show the plots of the optimized parametric fit to the data for $\hat{\varepsilon}_k$ for $k = 2$ for both time series. In particular, 100-year return level value and its 95% CI are estimated parametrically and plotted. For the wind speed data, the

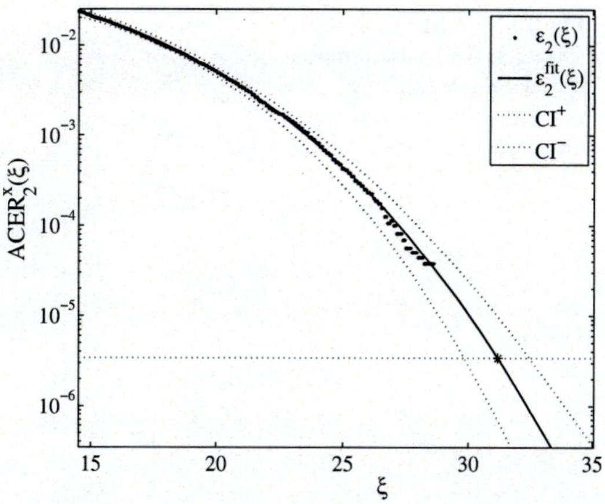

Fig. 10.13 Plot of $\hat{\varepsilon}_2(\xi)$ versus wind speeds ξ on a logarithmic scale for the optimized parameter values; $\xi_1 = 14.5$

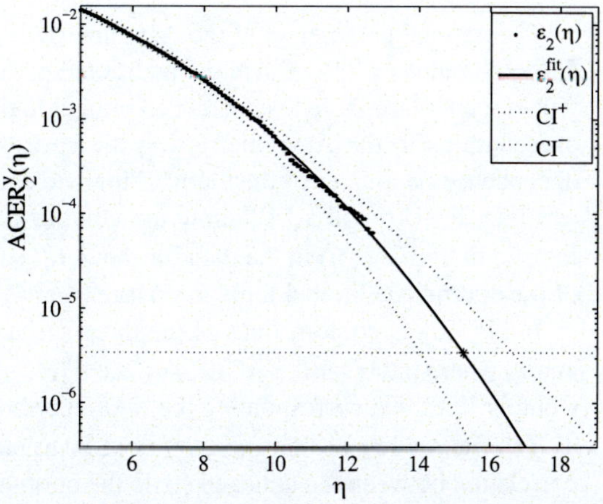

Fig. 10.14 Plot of $\hat{\varepsilon}_2(\eta)$ against wave heights η on a logarithmic scale for the optimized parameter values; $\eta_1 = 4.5$

optimal parameters are: $q = 0.05$, $b = 0.1$, $a = 1.9 \cdot 10^{-4}$, $c = 3.14$, while in case of Hs data, the parameters of the optimal curve are: $q = 0.04$, $b = -2.27$, $a = 0.02$, $c = 2.23$.

Figure 10.15 demonstrates the empirically estimated bivariate ACER surfaces $\hat{\mathscr{E}}_k(\xi, \eta)$ for different values of k on a logarithmic scale. $\hat{\mathscr{E}}_k(\xi, \eta)$ with $k = 1$ is the uppermost. As it is seen from the figure, the cross-section of the surfaces at a high level of the wave height η gives the univariate ACER functions of the wind speed data, while the cross-section at a high wind speed level represents the univariate ACER of the Hs time series, respectively.

The same arguments as in the univariate case are applied to make the decision about the bivariate ACER surface to be used in the analyses. That is, as long as the surfaces for $k \geq 2$ all converge in the tail and estimation of $\hat{\mathscr{E}}_2(\xi, \eta)$ is more accurate due to availability of more data, we would choose the surface with the degree of conditioning $k = 2$.

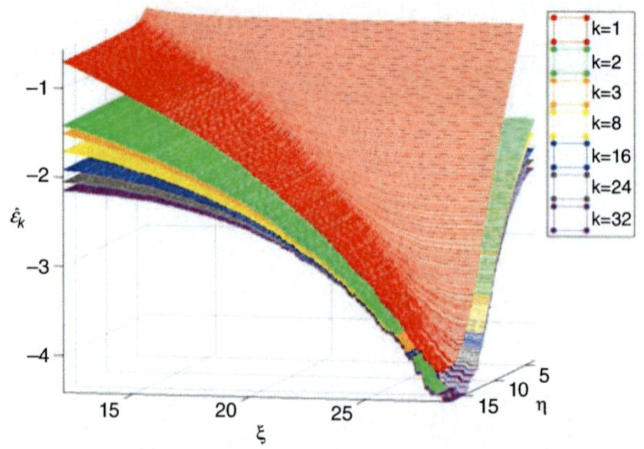

Fig. 10.15 Bivariate ACER surface estimates for different degrees of conditioning. $\hat{\mathscr{E}}_k(\xi, \eta)$ surfaces are plotted on a logarithmic scale

The optimal parameters of the Asymmetric logistic fit were found to be: $m_{AL} = 7$, $\theta = 1$ and $\phi = 0.91$. Since the additional parameters θ and ϕ are close to one, it would seem reasonable to consider a Gumbel logistic model. The same optimization procedure as in the Asymmetric logistic case, was applied to obtain the optimal dependence parameter of the Gumbel logistic copula $m_{GL} = 4.78$.

Figures 10.16 and 10.17 show the contour plots of the optimized Asymmetric logistic fit $\mathscr{A}\mathscr{L}_2(\xi, \eta)$ to the data for the $\hat{\mathscr{E}}_2(\xi, \eta)$ surface and also the contour plots of the optimized Gumbel logistic surface $\mathscr{G}\mathscr{L}_2(\xi, \eta)$.

In Fig. 10.16, contour lines of the three surfaces are plotted for those levels of ξ and η, where the bivariate ACER surface $\hat{\mathscr{E}}_2(\xi, \eta)$ have been empirically estimated. Contour lines that correspond to the return period levels are presented in Fig. 10.17.

The figures reveal that the empirical bivariate ACER surface $\hat{\mathscr{E}}_2$ captures high correlation between the data and so do the optimally fitted $\mathscr{G}\mathscr{L}_2$ and $\mathscr{A}\mathscr{L}_2$ surfaces. Note that the contour lines of the bivariate ACER surface of fully correlated data would show up as lines that consist of only horizontal and vertical line segments. This happens because for such data, $H_{xy}(\xi, \eta) = \min\big(F_x(\xi),\ G_y(\eta)\big)$, which implies that $\mathscr{E}_k(\xi, \eta) = \max\big(\varepsilon_k^x(\xi), \varepsilon_k^y(\eta)\big)$, cf. Eq. (10.18). It is seen that the level of agreement between the estimated bivariate ACER and optimized Asymmetric logistic and Gumbel logistic surfaces is equally significant. Thereby, the optimized Gumbel logistic model with asymptotically consistent marginals obtained from the optimized univariate ACER, can be used as the parametric representative of the bivariate ACER surface estimated from the given data set. The Matlab programs for ACER 2D analyses from Karpa (2014) were used to obtain the results presented in this section.

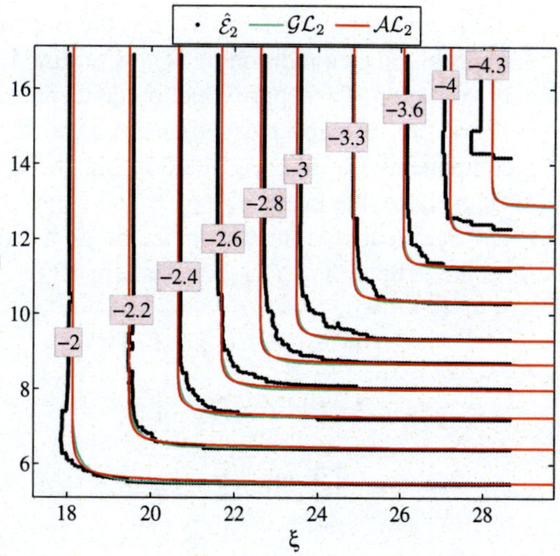

Fig. 10.16 Contour plot of the empirically estimated $\hat{\mathcal{E}}_2(\xi, \eta)$ surface (•), optimized Gumbel logistic $\mathcal{GL}_2(\xi, \eta)$ (∘) and optimized Asymmetric logistic $\mathcal{AL}_2(\xi, \eta)$ (—) surfaces. Boxes indicate levels on a logarithmic scale

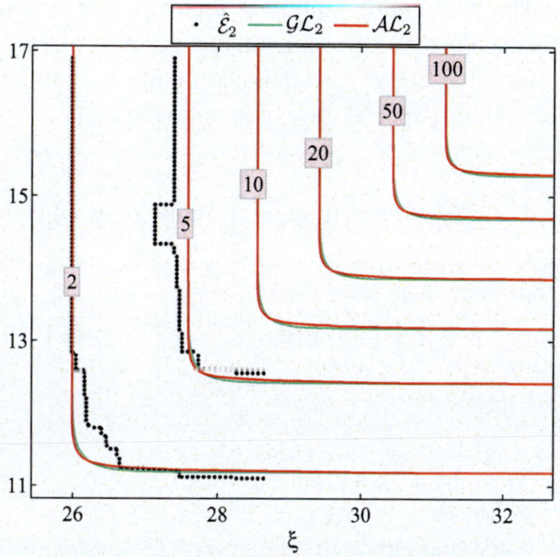

Fig. 10.17 Contour plot of the return period levels for $\hat{\mathcal{E}}_2(\xi, \eta)$ surface (•), optimized Asymmetric logistic $\mathcal{AL}_2(\xi, \eta)$ (—) and optimized Gumbel logistic $\mathcal{GL}_2(\xi, \eta)$ (∘) surfaces. Boxes indicate return period levels in years

Appendix 1: The Sequence of Conditioning Approximations

Consider a bivariate stochastic process $Z(t) = \big(X(t), Y(t)\big)$ with dependent component processes, which has been observed over a time interval, $(0, T)$ say. Assume that the sampled values $Z_1 = (X_1, Y_1), \ldots, Z_N = (X_N, Y_N)$ are allocated to the (usually equidistant) discrete times $t_1, \ldots, t_N$ in $(0, T)$. The goal is to determine the joint distribution function of the extreme value vector $M_N = \big(M_{x,N}, M_{y,N}\big)$, where $M_{x,N} = \max\big\{X_j\,;\, j = 1, \ldots, N\big\}$, and with a similar definition of $M_{y,N}$. Specifically, the goal is to estimate $P(\xi, \eta) = \text{Prob}\big(M_{x,N} \leq \xi, M_{y,N} \leq \eta\big)$ accurately for large values of ξ and η.

Following the pattern of Chap. 5, the implementation of a sequence of approximations based on conditioning is now outlined, where the first is a one-step memory approximation. This approximation concept is described by Naess (1985a, 1990a).

Hereafter, whenever expedient to ease the notation, $\zeta = (\xi, \eta)$ is used with a componentwise ordering relationship for Z_i, e.g., $Z_i \leq \zeta$ means $X_i \leq \xi$ and $Y_i \leq \eta$. Also, the event $\mathscr{C}_{kj}(\zeta) = \mathscr{C}_{kj}(\xi, \eta) = \{Z_{j-1} \leq \zeta, \ldots, Z_{j-k+1} \leq \zeta\}$ of $k-1$ consecutive componentwise non-exceedances ($k \geq 2$) is introduced. Then, from the definition of $P(\xi, \eta)$, it emerges that,

$$\begin{aligned} P(\xi,\eta) &= P(\zeta) = \text{Prob}\big(\mathscr{C}_{N+1,N+1}(\zeta)\big) \\ &= \text{Prob}\big(Z_N \leq \zeta \,|\, \mathscr{C}_{NN}(\zeta)\big) \cdot \text{Prob}\big(\mathscr{C}_{NN}(\zeta)\big) \\ &= \prod_{j=2}^{N} \text{Prob}\big(Z_j \leq \zeta \,|\, \mathscr{C}_{jj}(\zeta)\big) \cdot \text{Prob}\big(\mathscr{C}_{22}(\zeta)\big). \end{aligned} \tag{10.27}$$

The first approximation of the sequence is obtained by assuming that the observed data pairs are independent, that is, data points Z_i and Z_j are statistically independent for all $i, j, i \neq j$, so that all conditioning in Eq. (10.27) can be neglected.

In this special case, it is obtained that,

$$\begin{aligned} P(\zeta) \approx P_1(\zeta) &= \prod_{j=1}^{N} \text{Prob}\big(Z_j \leq \zeta\big) \\ &= \prod_{j=1}^{N} \big(1 - \text{Prob}(X_j > \xi) - \text{Prob}(Y_j > \eta) + \text{Prob}(X_j > \xi, Y_j > \eta)\big). \end{aligned} \tag{10.28}$$

Now, the designations $\alpha_{1j}(\xi;\eta) = \text{Prob}(X_j > \xi)$, $\beta_{1j}(\eta;\xi) = \text{Prob}(Y_j > \eta)$ and $\gamma_{1j}(\xi,\eta) = \gamma_{1j}(\zeta) = \text{Prob}(Z_j > \zeta)$ for $1 \leq j \leq N$ are introduced. It should be noted that although neither $\alpha_{1j}(\xi;\eta)$ depends on η nor $\beta_{1j}(\eta;\xi)$ depends on ξ, yet this notation is kept for the correct further derivations.

Equation (10.28) can now be rewritten as,

$$\begin{aligned} P(\zeta) \approx P_1(\zeta) &= \prod_{j=1}^{N} \{1 - \alpha_{1j}(\xi;\eta) - \beta_{1j}(\eta;\xi) + \gamma_{1j}(\zeta)\} \\ &\approx \exp\Big\{-\sum_{j=1}^{N} \big(\alpha_{1j}(\xi;\eta) + \beta_{1j}(\eta;\xi) - \gamma_{1j}(\zeta)\big)\Big\}; \end{aligned} \tag{10.29}$$

for large values of the components of ζ, where the approximation $1 - x \approx \exp(-x)$ for small values of x, has been applied, cf. Chap. 5.

In general, the variables Z_j are statistically dependent in the componentwise sense. In this case, clearly the first genuine conditioning approximation is obtained by neglecting all previous data except the immediate predecessor in Eq. (10.27). While this approximation sometimes captures the effect of dependence in the time series on the extreme value distribution, it is in general not sufficient. Consequently, the following $k-1$-step memory approximation is adopted,

$$P(\zeta) \approx P_k(\zeta) = \prod_{j=k}^{N} \text{Prob}\big(Z_j \leq \zeta \,|\, \mathcal{C}_{kj}(\zeta)\big) \cdot \text{Prob}\big(\mathcal{C}_{kk}(\zeta)\big)$$

$$= \prod_{j=k}^{N} \Big\{1 - \Big(1 - \text{Prob}\big(Z_j \leq \zeta \,|\, \mathcal{C}_{kj}(\zeta)\big)\Big)\Big\} \cdot \text{Prob}\big(\mathcal{C}_{kk}(\zeta)\big) \tag{10.30}$$

$$\approx \exp\Big\{-\sum_{j=k}^{N}\Big(1 - \text{Prob}\big(Z_j \leq \zeta \,|\, \mathcal{C}_{kj}(\zeta)\big)\Big)\Big\} \cdot \text{Prob}\big(\mathcal{C}_{kk}(\zeta)\big),$$

for large values of the components of ζ. Then, $k = 2$ means conditioning on only the previous observation.

By introducing the notations $\alpha_{kj}(\xi;\eta) = \text{Prob}\big(X_j > \xi \,|\, \mathcal{C}_{kj}(\zeta)\big)$, $\beta_{kj}(\eta;\xi) = \text{Prob}\big(Y_j > \eta \,|\, \mathcal{C}_{kj}(\zeta)\big)$ and $\gamma_{kj}(\xi,\eta) = \gamma_{kj}(\zeta) = \text{Prob}\big(Z_j > \zeta \,|\, \mathcal{C}_{kj}(\zeta)\big)$, for $k \leq j \leq N$, it can now be shown that,

$$\exp\Big\{-\sum_{j=k}^{N}\Big(1 - \text{Prob}\big(Z_j \leq \zeta \,|\, \mathcal{C}_{kj}(\zeta)\big)\Big)\Big\}$$

$$= \exp\Big\{-\sum_{j=k}^{N}\big(\alpha_{kj}(\xi;\eta) + \beta_{kj}(\eta;\xi) - \gamma_{kj}(\zeta)\big)\Big\}. \tag{10.31}$$

Similarly, it is found that,

$$\text{Prob}\big(\mathcal{C}_{kk}(\zeta)\big) \approx \exp\Big\{-\sum_{j=1}^{k-1}\big(\alpha_{jj}(\xi;\eta) + \beta_{jj}(\eta;\xi) - \gamma_{jj}(\zeta)\big)\Big\}; \tag{10.32}$$

for large values of the components of ζ. Under these conditions, this leads to the result,

$$P_k(\zeta) \approx \exp\Big\{-\sum_{j=k}^{N}\big(\alpha_{kj}(\xi;\eta) + \beta_{kj}(\eta;\xi) - \gamma_{kj}(\zeta)\big)$$

$$-\sum_{j=1}^{k-1}\big(\alpha_{jj}(\xi;\eta) + \beta_{jj}(\eta;\xi) - \gamma_{jj}(\zeta)\big)\Big\}. \tag{10.33}$$

Thereby, based on the definition of the extreme value distribution $P(\zeta)$ and the properties of conditional probability, a set $\{P_k(\zeta)\}_{k=1}^{N}$ of conditional probability distributions has been constructed, which converges to the target distribution $P(\zeta)$ of the extreme value M_N in the limit as k increases for large values of the components of ζ.

For most applications, and for practical significance, the following assumption on this sequence of approximations is made: there is an effective k_e satisfying $k_e \ll N$ such that $P(\zeta) = P_{k_e}(\zeta)$. Then, $P_1(\zeta) \le P_2(\zeta) \le \ldots \le P_{k_e}(\zeta) = P(\zeta)$. It may be noted that for a k-dependent stationary bivariate stochastic process $Z(t)$, that is, for data where Z_i and Z_j are independent componentwise whenever $|j - i| > k$, then $P(\zeta) = P_{k+1}(\zeta)$ exactly, as in the univariate case.

It will be verified that the property $k_e \ll N$ is indeed satisfied for the wind speed data analyzed in the present paper. Also, under this assumption, $\sum_{j=1}^{k-1}\big(\alpha_{jj}(\xi;\eta) + \beta_{jj}(\eta;\xi) - \gamma_{jj}(\zeta)\big)$ is generally negligible compared to $\sum_{j=k}^{N}\big(\alpha_{kj}(\xi;\eta) + \beta_{kj}(\eta;\xi) - \gamma_{kj}(\zeta)\big)$. This leads to the approximation ($k = k_e$),

$$P(\zeta) \approx P_k(\zeta) \approx \exp\Big\{-\sum_{j=k}^{N}\big(\alpha_{kj}(\xi;\eta) + \beta_{kj}(\eta;\xi) - \gamma_{kj}(\zeta)\big)\Big\}\,; \tag{10.34}$$

for large values of ξ and η, from which it emerges that for the estimation of the bivariate extreme value distribution, it is sufficient to estimate the sequence of functions $\{\alpha_{kj}(\xi;\eta) + \beta_{kj}(\eta;\xi) - \gamma_{kj}(\zeta)\}_{j=k}^{N}$.

Appendix 2: Empirical Estimation of the Bivariate ACER Functions

To get a more compact representation, it is expedient to introduce the concept of k'th order bivariate average conditional exceedance rate (ACER) function as follows,

$$\mathcal{E}_k(\zeta) = \frac{1}{N-k+1}\sum_{j=k}^{N}\big(\alpha_{kj}(\xi;\eta) + \beta_{kj}(\eta;\xi) - \gamma_{kj}(\zeta)\big)\,;\; k = 1, 2, \ldots \tag{10.35}$$

Hence, when $N \gg k$, and for large values of the components of ζ,

$$P_k(\zeta) \approx \exp\{-(N-k+1)\,\mathcal{E}_k(\zeta)\}\,. \tag{10.36}$$

Note that for higher dimensional cases, it is expedient to use an alternative formulation of the ACER functions, which is related to the problem of numerical calculation of the ACER functions, viz.,

$$\mathcal{E}_k(\zeta) = \frac{1}{N-k+1}\sum_{j=k}^{N}\Big(1 - \text{Prob}\big(Z_j \le \zeta \mid \mathcal{C}_{kj}(\zeta)\big)\Big)\;;\; k = 2, 3, \ldots \tag{10.37}$$

A few more details on the numerical estimation of the bivariate ACER functions are worth noting. It is useful to start by introducing a set of random functions. For $k = 2, \ldots, N$, and $k \le j \le N$, let,

$$\begin{aligned}
A_{kj}(\xi;\eta) &= \mathbf{1}\Big\{X_j > \xi \cap \mathcal{C}_{kj}(\zeta)\Big\},\\
B_{kj}(\eta;\xi) &= \mathbf{1}\Big\{Y_j > \eta \cap \mathcal{C}_{kj}(\zeta)\Big\},\\
G_{kj}(\xi,\eta) &= G_{kj}(\zeta) = \mathbf{1}\Big\{Z_j > \zeta \cap \mathcal{C}_{kj}(\zeta)\Big\},\\
C_{kj}(\xi,\eta) &= C_{kj}(\zeta) = \mathbf{1}\Big\{\mathcal{C}_{kj}(\zeta)\Big\},
\end{aligned} \tag{10.38}$$

where $\mathbf{1}\{\mathcal{A}\}$ denotes the indicator function of some event $\mathcal{A}$.

From these definitions it follows that, for instance,

$$\alpha_{kj}(\xi;\eta) = \frac{\mathbb{E}[A_{kj}(\xi;\eta)]}{\mathbb{E}[C_{kj}(\zeta)]}\,, \tag{10.39}$$

where $\mathbb{E}[\cdot]$ denotes the expectation operator. A similar equation holds for $\beta_{kj}(\eta;\xi)$ with $B_{kj}(\eta;\xi)$ instead of $A_{kj}(\xi;\eta)$ in the numerator, and for $\gamma_{kj}(\zeta)$ with $G_{kj}(\zeta)$ instead of $A_{kj}(\xi;\eta)$, by analogy.

Assuming ergodicity of the process $Z(t) = \big(X(t), Y(t)\big)$, then obviously $\mathcal{E}_k(\zeta) = \big(\alpha_{kk}(\xi;\eta) + \beta_{kk}(\eta;\xi) - \gamma_{kk}(\zeta)\big) = \ldots = \big(\alpha_{kN}(\xi;\eta) + \beta_{kN}(\eta;\xi) - \gamma_{kN}(\zeta)\big)$, and it may be assumed that for the bivariate time series at hand,

$$\mathcal{E}_k(\zeta) = \lim_{N\to\infty}\frac{\sum_{j=k}^{N}\big(\mathrm{a}_{kj}(\xi;\eta) + \mathrm{b}_{kj}(\eta;\xi) - \mathrm{g}_{kj}(\zeta)\big)}{\sum_{j=k}^{N}\mathrm{c}_{kj}(\zeta)}\,, \tag{10.40}$$

where $\mathrm{a}_{kj}(\xi;\eta)$, $\mathrm{b}_{kj}(\eta;\xi)$, $\mathrm{g}_{kj}(\zeta)$ and $\mathrm{c}_{kj}(\zeta)$ are the realized values of $A_{kj}(\xi;\eta)$, $B_{kj}(\eta;\xi)$, $G_{kj}(\zeta)$ and $C_{kj}(\zeta)$, respectively, for the observed time series.

While Eq. (10.40) clearly displays how the conditional exceedances are counted for the estimation of the bivariate ACER functions, this approach becomes very cumbersome for higher dimensional problems. Fortunately, this can be circumvented by invoking Eq. (10.37). It is observed that,

$$\mathrm{Prob}\big(Z_j \le \zeta \mid \mathscr{C}_{kj}(\zeta)\big) = \frac{\mathrm{Prob}\big(Z_j \le \zeta \cap \mathscr{C}_{kj}(\zeta)\big)}{\mathrm{Prob}\big(\mathscr{C}_{kj}(\zeta)\big)} = \frac{\mathbb{E}[C_{k+1,j+1}(\zeta)]}{\mathbb{E}[C_{kj}(\zeta)]}, \tag{10.41}$$

since clearly $\{Z_j \le \zeta\} \cap \mathscr{C}_{kj}(\zeta) = \mathscr{C}_{k+1,j+1}(\zeta)$. A consequence of this, is that Eq. (10.40) may be replaced by the following equation,

$$\mathscr{E}_k(\zeta) = 1 - \lim_{N\to\infty} \frac{\sum_{j=k}^{N} c_{k+1,j+1}(\zeta)}{\sum_{j=k}^{N} c_{kj}(\zeta)}, \tag{10.42}$$

which is a form particularly suitable for estimating $\mathscr{E}_k(\zeta)$ for high dimensional distributions.

Clearly, $\lim\limits_{\xi,\eta\to\infty} \mathbb{E}[C_{kj}(\zeta)] = 1$. Hence,

$$\lim_{\xi,\eta\to\infty} \tilde{\mathscr{E}}_k(\zeta) \,/\, \mathscr{E}_k(\zeta) = 1, \tag{10.43}$$

where

$$\tilde{\mathscr{E}}_k(\zeta) = \lim_{N\to\infty} \frac{\sum_{j=k}^{N} \Big(\mathbb{E}[A_{kj}(\xi;\eta)] + \mathbb{E}[B_{kj}(\xi;\eta)] - \mathbb{E}[G_{kj}(\zeta)]\Big)}{N-k+1}. \tag{10.44}$$

The advantage of using the modified bivariate ACER function $\tilde{\mathscr{E}}_k(\zeta)$ for $k \ge 2$ is that it is somewhat easier to use for non-stationary or long-term statistics than $\mathscr{E}_k(\zeta)$. This aspect is discussed in Chap. 5, see also Karpa and Naess (2013). Since our focus is on the values of the ACER functions at the extreme levels, any function may be used that provides correct predictions of the appropriate ACER function at these extreme levels.

Now, let us look at the problem of estimating confidence intervals for the bivariate ACER function. If several realizations of the time series $Z(t) = \big(X(t), Y(t)\big)$ are provided or the time series can be appropriately sectioned into several records, e.g., several annual, or other time span records, the sample estimate of $\mathscr{E}_k(\zeta)$ would be,

$$\hat{\mathscr{E}}_k(\zeta) = \frac{1}{R}\sum_{r=1}^{R} \hat{\mathscr{E}}_k^{(r)}(\zeta), \tag{10.45}$$

where R is the number of realizations (samples). $\hat{\mathscr{E}}_k^{(r)}(\zeta)$ is estimated using the result from Eq. (10.40) for the stationary time series, or using the result from Eq. (10.44) for non-stationary time series, where the index (r) refers to realization no. r. The sample standard deviation $\hat{s}_k(\zeta)$ then can be estimated by the standard formula,

$$\hat{s}_k(\zeta)^2 = \frac{1}{R-1}\sum_{r=1}^{R}\left(\hat{\mathscr{E}}_k^{(r)}(\zeta) - \hat{\mathscr{E}}_k(\zeta)\right)^2 . \tag{10.46}$$

Assuming that realizations are independent, Eq. (10.46) leads to a good approximation of the 95% confidence interval $\mathrm{CI} = \left(\mathrm{CI}^-(\zeta), \mathrm{CI}^+(\zeta)\right)$ for the value $\mathscr{E}_k(\zeta)$, where,

$$\mathrm{CI}^{\pm}(\zeta) = \hat{\mathscr{E}}_k(\zeta) \pm \tau \cdot \frac{\hat{s}_k(\zeta)}{\sqrt{R}} , \tag{10.47}$$

and $\tau = t^{-1}\big((1-0.95)/2,\ R-1\big)$ is the corresponding quantile of the Student's t-distribution with $R-1$ degrees of freedom.

Chapter 11
Space–Time Extremes of Random Fields

11.1 Introduction

The initial motivation for developing the methods presented in this chapter was the search for a practical solution to the air gap problem for offshore structures. Over the years several reports had been filed detailing damage of the deck structure of offshore platforms due to wave impacts. In fact, such damage seemed to occur more frequently than could be expected from predictions based on standard theory. These wave impacts are highly undesirable events as they may, in the worst case, compromise the structural integrity of the platform deck. Hence, considerable attention was being paid to this problem over some time, and now reasonably good predictive tools have become available, as will be demonstrated in this chapter.

It was eventually realized that the standard procedure for predicting the extreme wave crest height to be expected at a given platform location is largely based on extreme value statistics for a single point (Forristall, 2006). It is clear that if this is used as a basis for predicting the probability of wave impact on the deck structure, then the fact that there is also a significant area effect on the extreme crest height distribution over the deck area is neglected. Hence, for a proper solution of the air gap problem, it is necessary to model the ocean surface as a random field so that also the spatial aspect can be correctly dealt with (Forristall, 2006). For the case of a homogeneous Gaussian random wave field, this aspect is also discussed by Socquet-Juglard et al. (2005) and Baxevani and Rychlik (2006).

Recently, a new, simplified method for predicting the space–time extreme value statistics of homogeneous Gaussian random fields has been proposed (Naess and Batsevych, 2010). In the present chapter it is shown that this simplified method can be extended to deal with homogeneous non-Gaussian random fields and, in particular, with the special case of a second-order homogeneous ocean wave field (Naess and Batsevych, 2012). Additionally, a new semi-parametric representation of the space–time extreme value distribution for quite general homogeneous random fields over rectangular domains is proposed. It is demonstrated that both the

A. Naess, *Applied Extreme Value Statistics*,
https://doi.org/10.1007/978-3-031-60769-1_11

simplified method and the new semi-parametric method make it possible to make good predictions of the extreme crest heights over the deck area for jacket platforms, which are assumed to be transparent for the big waves. Even if the second-order ocean wave field is intrinsically non-Gaussian, its deviation from a Gaussian field is not very significant. Therefore, to implement a more stringent test of the two methods proposed, an example of a strongly non-Gaussian field has also been included. Both methods will be shown to provide accurate extreme value predictions for this case as well.

The proposed simplified approach, as opposed to the new semi-parametric method, can be easily modified to also provide good solutions of the air gap problem for a TLP or a semi-submersible platform, provided measured or simulated time series of the ocean surface elevation or air gap at a sufficient number of points below the deck structure are available, which properly account for the motions of the platform and the effect of the structure on the wave field. Due to the interaction effects between the wave field and the structure, the wave field is not homogeneous in space under the platform deck, even if the incoming wave field is of this type. Hence, a prediction method that can deal with a random field that is nonhomogeneous in space while stationary in time will be required. The simplified method proposed in this chapter can provide such a prediction tool.

Over the last 20 years or so, several approximations for the extreme value distributions of a real-valued random field $X(\mathbf{x}, t)$, where $\mathbf{x} = (x_1, \ldots, x_d) \in \mathscr{D} \subseteq \mathbb{R}^d$ and $t \in \mathscr{T} \subseteq \mathbb{R}$, have been proposed, of which a few are mentioned (Adler, 1981; Vanmarcke, 1983; Sun, 1993; Maes and Breitung, 1997; Ditlevsen, 2004). Of particular interest are two books which provide substantial theoretical results, mainly for Gaussian random fields (Piterbarg, 1996; Adler and Taylor, 2007).

11.2 Spatial–Temporal Extremes for Gaussian Random Fields

Let $X = X(\mathbf{x}, t)$ denote a zero mean random field defined on $\mathbb{R}^d \times \mathbb{R}$, and let

$$\hat{X} = \hat{X}(\mathscr{D} \times \mathscr{T}) = \max\{X(\mathbf{x}, t); \mathbf{x} \in \mathscr{D},\ t \in \mathscr{T}\}, \tag{11.1}$$

where $\mathscr{T}$ denotes a time interval, specifically, $\mathscr{T} = (0, T)$ for a given T.

In Chapter 14 of Adler and Taylor (2007), the following result is proved for a smooth Gaussian random field:

$$\left|\text{Prob}(\hat{X} \geq \xi) - \mathbb{E}[\phi(A_\xi(X, \mathscr{D} \times \mathscr{T}))]\right| < O\left(e^{-\alpha\xi^2/(2\sigma^2)}\right), \tag{11.2}$$

where $\phi(A_\xi(X, \mathscr{D} \times \mathscr{T}))$ denotes the Euler characteristic of $A_\xi(X, \mathscr{D} \times \mathscr{T})$, which is the excursion set of X over $\mathscr{D} \times \mathscr{T}$. That is, $A_\xi(X, \mathscr{D} \times \mathscr{T}) = \{(\mathbf{x}, t) \in \mathscr{D} \times \mathscr{T} : X(\mathbf{x}, t) \geq \xi\}$, σ^2 is the variance of X (assumed constant), and $\alpha > 1$ is a constant that can be identified as discussed by Adler and Taylor (2007). Certain restrictions apply to the geometry of the domain $\mathscr{D}$. Instead of going into detail about this formula, a largely equivalent result will be given, which is due to Piterbarg (1996). This result applies to a homogeneous Gaussian field, which has a simple geometry for $\mathscr{D}$, typically a nice convex domain, e.g., like a rectangle or a sphere. Any numerical representation of the homogeneous random field X by necessity leads to a truncation of the spectral density (assumed to exist). Hence, there is no practical limitation in assuming that the spectral density of X has bounded support, which leads to smoothness of any order. Piterbarg's (asymptotic) formula, cf. Chapter 1 of Piterbarg (1996), can then be written as ($\xi = u\sigma, u \to \infty$)

$$\text{Prob}(\hat{X} \leq \sigma u) \simeq P_A(u) = \exp\left\{-\varphi(u)(2\pi)^{(d+1)/2} H_d(u) V_d\right\}, \tag{11.3}$$

where $\simeq$ denotes asymptotically equal, $H_k(u) = (-1)^k \varphi(u)^{-1}(\mathrm{d}/\mathrm{d}u)^k \varphi(u)$ is a Hermite polynomial of order k, φ is the standard normal probability density, that is, $\varphi(u) = (\sqrt{2\pi})^{-1} e^{-u^2/2}$, $V_d = m(\mathscr{D} \times \mathscr{T})/m(\mathscr{V})$, $m(\mathscr{V}) = (2\pi\sigma)^{d+1}/\sqrt{\det \Lambda}$ (Λ is defined by Eq. (11.5) below), and $m(\mathscr{A})$ denotes the Lebesgue measure (volume) of the domain $\mathscr{A}$.

Before the calculation of V_d is discussed any further, a slight detour is made and some basic results are presented as detailed by Krogstad et al. (2004). Denoting the spectral density of X by $\Psi(\mathbf{k}, \omega)$, its covariance function $\gamma(\mathbf{x}, t)$ is given as

$$\gamma(\mathbf{x}, t) = \int_{\mathbb{R}^{d+1}} e^{\mathrm{i}(\mathbf{kx}' + \omega t)} \, \Psi(\mathbf{k}, \omega) \, d\mathbf{k} d\omega, \tag{11.4}$$

where $\mathbf{k}$ denotes the wave number vector.

Denote by $\Lambda = (\Lambda_{ij})$ the essentially nonsingular covariance matrix of the gradient $\nabla X(\mathbf{x}, t) = (\partial X/\partial x_1, \ldots, \partial X/\partial x_d, \partial X/\partial t)$ evaluated in the point $\mathbf{x} = \mathbf{0}$, $t = 0$

$$\begin{aligned} \Lambda_{ij} &= \text{Cov}\left(\nabla_i X(\mathbf{0}, 0), \nabla_j X(\mathbf{0}, 0)\right) \\ &= -\nabla_i \nabla_j \gamma(\mathbf{0}, 0) = \int_{\mathbb{R}^{d+1}} r_i r_j \Psi(\mathbf{r}) \, \mathrm{d}\mathbf{r}, \end{aligned} \tag{11.5}$$

where $\mathbf{r} = (r_1, \ldots, r_d, r_{d+1}) = (k_1, \ldots, k_d, \omega)$.

A linear space transformation $A : \mathscr{D} \times \mathscr{T} \to \tilde{\mathscr{D}} \times \tilde{\mathscr{T}}$ is now performed with the help of a nonsingular $(d + 1) \times (d + 1)$ matrix A

$$\tilde{\mathbf{y}} = A\tilde{\mathbf{x}}, \tag{11.6}$$

where $\tilde{\mathbf{x}} = (\mathbf{x}, t) = (x_1, \ldots, x_d, t)$, then a new Gaussian random field $\tilde{X}(\tilde{\mathbf{y}})$ defined as

$$\tilde{X}(\tilde{\mathbf{y}}) = X(A^{-1}\tilde{\mathbf{y}})/\sigma \tag{11.7}$$

has unit variance, $\tilde{\sigma} = 1$, and the covariance matrix of the gradient of $\tilde{X}$ is given by

$$\tilde{\Lambda} = \frac{1}{\sigma^2}(A^{-1})^T \Lambda A^{-1}\,. \tag{11.8}$$

Since Λ is a symmetrical matrix, the transformation $A = \sqrt{d+1}\sigma^{-1}\Lambda^{1/2}$ is well defined and results in $\tilde{\Lambda} = (d+1)^{-1}I$. For this special case of the Gaussian random field $\tilde{X}(\tilde{\mathbf{y}})$ corresponding to the mentioned covariance matrix $\tilde{\Lambda}$, a complementary extreme value distribution can be obtained for the particular case of a rectangular domain Δ with edges oriented along the coordinate axes, that is, $\Delta = \otimes_{i=1}^{d+1}[0, L_i]$, where $\otimes$ denotes a Cartesian product. This probability can be written as

$$\mathrm{Prob}(\hat{\tilde{X}} > u) \simeq \varphi(u)\sum_{k=1}^{d+1}\frac{H_{k-1}(u)}{[2\pi(d+1)]^{k/2}}\Sigma_k(\mathbf{L}) + \int_u^{\infty}\varphi(x)\mathrm{d}x\,, \tag{11.9}$$

where $\mathbf{L} = (L_1, \ldots, L_{d+1})$ and $\Sigma_k(\mathbf{L})$ is the kth elementary symmetric polynomial, defined as

$$\Sigma_k(\mathbf{L}) = \sum_{1\le j_1<j_2<\cdots<j_k\le d+1} L_{j_1}\cdot L_{j_2}\cdots L_{j_k}\,. \tag{11.10}$$

This result is obtained as a special case of Theorem 5.1 in the book by Piterbarg (1996).

Now, let us rewrite Eq. (11.9) for the case of the Gaussian field $X(\mathbf{x}, t)$ having variance σ and diagonal covariance matrix of general form, $\Lambda_X = \mathrm{diag}(\lambda_1, \ldots, \lambda_{d+1})$. This case is important since an arbitrary Gaussian field $Z(\mathbf{z}, t)$ with covariance matrix Λ_Z can be brought into this form by performing a rotation A ($A^T = A^{-1}$) of the coordinate axes to the principal directions. As a consequence, only rectangular domains with edges oriented along principal directions are considered.

It is now easy to check that for $X(\mathbf{x}, t)$

$$\mathrm{Prob}(\hat{X} > u\sigma) \simeq \varphi(u)\sum_{k=1}^{d+1}(2\pi)^{k/2}H_{k-1}(u)\Sigma_k(\mathbf{q}) + \int_u^{\infty}\varphi(t)\mathrm{d}t, \tag{11.11}$$

where $\mathbf{q} = (q_1, q_2, \cdots, q_{d+1})$, $q_i = L_i/\ell_i^0$, $\ell_i^0 = 2\pi\sigma/\sqrt{\lambda_i}$, $i = 1, \ldots, d+1$.

For high levels u and, correspondingly, small exceedance probabilities, one may write

$$\begin{aligned}\text{Prob}(\hat{X} \le u\sigma) &= 1 - \text{Prob}(\hat{X} > u\sigma) \simeq \exp\left\{-\text{Prob}(\hat{X} > u\sigma)\right\} \\ &\simeq P_B(u) = \exp\left\{-\varphi(u)\sum_{k=1}^{d+1}(2\pi)^{k/2}H_{k-1}(u)\Sigma_k(\mathbf{q})\right\}.\end{aligned} \tag{11.12}$$

This formula (referred to in the following as PB) is in agreement with the "conventional" Piterbarg's asymptotic formula Eq. (11.3) (referred to as PA) if only the leading term of the sum in Eq. (11.12) is taken into account, as $\Sigma_{d+1}(\mathbf{q}) = V_d$. The leading term in the exponent of PB (and the only one in PA) provides the probability of the exceedance from inside the volume of the domain $\mathscr{D} \times \mathscr{T}$, while other terms of PB represent probability of exceedance for the lower dimensional manifolds associated with the boundaries of the domain, like faces, edges, and vertices in the case of three-dimensional Δ. Therefore, if one wants to quantify the exceedance probability for a rectangle Δ, one of whose spatial sizes, say L_1, tends to zero, PA must be rewritten for the lower dimensional case. This demands in particular using V_{d-1} instead of V_d, as the last tends to zero, while the same formula PB will work for all ranges of Δ, that is, there is no need to change it upon a "squeeze" of one of the dimensions.

To check the validity of both PA and PB, a comparison of them will be made against numerical results for (1+1)- and (2+1)-dimensional Gaussian fields. For the (2+1)-dimensional Gaussian field, the explicit expression for PB is

$$\begin{aligned}P_B(u) = \exp\big\{ &- \exp(-u^2/2)\big[q_x + q_y + q_t \\ &+ \sqrt{2\pi}\, u\,(q_x\, q_y + q_x\, q_t + q_y\, q_t) + 2\pi\, q_x\, q_y\, q_t\,(u^2 - 1)\big]\big\}.\end{aligned} \tag{11.13}$$

Here $q_x = q_1 = L_1/\ell_1^0$, $q_y = q_2 = L_2/\ell_2^0$, $q_t = q_3 = L_3/\ell_3^0$, and $\ell_1^0 = 2\pi\sigma/\sqrt{\Lambda_{xx}}$, $\ell_2^0 = 2\pi\sigma/\sqrt{\Lambda_{yy}}$, $\ell_3^0 = 2\pi\sigma/\sqrt{\Lambda_{tt}} = T_z = 2\pi/\omega_z$, where

$$\begin{pmatrix}\Lambda_{xx}\\ \Lambda_{yy}\end{pmatrix} = \int \begin{pmatrix}k_x^2\\ k_y^2\end{pmatrix} \Psi(\mathbf{k},\omega)\, \mathrm{d}\omega\, \mathrm{d}\mathbf{k}\,, \tag{11.14}$$

$$\Lambda_{tt} = \omega_z^2\sigma^2 = \int \omega^2\, \Psi(\mathbf{k},\omega)\, \mathrm{d}\omega\, \mathrm{d}\mathbf{k}\,. \tag{11.15}$$

The last equality in Eq. (11.15) defines the zero-upcrossing (circular) frequency ω_z. A more general result corresponding to the special case of Eq. (11.13) with $q_t = 0$ was derived already by Ditlevsen (1971).

It can be shown that the rhs of Eq. (11.3) has the following Gumbel distribution as its max-stable asymptotic limit (Krogstad et al., 2004):

$$G(u\sigma) = \exp\big\{ - \exp\big[- h_V(u - h_V)\big]\big\} , \tag{11.16}$$

where the parameter h_V is obtained by solving the equation

$$V_d \, h_V^2 \, e^{-h_V^2/2} = 1 , \tag{11.17}$$

which has the approximate solution

$$h_V = \sqrt{2 \log V_d + 2 \log(2 \log V_d)} . \tag{11.18}$$

11.3 A Simplified Approach

The first step is rewriting Eq. (11.1) as follows:

$$\hat{X} = \max_{0 \le t \le T} \{\max_{\mathbf{x} \in \mathcal{D}} X(\mathbf{x}, t)\} . \tag{11.19}$$

A stochastic process consisting of spatial extremes is then introduced

$$\hat{X}_{\mathcal{D}}(t) = \max_{\mathbf{x} \in \mathcal{D}} X(\mathbf{x}, t) . \tag{11.20}$$

It is now assumed that a finite set of points $\{\mathbf{x}_i\}_{i=1}^{M}$ can be chosen so that

$$\hat{X}_{\mathcal{D}}(t) \approx \max_{1 \le i \le M} X(\mathbf{x}_i, t) , \tag{11.21}$$

for each $t \in (0, T)$, and that

$$\hat{X} \approx \max_{1 \le n \le N} \hat{X}_{\mathcal{D}}(t_n) \approx \max_{1 \le n \le N} \{ \max_{1 \le i \le M} X(\mathbf{x}_i, t_n)\} , \tag{11.22}$$

for a suitable discretization of the time interval $(0, T)$, $0 \le t_1 < \ldots < t_N \le T$. For each n, $n = 1, \ldots, N$, denote by $\mathbf{x}_{i_n}$ a point so that $\max_{1 \le i \le M} X(\mathbf{x}_i, t_n) = X(\mathbf{x}_{i_n}, t_n)$, and let $\tilde{X}_n = X(\mathbf{x}_{i_n}, t_n)$. Then

$$\hat{X} \approx \max_{1 \le n \le N} \tilde{X}_n . \tag{11.23}$$

If the sampling times t_n are sufficiently dense in $(0, T)$, then the time series $\tilde{X}_n$ can be considered as quasi-continuous in time and an accurate discrete representation of a smooth stochastic process $\tilde{X}(t)$ so that $\hat{X} \approx \max_{0 \le t \le T} \tilde{X}(t)$.

The derivations in this section apply also to the case of a nonhomogeneous field, and the extreme value distribution $F_{\hat{X}}(\xi)$ is then with good approximation given by the equation

$$F_{\hat{X}}(\xi) \approx \exp\left\{ - \int_0^T \nu^+_{\tilde{X}}(\xi; t)\, dt \right\}, \tag{11.24}$$

where $\nu^+_{\tilde{X}}(\xi; t)$ denotes the average upcrossing rate of the level ξ by $\tilde{X}(t)$ at time t. This approximation is contingent on the assumption that the probability of initial exceedance can be neglected. For estimation purposes, Eq. (11.24) is rewritten as

$$F_{\hat{X}}(\xi) \approx \exp\{-\overline{\nu}^+_{\tilde{X}}(\xi)\, T\}, \tag{11.25}$$

where the time average $\overline{\nu}^+_{\tilde{X}}(\xi)$ is given as

$$\overline{\nu}^+_{\tilde{X}}(\xi) = \frac{1}{T} \int_0^T \nu^+_{\tilde{X}}(\xi; t)\, dt, \tag{11.26}$$

which is in a form suitable for empirical estimation from time series. However, it should be clear that in the case of a nonhomogeneous field, the extreme value predictions can only refer to the initial space–time domain $\mathscr{D} \times \mathscr{T}$, if no additional modelling or structure is introduced.

The comparison between Eqs. (11.3), (11.12), and (11.25) reveals that $-\log(P_B(\xi/\sigma))$ is not a multiplicative function with respect to T (or, equivalently, q_t) in contrast to $-\log(P_A(\xi/\sigma))$ and $-\log(F_{\hat{X}}(\xi)) \approx \overline{\nu}^+_{\tilde{X}}(\xi)\, T$. More precisely,

$$P_A(\xi/\sigma) = \exp\{-\nu^+_A(\xi)\, T\}, \tag{11.27}$$

$$P_B(\xi/\sigma) = \exp\{-\nu^+_0(\xi) - \nu^+_B(\xi)\, T\}, \tag{11.28}$$

where, for the three-dimensional case,

$$\begin{aligned} \nu^+_A(u\sigma) &= 2\pi\, q_x\, q_y\, (u^2 - 1) \exp(-u^2/2)/T_z, \\ \nu^+_B(u\sigma) &= \nu^+_A(u\sigma) + \left[1 + \sqrt{2\pi}\, u(q_x + q_y)\right] \exp(-u^2/2)/T_z, \\ \nu^+_0(u\sigma) &= (q_x + q_y + \sqrt{2\pi}\, u\, q_x\, q_y) \exp(-u^2/2), \end{aligned} \tag{11.29}$$

as it immediately follows from Eq. (11.13) since $q_t = L_3/\ell_3^0 = T/T_z$. The term $\nu^+_0(\xi)$, the only one that survives in PB when $T \to 0$, gives the probability of an initial exceedance event on the "boundary" $\mathscr{D}$ of the $\mathscr{D} \times \mathscr{T}$ domain. Since the most interesting cases in practice would typically have domains where the time dimension is much greater than the spatial ones, which leads to $\max(q_x, q_y) \ll q_t$, $\nu_0^+(\xi)$ will be safely neglected for the numerical examples in this chapter. A similar approximation was also implemented in Eq. (11.24).

From Eqs. (11.3) and (11.16), it is seen that the mean upcrossing rate tail, say for $\xi \geq \xi_0$, behaves in a manner largely determined by a function of the form $\exp\{-a(\xi - b)^c\}$ $(\xi \geq \xi_0)$, where a, b and c are suitable constants. It is therefore assumed that the mean upcrossing rate function of $\tilde{X}(t)$ can be represented as

$$\nu^+(\xi) \approx q(\xi) \exp\{-a(\xi - b)^c\}\,, \; \xi \geq \xi_0\,, \tag{11.30}$$

for a suitable choice of ξ_0, where the function $q(\xi)$ is slowly varying compared with the exponential function $\exp\{-a(\xi - b)^c\}$ for tail values of ξ, cf. Naess and Gaidai (2008). Now, typically, the function $q(\xi)$ can be largely considered as a constant for tail values of ξ. This suggests an extrapolation strategy obtained by replacing $q(\xi)$ by a suitable constant value, q say.

The adopted procedure for identifying appropriate values for the parameters a, b, c, q, assuming a constant q, follows closely the optimization method detailed in Chap. 5. It is based on minimizing the following mean square error function with respect to the four arguments

$$F(q, a, b, c) = \sum_{j=1}^{N} w_j \left| \log \hat{\nu}^+(\xi_j) - \log q + a(\xi_j - b)^c \right|^2, \tag{11.31}$$

where $\hat{\nu}^+(\xi_j)$ is the empirical estimate of the upcrossing rate at the level ξ_j $(\xi_0 \leq \xi_1 \leq \ldots \leq \xi_N)$, and w_j denotes a weight factor that puts more emphasis on the more reliable estimates of $\hat{\nu}^+(\xi_j)$. The choice of weight factor is to some extent arbitrary. Here, a weight factor will be used based on the confidence interval associated with the empirical estimate of the upcrossing rate $\hat{\nu}^+(\xi_j)$.

11.4 Spatial–Temporal Extremes for Non-Gaussian Random Fields

In this section, the discussion is limited to random wave fields with two spatial dimensions. However, it should be quite obvious how to extend the methods discussed to higher dimensional spaces. Accordingly, let $X = X(\mathbf{x}, t)$ denote a zero mean random field defined on $\mathbb{R}^2 \times \mathbb{R}$, and let

$$\hat{X} = \hat{X}(\mathscr{D} \times \mathscr{T}) = \max\{X(\mathbf{x}, t); \mathbf{x} \in \mathscr{D}\,, \; t \in \mathscr{T}\}\,, \tag{11.32}$$

where $\mathscr{T}$ denotes a time interval, specifically, $\mathscr{T} = (0, T)$ for a given time T, and $\mathscr{D}$ is the area of interest.

Except for the case of homogeneous Gaussian fields and memoryless transformations of them, the asymptotic distribution function of $\hat{X}$ is unknown (Piterbarg, 1996; Adler and Taylor, 2007). In this section, the focus is on the case of second-order homogeneous wave fields. Since they are non-Gaussian and cannot be obtained by a simple transformation of a Gaussian field, the standard theory of Gaussian fields does not apply. However, even if no simple formulas can be derived for extremes of second-order wave fields, the simplified empirically based approach described in the previous section may still be applied, cf. Naess and Batsevych (2010).

The points $\{\mathbf{x}_i\}_{i=1}^{M}$ of Eq. (11.21) for the case of a wave field are typically chosen as the nodal points of a rectangular grid with a mesh size determined by the smallest significant wavelength of the field. A mesh size equal to one-tenth of this smallest wavelength would be sufficient. Then to ensure the approximation in Eq. (11.22), the (equidistant) discretization of the time interval $(0, T)$, $0 = t_0 < t_1 < \ldots < t_N \leq T$. $\Delta t = t_j - t_{j-1}$ can in most practical cases be chosen as one-tenth of the spectral peak period. The robustness of the results with respect to discretization can be verified by refining the mesh size in space and time.

Although the main focus of this section is on second-order ocean waves, the applicability of the methods discussed is far more wide ranging. However, for more general cases the validity of the approximation in Eq. (11.22) may require a more careful consideration. For a homogeneous random field which has a spectral density of bounded support, which can often be assumed in applications, then, the field is infinitely smooth and there will be uniform bounds on the rate of change of the field. Hence, a finite grid will always exist that secures the validity of the approximation in Eq. (11.22) within a given accuracy. For more general cases, assuming that the field has continuously differentiable realizations with uniformly bounded derivatives on bounded domains, then, again the validity of Eq. (11.22) within a specified level of accuracy is guaranteed for a suitable choice of the grid. Usually in practice, physical rather than mathematical arguments can be used to justify this approximation.

An alternative procedure is now introduced based on a detailed study of the extreme value statistics of Gaussian random fields described by Piterbarg (1996) and Adler and Taylor (2007). Consider a rectangular domain $\mathscr{D} = \mathscr{L}_x \times \mathscr{L}_y$, where $\mathscr{L}_x$ denotes an interval of length L_x, and similarly, $\mathscr{L}_y$ denotes an interval of length L_y. Neglecting the probability of initial exceedance, which is acceptable for long time intervals, it is then postulated that the distribution of $\hat{X}(\mathscr{D} \times \mathscr{T})$ for a homogeneous random field X can be written as

$$F_{\hat{X}}(\xi) = \exp\big\{ - \big(\nu_0^+(\xi) + \nu_x^+(\xi)L_x + \nu_y^+(\xi)L_y + \nu_{xy}^+(\xi)L_x L_y\big) T \big\}. \tag{11.33}$$

Here $\nu_0^+(\xi)$ denotes the one-point upcrossing rate of the level ξ. $\nu_x^+(\xi)$ $(\nu_y^+(\xi))$ represents corrections to the one-point upcrossing rate due to extensional effects in the x (y) direction, while $\nu_{xy}^+(\xi)$ corrects for area effects. In particular, $\nu_0^+(\xi)$ is the mean upcrossing rate of the stochastic process $X(\mathbf{x}, t)$ for an arbitrary, fixed point $\mathbf{x}$. $\nu_x^+(\xi)$ is determined by introducing the stochastic process $\hat{X}_{\mathscr{L}_x}(t) = \max_{\mathbf{x}\in\mathscr{L}_x\times\{y_0\}} X(\mathbf{x}, t)$, for a suitably fixed value y_0, and writing the mean upcrossing rate of this process as $\nu_0^+(\xi) + \nu_x^+(\xi)L_x$. This defines the mean normalized upcrossing rate $\nu_x^+(\xi)$. Similarly, $\nu_y^+(\xi)$ is determined by introducing the stochastic process $\hat{X}_{\mathscr{L}_y}(t) = \max_{\mathbf{x}\in\{x_0\}\times\mathscr{L}_y} X(\mathbf{x}, t)$, for a suitably fixed value x_0, and writing the mean upcrossing rate of this process as $\nu_0^+(\xi) + \nu_y^+(\xi)L_y$. This, then, defines the mean normalized upcrossing rate $\nu_y^+(\xi)$. Finally, $\nu_{xy}^+(\xi)$ is determined by writing the mean upcrossing rate of the process $\hat{X}_{\mathscr{D}}(t) = \hat{X}_{\mathscr{L}_x\times\mathscr{L}_y}(t) = \max_{\mathbf{x}\in\mathscr{L}_x\times\mathscr{L}_y} X(\mathbf{x}, t)$ as $\nu_0^+(\xi) + \nu_x^+(\xi)L_x + \nu_y^+(\xi)L_y + \nu_{xy}^+(\xi)L_xL_y$. Combining this with the previous relations provides a way to obtain $\nu_{xy}^+(\xi)$. These functions can be estimated in a manner similar to the procedure to be described next. The advantage of the representation provided by Eq. (11.33) is that the calibration of the right hand side needs to be done for only one rectangular domain. After that, the formula can then be used for any other rectangular domain (with the same orientation).

While the procedure for estimating the upcrossing rate functions described above is one way to achieve this, a somewhat different approach is used here. Instead of describing the procedure in full generality, it will be detailed as applied to the specific example in the section on numerical examples. The edges of the chosen rectangular domain for the calibration, which was 300×300 m, were divided into subintervals of length 25 m. An aggregate of rectangular subdomains $\mathscr{D}_{mn}$, $m, n = 0, 1, \ldots, 12$, where $\mathscr{D}_{mn} = (0,\ 25\cdot m) \times (0,\ 25\cdot n)$, is then created. Note that this includes the degenerate domains $(0, 0)$, $(0,\ 25\cdot m)\times(0, 0)$, and $(0, 0)\times(0,\ 25\cdot n)$. A linear regression approach based on the upcrossing rate of the area extremes process for each of the resulting rectangular domains $\mathscr{D}_{mn}$ obtained for $m, n = 0, 1, \ldots, 12$ was used to estimate $\nu_0^+(\xi_j)$, $\nu_x^+(\xi_j)$, $\nu_y^+(\xi_j)$, $\nu_{xy}^+(\xi_j)$ for each value of ξ_j, $j = 1, \ldots, J$, which denotes a preassigned range and number of ξ-levels leading to meaningful estimates of various upcrossing rates from the data.

In the following, $\nu^+(\xi)$ will be used as a generic notation denoting either the mean or the time averaged upcrossing rate, as the case may be. A key issue for prediction of extreme values is the estimation of $\nu^+(\xi)$. Also for the case considered in this section, it is assumed that the mean upcrossing rate tail, say for $\xi \geq \xi_0$, behaves in a manner largely determined by a function of the form $\exp\{-a(\xi - b)^c\}$ $(\xi \geq \xi_0)$, where a, b, and c are suitable constants. Consequently, it is assumed that the mean upcrossing rate function of $\tilde{X}(t)$ can be represented as

$$\nu^+(\xi) \approx q(\xi)\,\exp\{-a(\xi - b)^c\}\,,\ \xi \geq \xi_0\,, \tag{11.34}$$

for a suitable choice of ξ_0, where the function $q(\xi)$ is slowly varying compared with the exponential function $\exp\{-a(\xi - b)^c\}$ for tail values of ξ, cf. Naess and Gaidai (2008). The function $q(\xi)$ can be largely considered as a constant for tail values of ξ. This points to an extrapolation strategy based on replacing $q(\xi)$ by a suitable constant value, q say.

The adopted procedure for identifying appropriate values for the parameters a, b, c, q, assuming a constant q, is largely identical to that of the previous section.

11.5 Empirical Estimation of the Mean Upcrossing Rate

In the previous section it was shown that the key to providing estimates of the extreme values of the response process $X(t)$ on the basis of simulated response time histories is the estimation of the mean upcrossing rate. By assuming the requisite ergodic properties of the response process for a short-term condition, the mean upcrossing rate is conveniently estimated from the ergodic mean value. That is, it may be assumed that

$$\nu^+(\xi) = \lim_{t\to\infty} \frac{1}{t}\, n^+(\xi; 0, t)\,, \tag{11.35}$$

where $n^+(\xi; 0, t)$ denotes a realization of $N^+(\xi; 0, t)$, that is, $n^+(\xi; 0, t)$ denotes the counted number of upcrossings during time t from a particular simulated time history for which the starting point $t = 0$ is suitably chosen. In practice, k time histories of a specified length, T_0 say, are simulated. The appropriate ergodic mean value estimate of $\nu^+(\xi)$ is then

$$\hat{\nu}^+(\xi) = \frac{1}{k\,T_0} \sum_{j=1}^{k} n_j^+(\xi; 0, T_0)\,, \tag{11.36}$$

where $n_j^+(\xi; 0, T_0)$ denotes the counted number of upcrossings of the level ξ by time history no. j. This will be the approach to the estimation of the mean upcrossing rate adopted in this chapter.

For a suitable number k, e.g., $k \geq 20$, and provided that T_0 is sufficiently large, a fair approximation of the 95% confidence interval for the value $\nu^+(\xi)$ can be obtained as $\mathrm{CI}_{0.95}(\xi) = \big(C^-(\xi)\,,\ C^+(\xi)\big)$, where

$$C^\pm(\xi) = \hat{\nu}^+(\xi) \pm 1.96\,\frac{\hat{s}(\xi)}{\sqrt{k}}\,, \tag{11.37}$$

and the empirical standard deviation $\hat{s}(\xi)$ is given as

$$\hat{s}(\xi)^2 = \frac{1}{k-1} \sum_{j=1}^{k} \left(\frac{n_j^+(\xi; 0, T_0)}{T_0} - \hat{\nu}^+(\xi) \right)^2. \tag{11.38}$$

Note that k and T_0 may not necessarily be the number and length of the actually simulated response time series. Rather, they may be chosen to optimize the estimate of Eq. (11.38). If initially $\tilde{k}$ time series of length $\tilde{T}$ are simulated, then $k = \tilde{k}k_0$ and $\tilde{T} = k_0 T_0$. That is, each initial time series of length $\tilde{T}$ has been divided into k_0 time series of length T_0, assuming, of course, that $\tilde{T}$ is large enough to allow for this in an acceptable way. The consistency of the estimates obtained by Eq. (11.38) can be checked for large values of ξ by the observation that $\mathrm{Var}[N^+(\xi; 0, t)] = \nu^+(\xi)t$ since $N^+(\xi; 0, t)$ is then a Poisson random variable by assumption. This leads to the equation

$$\hat{s}(\xi)^2 = \frac{1}{k}\mathrm{Var}\left[\sum_{j=1}^{k} \frac{N_j^+(\xi; 0, T_0)}{T_0}\right] = \frac{\nu^+(\xi)}{T_0}, \tag{11.39}$$

where $\{N_1^+(\xi; 0, T_0), \ldots, N_k^+(\xi; 0, T_0)\}$ denotes a random sample with a possible outcome $\{n_1^+(\xi; 0, T_0), \ldots, n_k^+(\xi; 0, T_0)\}$. Hence, $\hat{s}(\xi)^2/k \approx \nu^+(\xi)/kT_0$. Since this last relation is consistent with the adopted assumptions, it could have been used as the empirical estimate of the sample variance in the first place. It is also insensitive to the blocking of data discussed above since $kT_0 = \tilde{k}\tilde{T}$. However, the advantage of Eq. (11.38) is that it applies whatever the value of ξ, and it does not rely on any specific assumptions about the statistical distributions involved.

11.6 Numerical Examples for Gaussian Random Fields

Since a major motivation for developing the approach discussed in this chapter is application to random ocean wave fields, the numerical examples presented will be for zero mean, homogeneous Gaussian fields specified by spectral densities defined in terms of a so-called JONSWAP spectrum $S(\omega)$, which is a one-sided spectrum, and a directional spreading function $D(\theta)$ (Sarpkaya and Isaacson, 1981):

$$S(\omega) = \frac{\alpha g^2}{\omega^5}\exp\left\{-\frac{5}{4}\left(\frac{\omega_p}{\omega}\right)^4 + \ln\gamma\,\exp\left[-\frac{1}{2\sigma^2}\left(\frac{\omega}{\omega_p}-1\right)^2\right]\right\}, \omega > 0, \tag{11.40}$$

where $g = 9.81\,\mathrm{ms}^{-2}$, ω_p denotes the peak frequency in rad/s, and α, γ, and σ are parameters related to the spectral shape. $\sigma = 0.07$ when $\omega \le \omega_p$, and $\sigma = 0.09$ when $\omega > \omega_p$. The parameter γ is chosen to be equal to 3.0. The parameter α is determined from the following empirical relationship (Naess et al., 2007):

$$\alpha = 5.06\left(\frac{H_s}{T_p^2}\right)^2(1 - 0.287\ln\gamma), \tag{11.41}$$

H_s = significant wave height and $T_p = 2\pi/\omega_p$ = spectral peak wave period. H_s = 14.0 m and T_p = 16 s are chosen for all the following examples. The directional spreading function is expressed as

$$D(\theta) = \frac{2^{2s-1}\Gamma^2(s+1)}{\pi\Gamma(2s+1)} \cos^{2s}\frac{(\theta-\theta_0)}{2}, \tag{11.42}$$

where θ_0 is the main wave direction, chosen so that $\theta_0 = 0$, and then $-\pi < \theta \le \pi$. The choice of spreading parameter is $s = 8$, which is a typical value often used.

The extreme value distributions PA and PB may be considered as having been derived under the Poisson assumption, i.e., when independence of individual upcrossings or exceedances is assumed, supplemented by Rice-like formulas for the exceedance rates $\nu_A^+(\xi)$ and $\nu_B^+(\xi)$. Thus PA and PB are liable to work in case of a wideband spectrum $\Psi(\mathbf{k}, \omega)$, as they do in the one-dimensional case of a random process. It is therefore to be expected that for the narrow-band case, there will be some discrepancy between the exact values and the predictions provided by PA and PB. Also the simplified approach proposed in this chapter is based on the use of the mean upcrossing rate function. However, this approach can easily be amended to cope with statistical dependence between the peak values in the extracted time series by using the ACER method, cf. Chap. 5.

Another aspect that comes into question in the case of a random ocean wave field is the effective dimension of the field. Due to the dispersion relation, which implies a strong coupling between frequency and wave number, the time and space dimensions of the wave field will likewise be strongly coupled, indicating a degeneracy. In fact, the dispersion relation implies that all the mass of the spectral density function $\Psi(\mathbf{k}, \omega)$ will be localized on a surface in $(\omega, \mathbf{k})$-space.

Numerical results obtained for three examples will now be presented for $d \le 2$. The simulations of the example random fields were carried out on a standard desktop computer, and the required CPU time was of the order of 1 hour for each case. For each of the examples, predictions of the 99.9% quantile of a 3-hour extreme value distribution by the proposed method (marked by an asterisk in the relevant figures) and the analytical method are also provided.

11.6.1 1+1-Dimensional Gaussian Field

In the first example, a wideband 1+1-dimensional Gaussian field is considered. It has a spectral density given as

$$\Psi(\omega, k) = S(|\omega|)S(|k|), \tag{11.43}$$

where $S(\cdot)$ is defined by Eq. (11.40). This is clearly an artificial example, but it serves the purpose of securing no interaction between the time and space dimensions of the random field. It is therefore to be expected that the theoretical formulas with

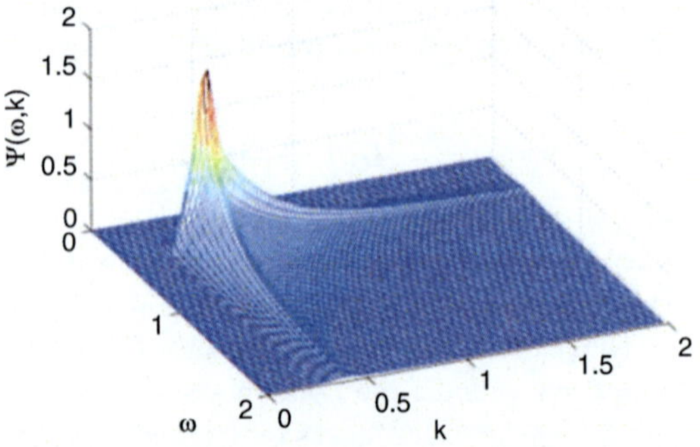

Fig. 11.1 The spectral density $\Psi(\omega, k)$ of the 1+1-dimensional Gaussian field

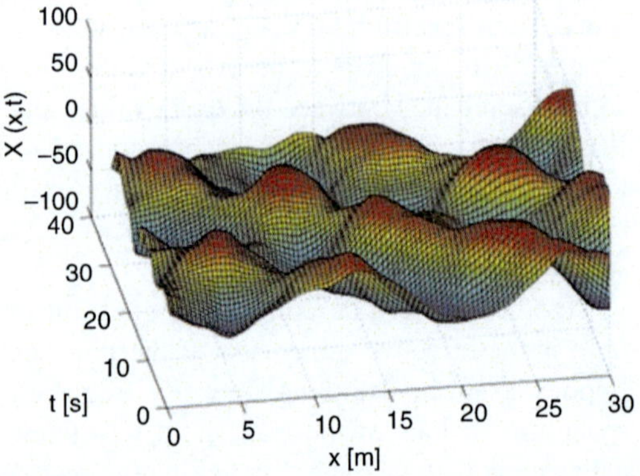

Fig. 11.2 Part of a realization of the 1+1-dimensional Gaussian random field

$d = 1$ should apply. A plot of the spectral density is shown in Fig. 11.1, while part of a realization of the associated Gaussian random field is given in Fig. 11.2. Plots of the empirical and analytical results are presented in Figs. 11.3 and 11.4 for one single point ($L_1 = 0$), which corresponds to an ordinary Gaussian temporal process, and for an interval ($L_1 = 100$), respectively. It is seen that the empirical and analytical results are in very good agreement in the tail for both cases and that the predictions of the 99.9% quantile in the 3-hour extreme value distributions are in good agreement.

11.6.2 1+1-Dimensional Gaussian Sea

The second example is a 1+1-dimensional Gaussian sea, which corresponds to the limiting case of long-crested waves. In this case the power spectral density assumes the form

$$\Psi(\omega, k) = \frac{1}{4} S(|\omega|)\, \delta\left(k - \frac{\omega|\omega|}{g}\right), \tag{11.44}$$

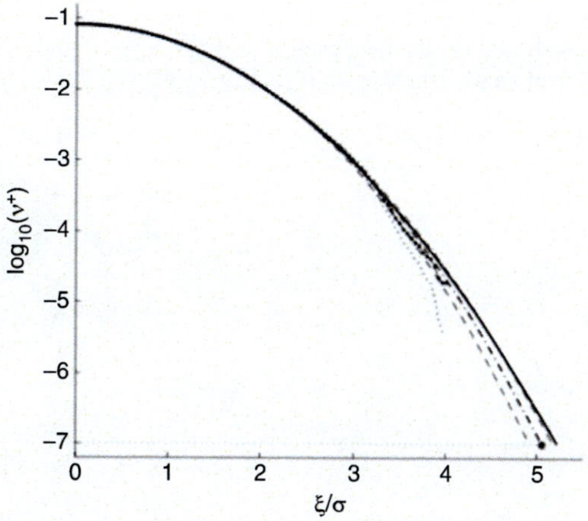

Fig. 11.3 Plot of the empirical $\log_{10}(\nu^+(\xi))$ (• • •), analytical $\log_{10}(\nu_B^+(\xi))$ (thick solid line), optimal fitted curve (– · –) with both empirical (dotted lines) and fitted (dashed lines) 95% confidence band for a point $L_1 = 0$ for Example 1. Predictions of 99.9% quantile: 5.22 (analytical); 5.06 and 95% CI = (4.93, 5.18) (proposed method)

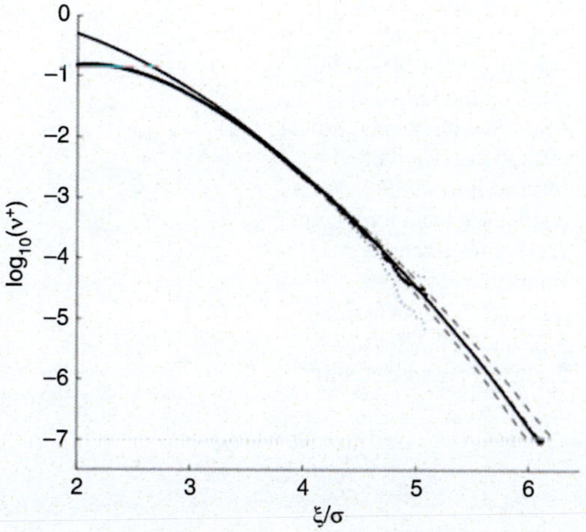

Fig. 11.4 Plot of the empirical $\log_{10}(\nu^+(\xi))$ (• • •), analytical $\log_{10}(\nu_B^+(\xi))$ (thick solid line), optimal fitted curve (– · –) with both empirical (dotted lines) and fitted (dashed lines) 95% confidence band for an interval $L_1 = 100$ m for Example 1. Predictions of 99.9% quantile: 6.09 (analytical); 6.11 and 95% CI = (6.03, 6.18) (proposed method)

where the dispersion relation for deep water waves is implemented in the form of a delta function. The resulting degeneracy would expectedly have some influence on the extreme value distribution. In the present case, the wave field can be represented as

$$X(t,x) = \mathrm{Re}\left\{\sum_{j=0}^{N}\sqrt{S(\omega_j)\Delta\omega}\cdot C_j\, e^{\mathrm{i}\omega_j t-\mathrm{i}\frac{\omega_j^2}{g}x}\right\}, \tag{11.45}$$

where C_j are complex $N(0,1)$-distributed random variables, that is, $C_j = R_j + \mathrm{i}S_j$ with R_j and S_j two independent $N(0,1/2)$-distributed variables.

Part of a realization of the Gaussian random wave field generated by Eq. (11.45) is plotted in Fig. 11.5. It is clearly seen how the strong coupling between frequency and wave number manifests itself.

Fig. 11.5 Part of a realization of the long-crested random wave field

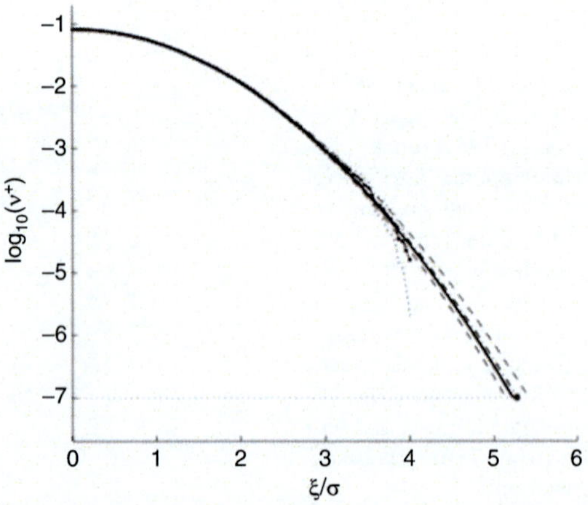

Fig. 11.6 Plot of the empirical $\log_{10}(\nu^+(\xi))$ (• • •), analytical $\log_{10}(\nu_B^+(\xi))$ (thick solid line), optimal fitted curve (− · −) with both empirical (dotted lines) and fitted (dashed lines) 95% confidence band for a point $L_1 = 0$ for Example 2. Predictions of 99.9% quantile: 5.22 (analytical); 5.27 and 95% CI = (5.14, 5.41) (proposed method)

Plots of the empirical and analytical results are presented in Figs. 11.6 and 11.7 for a point ($L_1 = 0$), which again corresponds to an ordinary Gaussian temporal process, and for an interval ($L_1 = 100$ m). It is seen that the empirical and analytical results are still in complete agreement in the tail for the one-point case, while there is a significant discrepancy between the results for $L_1 = 100$ m. In fact, the results for $L_1 = 100$ m are only slightly higher than the corresponding results for the one-point case, illustrating the effect of the degeneracy on the extreme values. It is interesting to observe that if the normalization constant ℓ_1^0 is calibrated to the empirical results, very accurate predictions are obtained, and they remain accurate for other domains without recalibration. This will be illustrated in the next example.

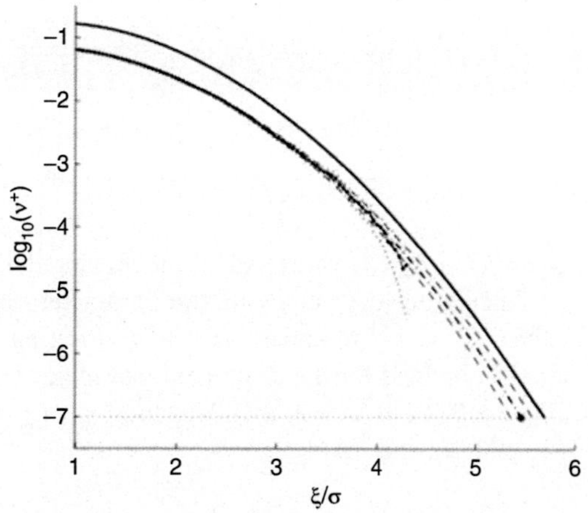

Fig. 11.7 Plot of the empirical $\log_{10}(\nu^+(\xi))$ (• • •), analytical $\log_{10}(\nu_B^+(\xi))$ (thick solid line), optimal fitted curve (– · –) with both empirical (dotted lines) and fitted (dashed lines) 95% confidence band for an interval $L_1 = 100$ m for Example 2. Predictions of 99.9% quantile: 5.70 (analytical); 5.47 and 95% CI = (5.36, 5.58) (proposed method)

11.6.3 A Short-Crested Gaussian Sea

In this example a model of a short-crested Gaussian random sea way is simulated. It corresponds to a case where the main direction of propagation of the waves is the x-direction. Having assumed deep water, the relation between the wave number vector and the spreading angle will be $\mathbf{k} = (k_x, k_y) = (\omega^2/g)\big(\cos\theta, \sin\theta\big)$. The spectral density of the random field X can then be expressed as

$$\Psi(\mathbf{k}, \omega) = \int S(\omega)\, D(\theta)\, \delta\!\left(k_x - \frac{\omega^2}{g}\cos\theta\right)\, \delta\!\left(k_y - \frac{\omega^2}{g}\sin\theta\right) \mathrm{d}\theta\,. \tag{11.46}$$

This leads to the expressions

$$\begin{aligned}
\Lambda_{xx} &= \int k_x^2\, \Psi(\mathbf{k}, \omega)\, \mathrm{d}\omega\, \mathrm{d}\mathbf{k} = D_c\, \omega_c^4\, \sigma^2/g^2\,, \\
D_c &= \int D(\theta)\, \cos^2(\theta)\, \mathrm{d}\theta\,, \\
\omega_c^4 &= \frac{1}{\sigma^2}\int \omega^4 S(\omega)\mathrm{d}\omega = (2\pi/t_c)^4\,, \\
\Lambda_{yy} &= \int k_y^2\, \Psi(\mathbf{k}, \omega)\, \mathrm{d}\omega\, \mathrm{d}\mathbf{k} = (1 - D_c)\, \omega_c^4\, \sigma^2/g^2\,, \\
\Lambda_{tt} &= \int \omega^2\, \Psi(\mathbf{k}, \omega)\, \mathrm{d}\omega\, \mathrm{d}\mathbf{k} = \omega_z^2 \sigma^2\,, \qquad (11.47)\\
\ell_1^0 &= 2\pi\sigma/\sqrt{\Lambda_{xx}} = \frac{2\pi g}{\sqrt{D_c}\omega_c^2} = \frac{g\, t_c^2}{2\pi\sqrt{D_c}}\,,
\end{aligned}$$

$$\ell_2^0 = 2\pi\sigma/\sqrt{\Lambda_{xx}} = \frac{2\pi g}{\sqrt{1 - D_c\omega_c^2}} = \frac{g\,t_c^2}{2\pi\sqrt{1 - D_c}},$$

$$\ell_3^0 = 2\pi\sigma/\sqrt{\Lambda_{tt}} = 2\pi/\omega_z = T_z,$$

$$q_x = L_1/\ell_1^0, \quad q_y = L_2/\ell_2^0 \quad q_t = L_3/\ell_3^0 = T/T_z,$$

where (L_1, L_2, T) is the size of the rectangular domain of interest.

Calculating the values of the parameters using the formulas above gives the values $\ell_1^0 = 135$ m and $\ell_2^0 = 280$ m. Plotting the empirical against the analytical results obtained for the calculated parameter values for $L_1 = L_2 = 0$, $L_1 \times L_2 = 100\,\text{m} \times 0$, $L_1 \times L_2 = 0 \times 100\,\text{m}$, and $L_1 \times L_2 = 100\,\text{m} \times 100\,\text{m}$ leads to Figs. 11.8, 11.9, 11.10, and 11.11.

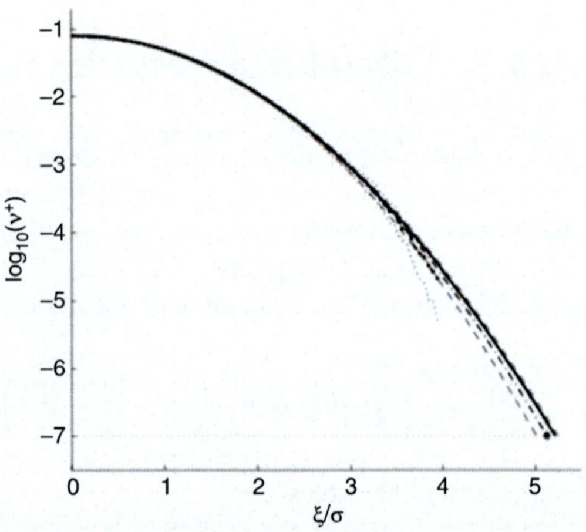

Fig. 11.8 Plot of the empirical $\log_{10}(\nu^+(\xi))$ (• • •), analytical $\log_{10}(\nu_B^+(\xi))$ (thick solid line), optimal fitted curve (− · −) with both empirical (dotted lines) and fitted (dashed lines) 95% confidence band for $L_1 = L_2 = 0$ for Example 3. Predictions of 99.9% quantile: 5.22 (analytical); 5.13 and 95% CI = (5.01, 5.25) (proposed method)

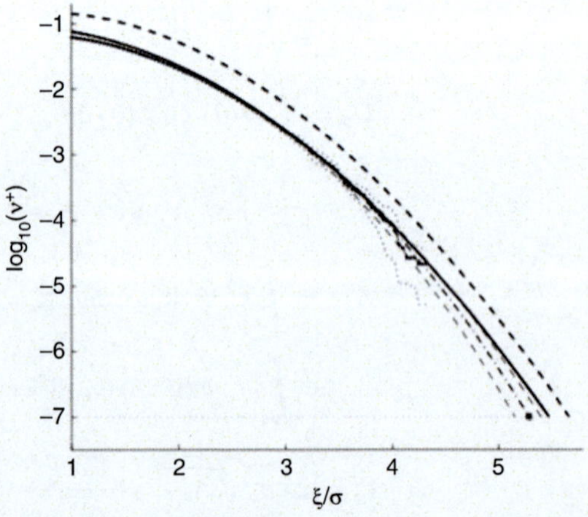

Fig. 11.9 Plot of the empirical $\log_{10}(\nu^+(\xi))$ (• • •), analytical $\log_{10}(\nu_B^+(\xi))$ (original: thick dashed line; tuned: thick solid line), optimal fitted curve (− · −) with both empirical (dotted lines) and fitted (dashed lines) 95% confidence band for the interval $L_1 \times L_2 = 100\,\text{m} \times 0$ for Example 3. Predictions of 99.9% quantile: 5.47 (tuned analytical); 5.33 and 95% CI = (5.26, 5.40) (proposed method)

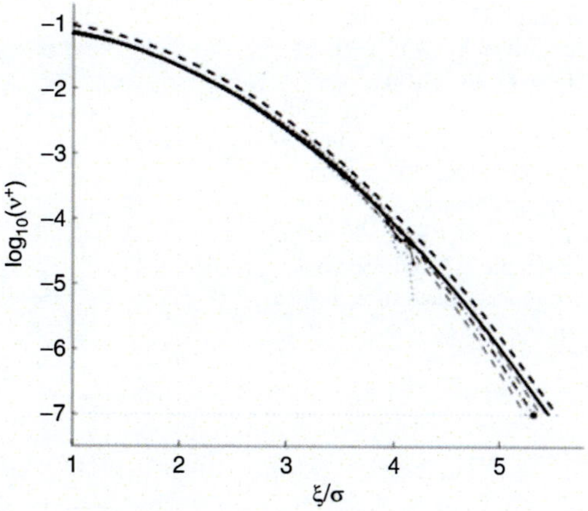

Fig. 11.10 Plot of the empirical $\log_{10}(\nu^+(\xi))$ (• • •), analytical $\log_{10}(\nu_B^+(\xi))$ (original: thick dashed line; tuned: thick solid line), optimal fitted curve (− · −) with both empirical (dotted lines) and fitted (dashed lines) 95% confidence band for the interval $L_1 \times L_2 = 0 \times 100$ m for Example 3. Predictions of 99.9% quantile: 5.47 (tuned analytical); 5.31 and 95% CI = (5.18, 5.41) (proposed method)

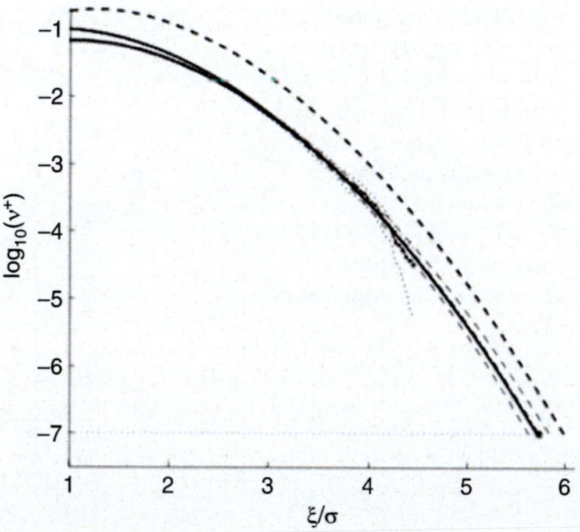

Fig. 11.11 Plot of the empirical $\log_{10}(\nu^+(\xi))$ (• • •), analytical $\log_{10}(\nu_B^+(\xi))$ (original: thick dashed line; tuned: thick solid line), optimal fitted curve (− · −) with both empirical (dotted lines) and fitted (dashed lines) 95% confidence band for the square domain $L_1 \times L_2 = 100 \times 100$ m for Example 3. Predictions of 99.9% quantile: 5.72 (tuned analytical); 5.74 and 95% CI = (5.56, 5.89) (proposed method)

It is seen from these figures that there is also in this case a significant effect of dimensional degeneracy. Since the main wave direction is along the x-axis, the strongest effect of dimensional degeneracy is to be expected for the cases $L_1 \times L_2 = 100\,\text{m} \times 0$ and $L_1 \times L_2 = 100 \times 100$ m, which is also corroborated by the plots.

As mentioned in the previous example, by tuning the parameters ℓ_1^0 and ℓ_2^0 to the empirical results, the tuned theoretical predictions become very accurate. The tuning can be done for each dimension separately. That is, ℓ_1^0 is calibrated to the results for the case $L_1 \times L_2 = 100\,\text{m} \times 0$, while ℓ_2^0 is calibrated to the results for the case $L_1 \times L_2 = 0 \times 100$ m. It has been verified that the calibrated values apply also to other domain sizes. For the particular example at hand, it was found that the tuned parameters are $\ell_1^0 = 474$ m, $\ell_2^0 = 477$ m. The results obtained by using the tuned parameters have also been plotted in Figs. 11.8, 11.9, 11.10, and 11.11 together with the predictions of the 99.9% quantiles. It is seen that fairly good agreement is now achieved for all cases.

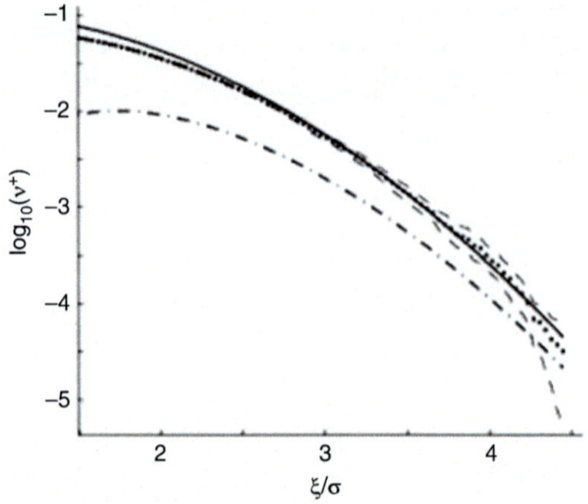

Fig. 11.12 Plot of the empirical $\log_{10}(\nu^+(\xi))$ (• • •), analytical $\log_{10}(\nu_A^+(\xi))$ (original: thick dash dot line; tuned: thick solid line) results for the square domain $L_1 \times L_2 = 100 \times 100$ m for Example 3. Empirical 95% confidence band (thin dashed lines)

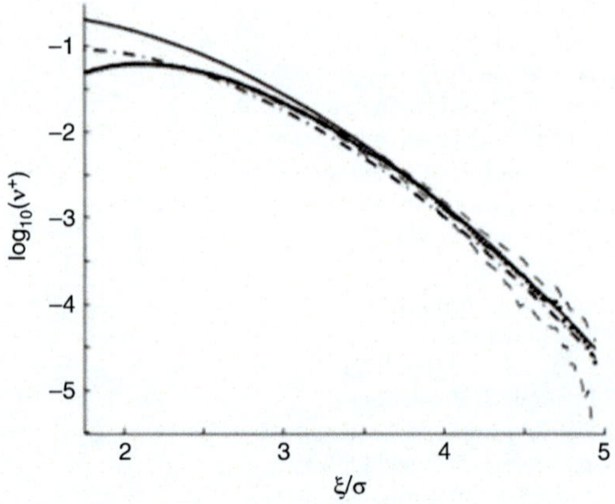

Fig. 11.13 Plot of the empirical $\log_{10}(\nu^+(\xi))$ (• • •), analytical $\log_{10}(\nu_A^+(\xi))$ (original: thick dash dot line; tuned: thick solid line) results for the square domain $L_1 \times L_2 = 300 \times 300$ m for Example 3. Empirical 95% confidence band (thin dashed lines)

It has been mentioned already that the tuned version of PB seems to give accurate predictions for practically any size. This has been verified for a range of domain sizes providing substantial support for this assertion. It is also of interest to investigate to what extent PA can give accurate predictions. If tuned to each particular size, it could perhaps provide reasonable predictions, but that is impractical. Let us therefore illustrate how it performs using the same tuned parameters as previously established, that is, $\ell_1^0 = 474$ m, $\ell_2^0 = 477$ m. The results for the two square domains of size 100×100 m and 300×300 m have been plotted in Figs. 11.12 and 11.13. It is seen that there is a significant discrepancy between the tuned PA results and the empirical results for the smaller domain, while there is good agreement for the large one, giving a good indication of the importance of the neglected terms in PA. If the domain is too small, the leading term is not sufficient to obtain accurate estimates.

The value of the 99% fractile of a 3-hour extreme value distribution, which corresponds to $\nu^+ = 10^{-6}$, as a function of the size of a square domain has been plotted in Fig. 11.14. It is seen that the extreme values show a fairly strong

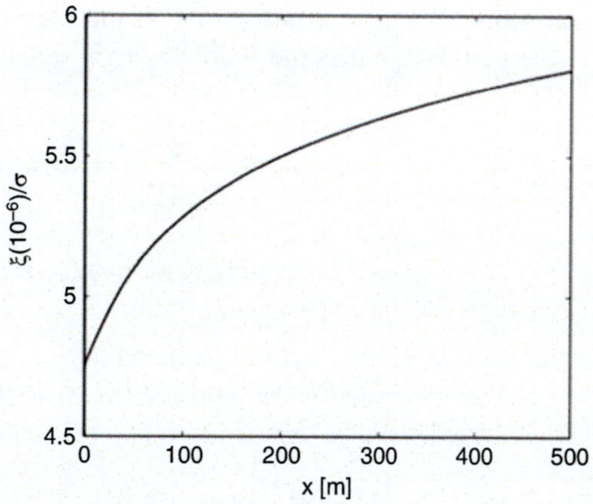

Fig. 11.14 The 99% fractile value of a 3-hour extreme value distribution as a function of the size of a square domain with $x = L_1 = L_2$ for the short-crested Gaussian sea

dependence on the area of the domain. For prediction of extremes over, e.g., the deck area of an offshore structure, which would typically be about 100×100 m, it is seen that the area effect amounts to 10–15% larger extremes than predicted by a one-point estimate.

11.7 Numerical Examples for Non-Gaussian Random Fields

11.7.1 A Second-Order Wave Field

Since a major motivation for developing the approach discussed in this chapter is application to random ocean wave fields, the numerical example presented will be for a zero mean, homogeneous second-order short-crested random wave field. It has been shown, cf., e.g., Toffoli et al. (2008), that such a wave field can be built up in the following way. First, let $\eta(t, r, \theta)$ denote the wave elevation of a long-crested sea propagating in the direction specified by the angle θ, where $r = x \cos\theta + y \sin\theta$. Then $\eta(t, r, \theta) = \eta_1(t, r, \theta) + \eta_2(t, r, \theta)$. Here the linear, first-order part $\eta_1(t, r, \theta)$ is given by the relation

$$\eta_1(t, r, \theta) = \sum_{k=1}^{N} |C_k| \cos(\chi_k), \tag{11.48}$$

where $C_k = |C_k| e^{i\varepsilon_k}$, $k = 1, \ldots, N$, is a set of independent complex random variables, whose real and imaginary parts are independent and normally distributed as $N(0, S(\omega_k)\Delta\omega)$. $\chi_k = r_k - \omega_k t + \varepsilon_k$, where $r_k = \omega_k^2 r/g$. $S(\omega)$ is the JONSWAP spectrum, which was introduced in the previous section. The nonlinear, second-order part $\eta_2(t, r)$ has the following representation:

$$\eta_2(t,r,\theta) = \sum_{k,l=1}^{N} \Big(m_{kl}|C_k C_l| \cos(\chi_k) \cos(\chi_l) - M_{kl}|C_k C_l| \sin(\chi_k) \sin(\chi_l) \Big), \qquad (11.49)$$

where $m_{kl} = (2g)^{-1} \min(\omega_k^2, \omega_l^2)$ and $M_{kl} = (2g)^{-1} \max(\omega_k^2, \omega_l^2)$.

The short-crested wave field $X(x, y, t)$ is then obtained by superposition of m long-crested wave fields as follows:

$$X(x, y, t) = \sum_{k=1}^{m} \sqrt{D(\theta_k)\Delta\theta}\, \eta(t, x\cos\theta_k + y\sin\theta_k, \theta_k), \qquad (11.50)$$

where $D(\theta)$ denotes the directional spreading function.

In the numerical example, a specific model of a short-crested, second-order random sea way is simulated. It corresponds to a case where the main direction of propagation of the waves is the x-direction. Having assumed deep water, the relation between the wave number vector and the spreading angle will be $\mathbf{k} = (k_x, k_y) = (\omega^2/g)\big(\cos\theta, \sin\theta\big)$. The spectral density in terms of wave numbers and frequency of the first-order part of the random field X can then be expressed as

$$\Psi(\mathbf{k}, \omega) = \int_{-\pi}^{\pi} S(\omega)\, D(\theta) \cdot \delta\left(k_x - \frac{\omega^2}{g}\cos\theta\right) \delta\left(k_y - \frac{\omega^2}{g}\sin\theta\right) d\theta\,. \qquad (11.51)$$

Part of a realization of the short-crested, second-order random wave field is shown in Fig. 11.15. The standard deviation of the wave field was estimated to

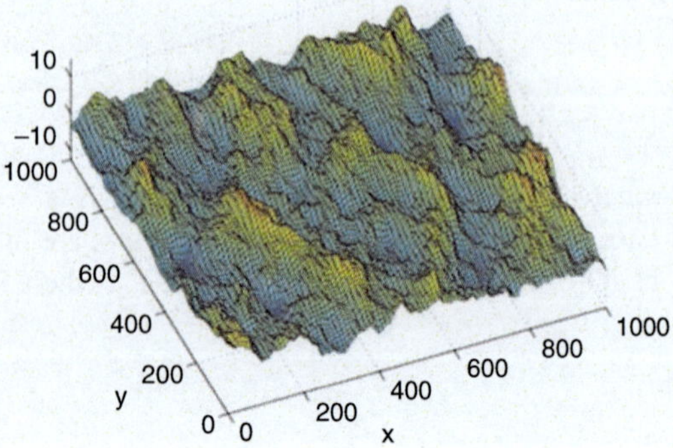

Fig. 11.15 Part of a realization of the short-crested, second-order random wave field

Fig. 11.16 The one-point upcrossing rate $\nu_0^+(\xi)$, cf. Eq. (11.33)

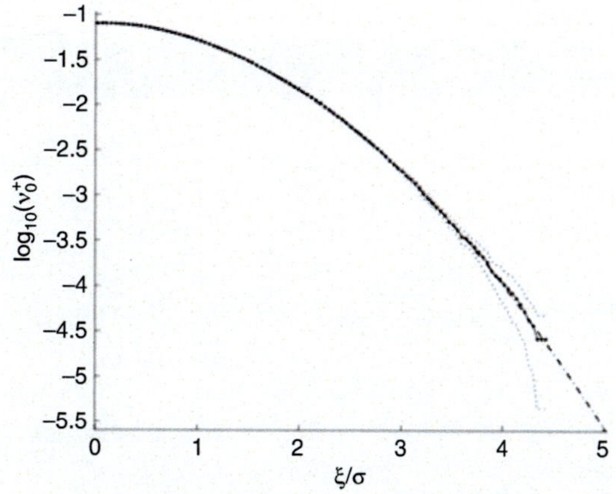

Fig. 11.17 The upcrossing rate $\nu_x^+(\xi)$, cf. Eq. (11.33)

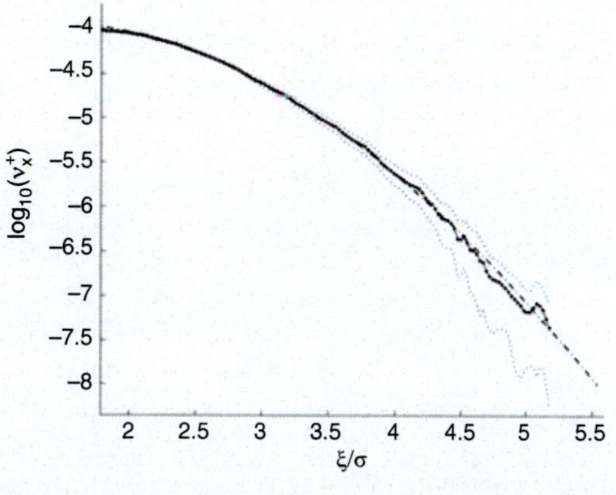

be $\sigma = 3.53$m. From the simulated wave field, the four upcrossing rate functions of Eq. (11.33) have been estimated by the linear regression approach described in Sect. 11.4 on space–time extremes. The results are shown in Figs. 11.16, 11.17, 11.18, and 11.19 together with the optimally fitted curves obtained by the point process procedure.

The prediction results obtained by the simplified empirical approach in combination with the point process procedure have been plotted in Figs. 11.20, 11.21, 11.22, and 11.23. Similarly, the results obtained by the parametric approach of Eq. (11.33) are also plotted. In the figures the predicted 99.9% fractile value of the 3-hour extreme value distribution obtained by the simplified approach has been indicated by an asterisk. This fractile value is achieved at $\nu^+ = 10^{-7}$. It is seen that the agreement between the two approaches proposed in this chapter is very good for the cases studied.

The value of the 99.9% fractile of a 3-hour extreme value distribution as a function of the size of a square domain has been plotted in Fig. 11.24. It is seen that the extreme values show a fairly strong dependence on the area of the domain. For prediction of extremes over, e.g., the deck area of an offshore structure, which

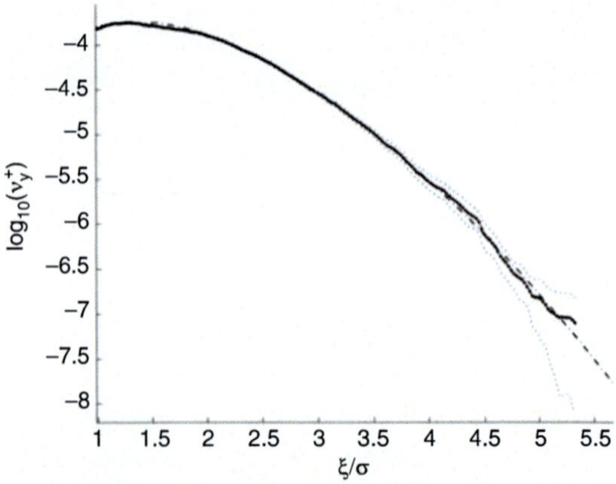

Fig. 11.18 The upcrossing rate $\nu_y^+(\xi)$, cf. Eq. (11.33)

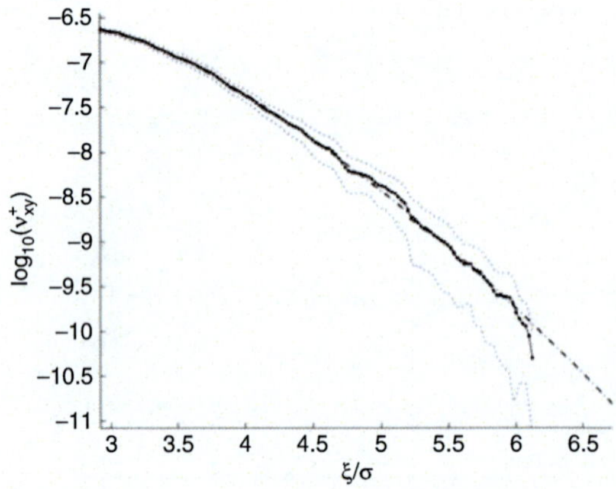

Fig. 11.19 The upcrossing rate $\nu_{xy}^+(\xi)$, cf. Eq. (11.33)

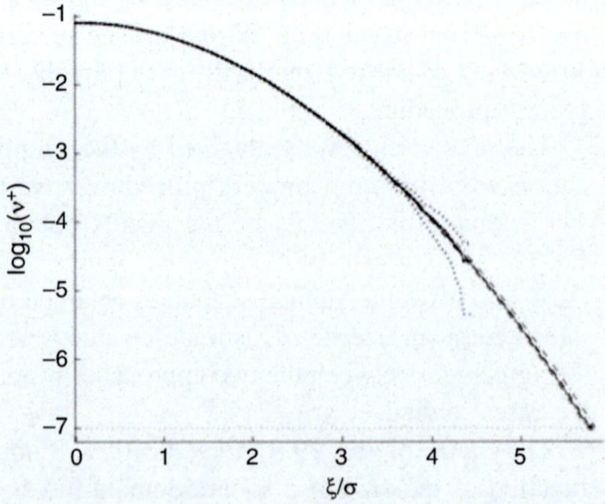

Fig. 11.20 Case 1: Single point. Plot of the empirical upcrossing rate $\log_{10}(\nu^+(\xi))$ (• • •), of the optimal fitted curve for the simplified approach (— — —), of $\log_{10}(\nu_P^+(\xi))$ (solid line) for the parametric approach. Both empirical (dotted lines) and fitted (dash-dotted lines) 95% confidence bands are shown. Predictions of 99.9% quantile: 5.80 and 95% CI = (5.75, 5.84) (simplified method); 5.77 (calibrated parametric)

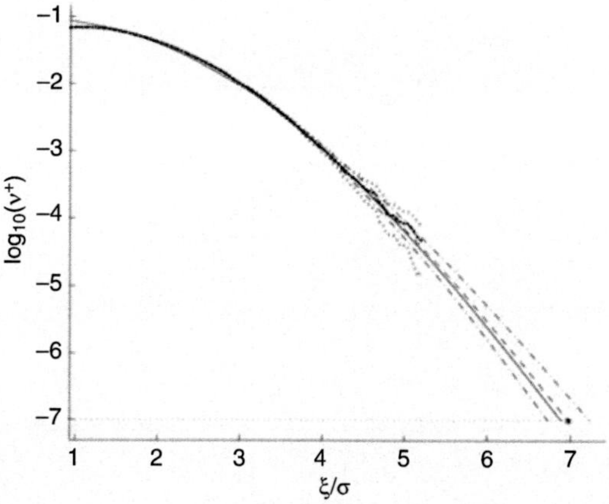

Fig. 11.21 Case 2: $100 \times 100\,\mathrm{m}^2$. Plot of the empirical upcrossing rate $\log_{10}(\nu^+(\xi))$ (• • •), of the optimal fitted curve for the simplified approach (— — —), of $\log_{10}(\nu_P^+(\xi))$ (solid line) for the parametric approach. Both empirical (dotted lines) and fitted (dash-dotted lines) 95% confidence bands are shown. Predictions of 99.9% quantile: 6.98 and 95% CI = (6.73, 7.23) (simplified method); 6.89 (calibrated parametric)

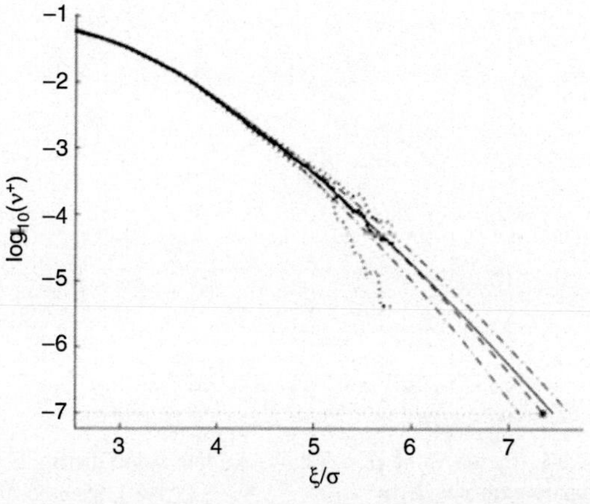

Fig. 11.22 Case 3: $300 \times 300\,\mathrm{m}^2$. Plot of the empirical upcrossing rate $\log_{10}(\nu^+(\xi))$ (• • •), of the optimal fitted curve for the simplified approach (— — —), of $\log_{10}(\nu_P^+(\xi))$ (solid line) for the parametric approach. Both empirical (dotted lines) and fitted (dash-dotted lines) 95% confidence bands are shown. Predictions of 99.9% quantile: 7.35 and 95% CI = (7.11, 7.62) (simplified method); 7.45 (calibrated parametric)

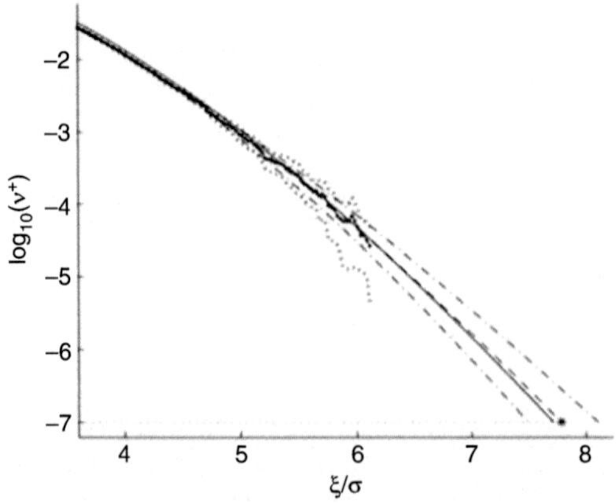

Fig. 11.23 Case 4: 500 × 500 m^2. Plot of the empirical upcrossing rate $\log_{10}(\nu^+(\xi))$ (• • •), of the optimal fitted curve for the simplified approach (— — —), of $\log_{10}(\nu_P^+(\xi))$ (solid line) for the parametric approach. Both empirical (dotted lines) and fitted (dash-dotted lines) 95% confidence bands are shown. Predictions of 99.9% quantile: 7.79 and 95% CI = (7.50, 8.10) (simplified method); 7.72 (calibrated parametric)

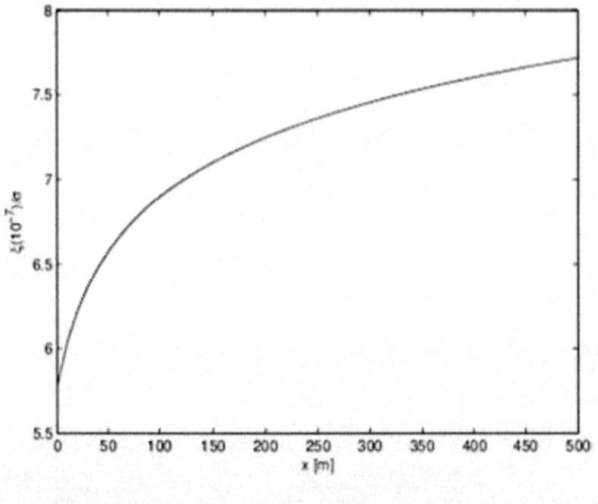

Fig. 11.24 The 99.9% fractile value of a 3-hour extreme value distribution as a function of the size of a square domain with $x = L_x = L_y$

would typically be about 100 × 100 m, it is seen that the area effect may amount to as much as 15–20% larger extreme crest heights than predicted by one-point estimates.

11.7.2 A Student's t Random Field

To test the proposed methods on a strongly non-Gaussian random field, what shall be referred to as a Student's t field has been constructed. This was done in the

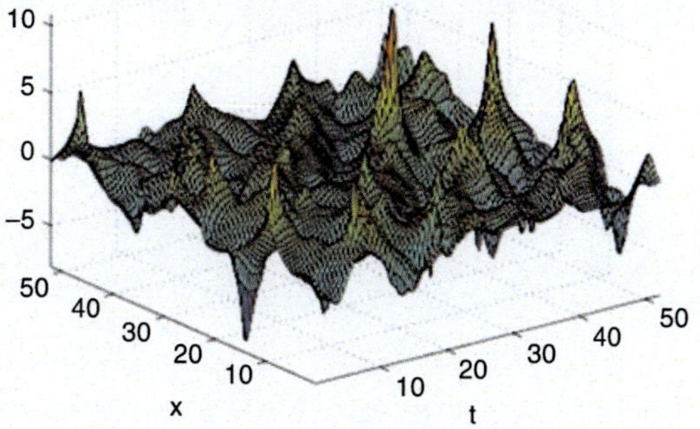

Fig. 11.25 Part of a realization of the Student's t random field

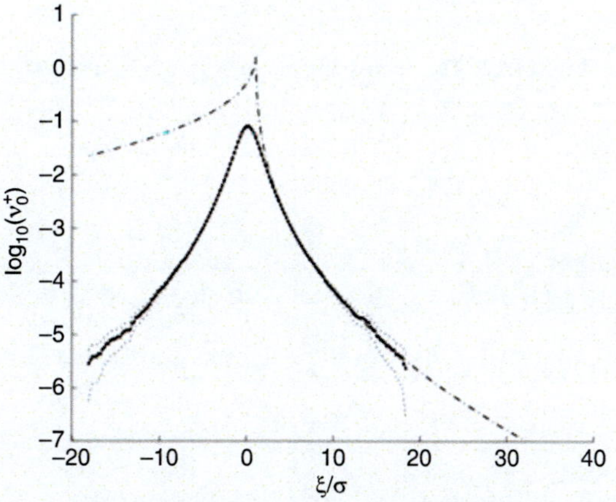

Fig. 11.26 The one-point upcrossing rate $\nu_0^+(\xi)$, cf. Eq. (11.33)

following manner. Let $X(x,t) = \eta_1(t,x,0)$, where $\eta_1(t,x,0)$ is as defined in the previous example. $X(x,t)$ is then a homogeneous Gaussian random field of zero mean with one spatial dimension. Let $X_j(x,t)$, $j = 1,\dots,4$, denote independent copies of $X(x,t)$. The random field $Z(x,t)$ is now constructed as follows:

$$Z(x,t) = \frac{X(x,t)}{\sqrt{\frac{1}{4}\sum_{k=1}^{4} X_j(x,t)^2 + \frac{\sigma}{20}}}\,, \tag{11.52}$$

where σ^2 denotes the variance of $X(x,t)$. The added term $\sigma/20$ in the denominator is introduced to avoid near singularities in the generated fields and make sure that it satisfies the conditions discussed after Eq. (11.22). Part of a realization of the resulting random field is shown in Fig. 11.25.

In Figs. 11.26 and 11.27 are shown the obtained results for $\nu_0^+(\xi)$ and $\nu_x^+(\xi)$, respectively. These figures clearly display the strongly non-Gaussian characteristics of this random field.

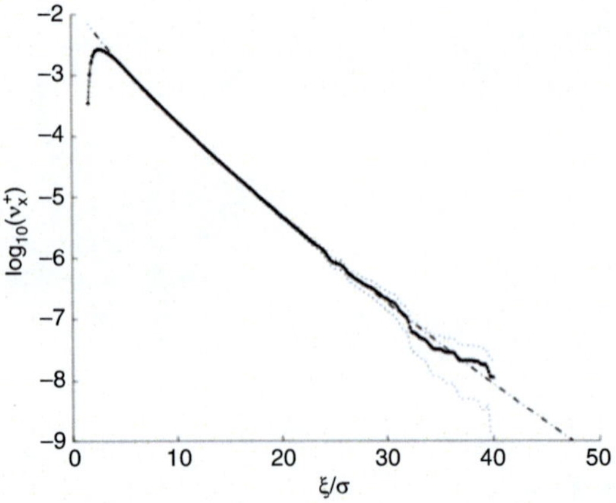

Fig. 11.27 The upcrossing rate $\nu_x^+(\xi)$, cf. Eq. (11.33)

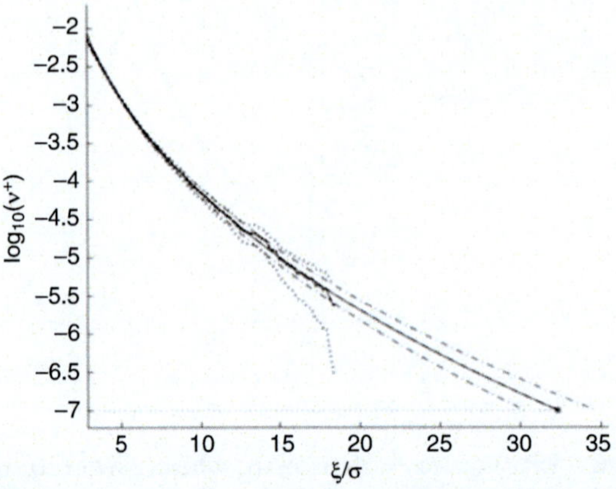

Fig. 11.28 Single point. Plot of the empirical upcrossing rate $\log_{10}(\nu^+(\xi))$ ($\bullet\bullet\bullet$), of the optimal fitted curve for the simplified approach (— — —), of $\log_{10}(\nu_P^+(\xi))$ (solid line) for the parametric approach. Both empirical (dotted lines) and fitted (dash-dotted lines) 95% confidence bands are shown. Predictions of 99.9% quantile: 32.33 and 95% CI = (30.25, 34.69) (simplified method); 32.33 (calibrated parametric)

Figure 11.28 shows the one-point prediction result, while Fig. 11.29 shows the prediction results for an interval of size 100 m. It is seen that there is a strong extensional effect of about 50% on the predicted extreme values. It is also clearly demonstrated that the predictions obtained by the two proposed methods are in excellent agreement.

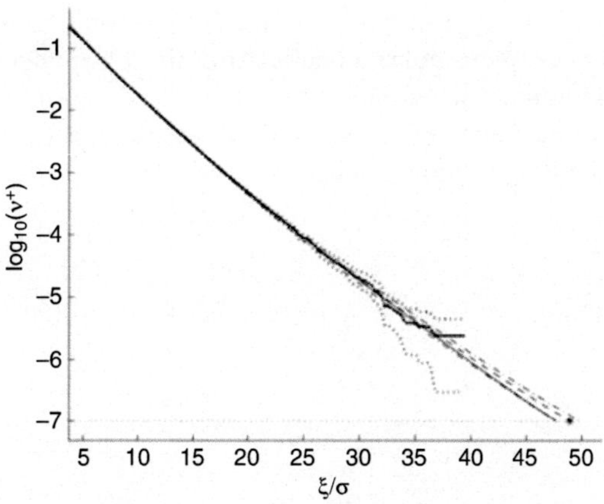

Fig. 11.29 $x = 100$ m. Plot of the empirical upcrossing rate $\log_{10}(\nu^+(\xi))$ (• • •), of the optimal fitted curve for the simplified approach (— — —), of $\log_{10}(\nu_P^+(\xi))$ (solid line) for the parametric approach. Both empirical (dotted lines) and fitted (dash-dotted lines) 95% confidence bands are shown. Predictions of 99.9% quantile: 48.92 and 95% CI = (47.73, 49.78) (simplified method); 47.70 (calibrated parametric)

11.8 Comments

Analytical formulas for the extreme value distribution of a homogeneous Gaussian random field have been discussed at some length, and some properties of these formulas have been highlighted. It has been shown that by proper calibration, the analytical formulas provide good results also for the case of a Gaussian random sea.

Provided there is no degeneracy in the dimensionality of the Gaussian random field, the extreme value distribution PA is asymptotically accurate with respect to the level ξ as well as to the size of the domain, while PB is asymptotically accurate with respect to the level ξ for any size.

The region of applicability of the approximation PB is suggested to be $\nu_B^+(\xi) \leq 10^{-2}$. It is then formulated in terms of the value of ν_B^+ rather than for the level ξ, and it appears to be valid for a domain of arbitrary size $L_1 \times L_2$.

The accuracy of the proposed prediction procedure obtained by optimal fitting to the simulated data has also been amply demonstrated. This leaves the door open for accurate prediction of extremes of non-Gaussian random fields. While the analytical formulas are restricted to homogeneous Gaussian fields, no such restriction applies to the proposed prediction method.

A simplified, empirically based procedure for prediction of space–time extremes of homogeneous non-Gaussian random fields has been presented. The advantage of this method is its flexibility and general applicability, while it appears to be accurate and robust. A drawback of the method is that if the spatial domain is changed, the whole analysis has to be repeated. In an effort to ameliorate this situation, a quasi-parametric representation of the extreme value distribution for rectangular domains is proposed. When this approach applies, it has the advantage that it only needs to be calibrated for one rectangular domain. After that, it applies to any rectangular domain (with the same orientation). The case studies presented show that the two methods proposed seem to provide very good practical tools for the prediction of space–time extremes of random fields.

Chapter 12
A Case Study—Extreme Water Levels

12.1 Introduction

As part of the efforts to reduce the vulnerability to flooding, it is of paramount importance to have available flooding charts that spell out the risk of flooding for a given location. This risk is typically expressed as the flooding levels associated with various return periods, e.g., the 100 year flooding level. In Norway the concerns about flooding events are primarily connected with two types of flooding: On the one hand, in rivers and lakes, which are mainly related to melting snow; and on the other, along the coast, which are mainly due to a combination of offshore storm surges and tides. In this chapter the attention is limited to coastal areas. Specifically, the extreme value statistics of sea levels measured at three stations along the Norwegian coastline will be investigated: Oslo, Heimsjø and Honningsvåg, see Fig. 12.1. This provides an excellent opportunity to compare the performance of the following four methods for extreme value estimation: The Annual Maxima (AM) method (Chap. 2), the Peaks-Over-Threshold (POT) method (Chap. 3), the ACER method (Chap. 5), and also the Revised Joint Probabilities (RJP) method (Tawn and Vassie 1989; Tawn 1992) (see also Batstone et al. (2009) for modifications of the RJP method). Since the RJP method has not been discussed previously in this book, it will be explained in some detail here. This chapter largely follows the work presented by Skjong et al. (2013).

12.2 Data Sets

The data sets used in this chapter are water level measurements and tidal predictions for three locations on the Norwegian coast. Water level measurements are collected by automated equipment, while the tidal predictions are based on standard numerical models.

A. Naess, *Applied Extreme Value Statistics*,
https://doi.org/10.1007/978-3-031-60769-1_12

Fig. 12.1 Map with locations of the sea level measuring stations considered in this paper

There is one data set from the south of Norway, Oslo. Tidal variations are not as large an influence on the water level here as it is further north. The location is referred to as *surge-dominant*. The total height of sea levels are also lower here than further north. There is one data set from the middle of Norway, Heimsjø, and one from the far north, Honningsvåg. The sea levels are much more influenced by tides here; the locations are *tide-dominant*. The data sets are not in the public domain, but were provided by the Norwegian Hydrographic Service, a division of the Norwegian Mapping Authority.

The sea level measurements are done either hourly or every 10 minutes. For the 10 minute data, hourly measurements have been extracted for use with the POT, RJP and ACER methods. There are several reasons for this: Firstly, the RJP method uses tidal predictions which were provided as hourly sea levels. Secondly, many locations have hourly data for older periods and ten minute measurements for the recent years. In order to use both data sets, values every hour are extracted from the more frequent observations. Thirdly, the literature used in the theoretical studies have hourly measurements. Methodological comparisons are therefore made simpler.

If not otherwise noted, all references to the height of the sea level are in centimeters relative to a mean sea level. All return periods are, if not otherwise noted, given in years.

12.2.1 Oslo

For the measurement station in Oslo, there are available hourly measured sea levels from December 10, 1914, to December 31, 1991, and 10 minute interval data from October 1, 1991, to September 16, 2010, which was the cutoff date for our analyses. These data are uncorrected for post-glacial rebound, which is an important factor in the Oslo area. To correct for this, 4 mm/year have been added to data points for years after 1988 and the same amount has been subtracted per year before 1988, according to the formula $currentValue + 0.4 \cdot (currentYear - 1988)$. This was found in a previous report (Hansen and Roald 2000), and is based on the fact that 1988 was the base year to calculate mean sea level (MSL) in Oslo.

For the same periods, tidal sea level predictions are available. These are based on a numerical model, the details of which are not publicly available.

12.2.2 Heimsjø

Heimsjø is located on the coast of Sør-Trøndelag, and the measurement station is found at latitude 63°26′ N and longitude 09°07′ E. Data are available as hourly measurements from November 1, 1928, to December 31, 1990, and as 10-minute interval measurements from November 1, 1990, to September 16, 2010. Tide predictions are also available for the same periods.

12.2.3 Honningsvåg

Honningsvåg is found at the very north of Norway, in Finnmark. The location of the measurement station is latitude 70°59′ N and longitude 25°59′ E. Measured values exist in hourly form from June 5, 1970, to December 31, 1988, and as 10 minute data from June 1, 1988, to September 16, 2010. Corresponding tidal predictions are also available.

12.3 Annual Maxima Method

As we know from Chap. 2, the annual maxima (AM) method is based on the assumption that $M_N = \max\{X_1, \ldots, X_N\}$, where $X_1, \ldots, X_N$ are independent observations using a block size of one year, is distributed according to the generalized extreme value (GEV) distribution.

For the published work done on Norwegian sea levels, the shape parameter of the GEV distribution is assumed to be fixed at zero. This means that the

Gumbel distribution is adopted, which can be ascertained by studying the underlying statistics and their domain of attraction. However, results for the case of a nonzero shape parameter will also be presented below. This is done because it has been argued that the GEV distribution to be used should be decided on the basis of the extreme value data (Coles 2001). It will be shown that this can be misguided advice.

12.3.1 Application to Water Level Measurements

Oslo

Oslo has data available from all years from 1914 to 2010—except 1939, from which there are no available measurements. In addition, some of the years with available data have important data missing. For instance, there are only 312 data points in 1914, all from the month of December. This means that there is a very real possibility that the real annual maximum is excluded. In 1915, there is much missing data from the important autumn and winter months, where annual maxima are often found. This year is therefore also excluded from consideration. In 1972, there are no measurements in February, July, August, September and October, which means that so much data are missing that the probability that the year's maximum is left out is large. 1974 lacks data for July, August and September, and 1991 only has data for October, November and December. Both these are excluded. Finally, measurements for 2010 only go to September 16, so its maximum is also dropped from consideration.

For the Gumbel model, the parameter estimates are $\hat{\mu} = 167.31$ (2.074) and $\hat{\sigma} = 18.630$ (1.497). For the GEV model, $\hat{\mu} = 168.21$ (2.233), $\hat{\sigma} = 19.030$ (1.579) and $\hat{\gamma} = -0.089$ (0.070). Standard errors are shown in parentheses.

With the data as specified above, estimates of return period levels together with maximum likelihood 95% confidence intervals are published in Table 12.1 for the Gumbel and GEV models. The obtained results agree fairly well with published values collected from a report by the Norwegian Map Authority (Hansen and Roald 2000), which were also based on the AM method.

The return level plot in Fig. 12.2 shows that all points stay within the 95% confidence intervals for both models. The two curves follow slightly different paths; the Gumbel model in Fig. 12.2a fits fairly well to all points, while the GEV model

Table 12.1 Estimates of return levels and 95% confidence intervals for Oslo

	Gumbel model		GEV model	
$R = 1/p$	$\hat{z}_p$	CI	$\hat{z}_p$	CI
5	195.2	(188.4, 202.1)	194.9	(188.7, 201.1)
10	209.2	(200.4, 218.0	207.0	(199.1, 214.9)
20	222.6	(211.9, 233.4)	217.9	(207.4, 228.3)
100	253.0	(237.7, 268.2)	240.1	(220.5, 259.7)
200	265.9	(248.7, 283.2)	248.8	(223.9, 273.8)

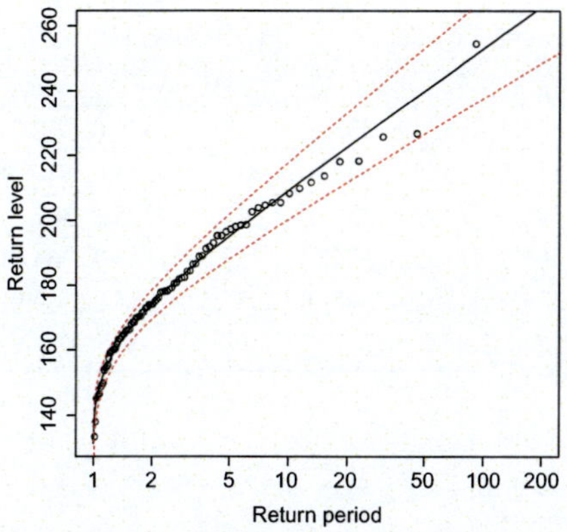

(a)

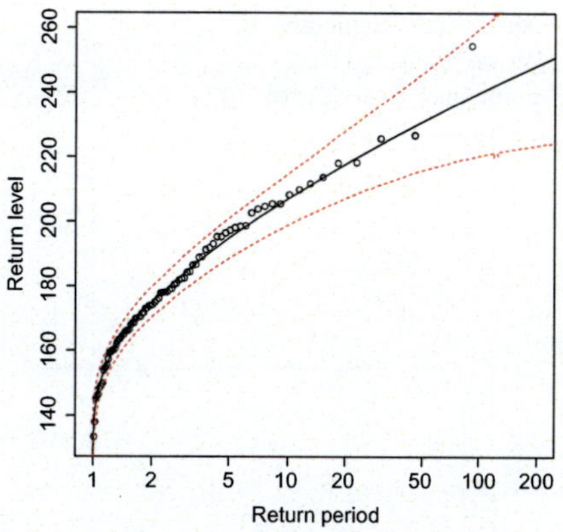

(b)

Fig. 12.2 Return level plots for the AM methods, Oslo. (**a**) Gumbel model. (**b**) GEV model

in Fig. 12.2b fits generally better to most points but poorly to the rightmost point. Since the estimated shape parameter is negative, the return levels of the GEV model are bounded at $\hat{\mu} - \hat{\sigma}/\hat{\gamma} = 381.7$ (cm), while the Gumbel model return levels are unbounded.

Heimsjø

Heimsjø has data for all years from 1928 to 2010. As with Oslo, some years are missing so much data that it is likely that the true annual maximum has been left out. One example is 1934, where data for all the first seven months are missing. This year is left out of the analysis. 1938 has missing data for January, February and December; important months where the real annual maximum is likely to be. 1943 lacks data for months January to April. 1959 lacks any data from September to December and has little from February and August. Finally, 2010 lacks data for the last months of the year. All these years are therefore left out.

For the Gumbel model, the parameter estimates are $\hat{\mu} = 311.84$ (1.399) and $\hat{\sigma} = 11.685$ (0.978). For the GEV model, $\hat{\mu} = 312.69$ (1.480), $\hat{\sigma} = 11.889$ (1.022) and $\hat{\gamma} = -0.137$ (0.066). Estimates of return levels are presented in Table 12.2.

The return level plots in Fig. 12.3 show the curves of the Gumbel and GEV models, together with confidence intervals and observed data points. The Gumbel model in Fig. 12.3a seems to slightly overestimate the points at high levels, while the GEV model in Fig. 12.3b perhaps underestimates them. All points are within confidence intervals, however. The GEV model return levels are bounded at 399.7 cm.

Table 12.2 Estimates of return levels and 95% confidence intervals for Heimsjø

	Gumbel model		GEV model	
$R = 1/p$	$\hat{z}_p$	CI	$\hat{z}_p$	CI
5	329.4	(324.8, 333.9)	328.8	(325.0, 332.7)
10	338.1	(332.3, 343.9)	335.7	(331.0, 340.4)
20	346.5	(339.5, 353.6)	341.7	(335.8, 347.6)
100	365.5	(355.5, 375.6)	353.3	(343.2, 363.3)
200	373.7	(362.4, 385.0)	357.5	(345.2, 369.9)

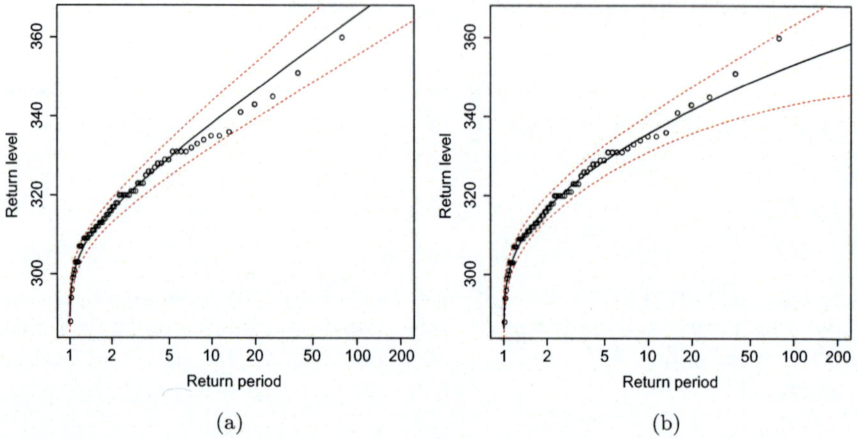

Fig. 12.3 Return level plots for the annual maxima methods, Heimsjø. (**a**) Gumbel model. (**b**) GEV model

Table 12.3 Estimates of return levels and 95% confidence intervals for Honningsvåg

	Gumbel model		GEV model	
$R = 1/p$	$\hat{z}_p$	CI	$\hat{z}_p$	CI
5	348.8	(342.0, 355.6)	348.0	(343.2, 352.8)
10	357.7	(349.0, 366.3)	353.5	(348.3, 358.7)
20	366.2	(355.6, 376.7)	357.8	(351.7, 363.9)
100	385.4	(370.4, 400.4)	365.0	(355.3, 374.7)
200	393.6	(376.7, 410.6)	367.2	(355.6, 378.7)

Honningsvåg

Honningsvåg has data for all years from 1970 to 2010, except 1985. Most of the years have acceptable amounts of data, but four years are missing important data: 1970 has no data from January to May, 1988 nothing in February and very little in March, 1989 nothing in August and September and 2010 nothing from October to December. All these are excluded from the model fitting.

For the Gumbel model, parameter estimates are $\hat{\mu} = 331.08(2.087)$ and $\hat{\sigma} = 11.812(1.458)$. For the GEV model, $\hat{\mu} = 332.85(2.285)$, $\hat{\sigma} = 12.344(1.626)$ and $\hat{\gamma} = -0.277(0.115)$. Estimates of return levels are given in Table 12.3.

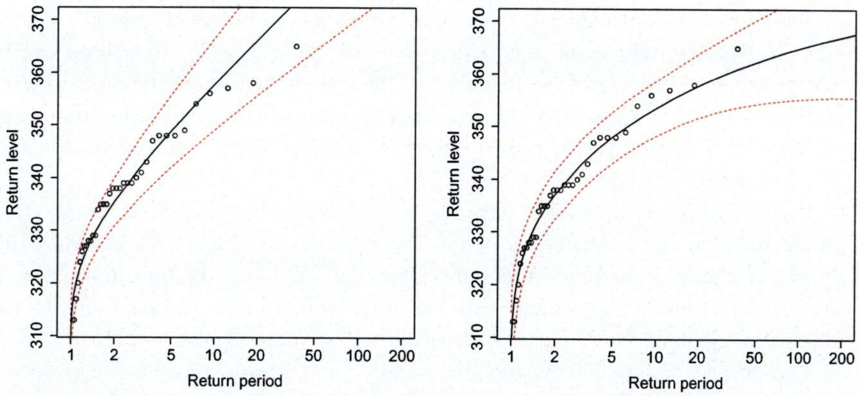

(a) (b)

Fig. 12.4 Return level plots for the annual maxima methods, Honningsvåg. (**a**) Gumbel model. (**b**) GEV model

The probability and quantile plots show that there are fewer points available for the model estimation than for Oslo and Heimsjø. The fit is still quite good overall, and similar to the other two locations—even though only a few points are on the line, the points seem to follow a sort of oscillating S shape around the optimal straight line.

The return level plots in Fig. 12.4 show the trend of the Gumbel and GEV models. Again it can be seen that the fit is not impressive, although the data points stay within the confidence bounds. As for Oslo and Heimsjø, the estimated shape parameter is negative, $\hat{\gamma} = -0.2771$, meaning that the Gumbel model gives higher estimates for the return levels for high return periods. Thus, it would appear that the higher return levels might be overestimated by the Gumbel model in Fig. 12.4a and somewhat underestimated by the GEV in Fig. 12.4b. The GEV return levels are bounded at 377.4 cm, significantly below even the 100-year return level of the Gumbel model. This is clearly not a sensible situation, and points to the hazard of using essentially curve fitting to decide on which GEV to use for estimating long return period sea levels.

12.4 The Peaks-Over-Threshold Method

For many physical processes, the assumption of temporally independent observations is unrealistic. Stationarity is usually a more plausible assumption, and it says that even though observations or data may be dependent, their stochastic properties are temporally homogeneous. A GEV distribution remains an appropriate model for block maxima of stationary series, and a GP distribution can also be shown to remain appropriate for threshold excesses, see Beirlant et al. (2004).

The degree of dependence between data points is in some sense quantified in the *extremal index* θ, which satisfies $0 < \theta \leq 1$, cf. Sect. 2.9. This parameter can be interpreted as a measure of the tendency of the process to cluster at extreme levels, and one can informally say that the inverse of the extremal index is the limiting mean cluster size. This means that if $\theta = 0.5$, then extreme values would approximately arrive in groups of two.

For the block maxima case, dependence is mostly absorbed in the parameters, which have to be estimated anyway. However, some change is necessary for threshold excess modeling. A common method used to overcome this issue is *declustering*, where the generalized Pareto distribution is instead fitted to the maxima of clusters. These clusters are identified by some empirical rule.

The easiest method of cluster identification is taking a threshold and saying that subsequent observations must be above this threshold to be part of the current cluster. A modification can be made by allowing one or more subsequent observations to be below the threshold before the current cluster is left. Selecting the amount of subsequent observations allowed before a cluster is left, is done by selecting a number for r, the allowed distance between observations above the threshold. For instance, this means that if observation number 50 and observation number 55 are both above the threshold u, they would be in the same cluster for $r \geq 5$ since the distance between the observations is 5. For $r < 5$, they would be considered to be from different clusters.

To be able to use a POT model in practice, the threshold u must be selected, as discussed in Chap. 3. In the mean residual life plots, cf. Eq. (3.11), for the three data sets that are analyzed in this chapter, the choice of u_0 has been clearly marked. Another method is available, based on fitting the model to a wide range of thresholds. In Eq. (3.10) it was used that there is a linear relationship between scale parameters σ_u above a valid threshold u_0. Furthermore, shape parameters γ_u should be constant. By reparametrizing the scale parameter to $\sigma_u^* = \sigma_u - \gamma_u u$, one obtains a parameter which should also be constant above u_0. The plot of (u, γ_u) and (u, σ_u^*) is hereby called the *stability plot*. Stability plots are also presented for the three available data sets.

Having chosen the reference level u_0, one would typically calculate the return period levels for several values of $u \geq u_0$. This is obtained by using the following formulas:

$$x_N = u + \sigma[(N n_y \zeta_u \theta)^\gamma - 1]/\gamma, \tag{12.1}$$

for the case of $\gamma \neq 0$, and

$$x_N = u + \sigma \ln(N n_y \zeta_u \theta), \tag{12.2}$$

for $\gamma = 0$. Here N = number of years, n_y = the average number of observations per year, $\zeta_u = \text{Prob}(X > u)$, θ = the extremal index.

The parameters of the Generalized Pareto distributions in this section are estimated by maximum likelihood. This was done with the maximum likelihood procedure in the fpot function in the evd library in R. Standard errors for the GP parameters come from this R function as well, and are extracted from a numerical approximation of the observed information. Confidence intervals for the parameters are then calculated by using the approximate normality of the maximum likelihood estimator. The uncertainty in ζ_u is ignored since it is usually small compared to the errors of the other parameters (Coles 2001).

12.4.1 Application to Water Level Measurements

Oslo

Figure 12.5a shows the mean residual life plot for Oslo. To select an appropriate threshold, one would search for approximate linearity within the confidence bounds. The vertical line shows the choice made at 134 cm. The figure does not paint a completely clear picture, as is the case with most real data sets, but from around the indicated spot and up to about 200 cm there is a certain level of linearity.

Figure 12.5b shows the effect of threshold selection on the model parameters. Ideally, the shape and modified scale parameters should be invariant to threshold change as long as the data are above the minimum threshold. One can see that this is basically the case from the beginning of the plot and up to about 170–175 cm. A threshold of 134 cm is within this range and cannot be rejected based on Fig. 12.5.

Having selected a threshold, one would go on to select the allowed distance between points in a cluster, r. Figure 12.6a shows the behaviour of the extremal index as the allowed distance between observations in a cluster is increased. The extremal index declines quite sharply until about $r = 12$, then almost flattens out

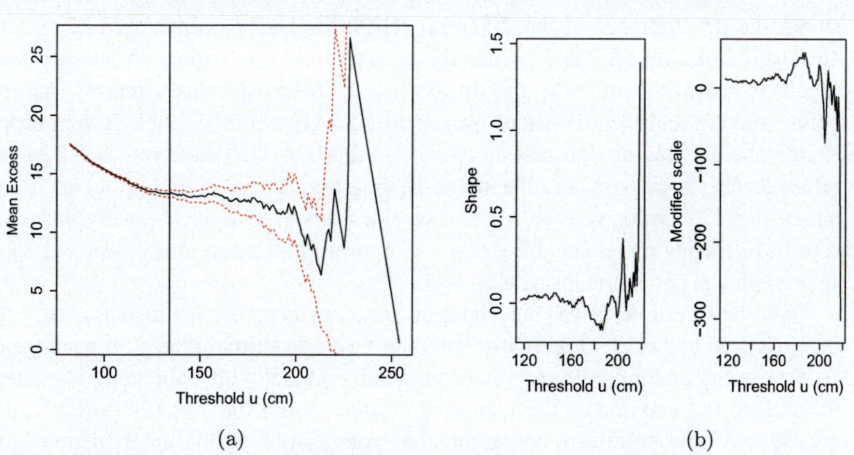

Fig. 12.5 Threshold selection plots for the POT model for Oslo. (**a**) Mean of exceedances of increasing thresholds (cm). (**b**) Stability of parameters for increasing threshold (cm)

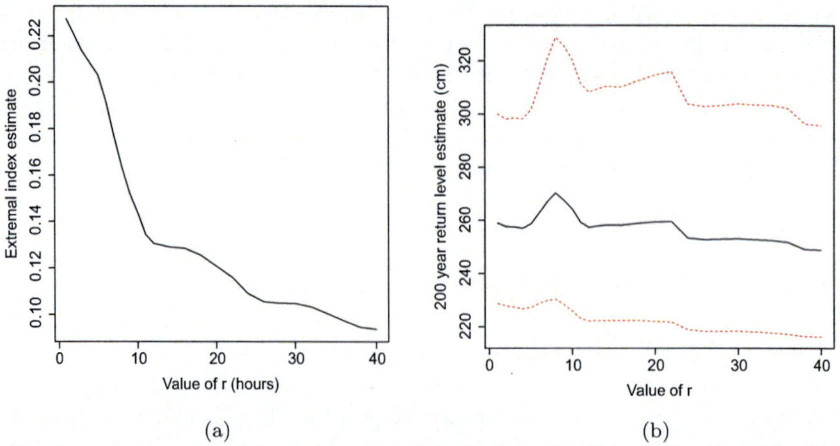

Fig. 12.6 Change in extremal index and 200 year return level as calculated by the POT model for Oslo. (**a**) Extremal index for r between 1 and 40. (**b**) 200 year return level for r between 1 and 40

before declining faster again. A similar tapering in the steepness of the curve is seen at about $r = 24$, and possibly at about $r = 36$.

With hourly observations, these changes in steepness coincide with the 12 hour cycle of the lunar tidal component. It is therefore safe to assume that extreme storms above $u = 134$ may last from one high tide to another. This suggests selecting a value of r larger than 12 to encapsulate the full length of such storms. But it does not seem plausible to have r much larger, since if the tide was the dampening factor, then the subsequent rising tide should bring the storm with it up to extreme levels again.

More insight into the choice of r is granted by looking at Fig. 12.6b, which shows the development of the 200 year return level as r is increased from 1 to 40. The return level stays approximately constant from $r = 1$ to $r = 5$ but increases relatively sharply from there and up to $r = 8$. It then decreases just as sharply, before staying relatively constant from $r = 12$ up to about $r = 22$. It then drops sharply, before staying constant up to $r = 36$. For $r \in (1, 5)$ and $r \in (12, 22)$, the return level is approximately the same. Having argued that r should not be much larger than 12, a value of $r = 1$ is chosen since this gives approximately the same result while using more data. Note that $r = 1$ means that a new cluster starts as soon as one value is below the threshold.

With the threshold at 134 cm, the model parameters were estimated to be $\hat{\gamma} = 0.019(0.025)$ and $\hat{\sigma} = 13.993(0.494)$ by the `fpot` function in the `evd` package in R. Probability and quantile plots indicate that the model with estimated parameters fits well to the maxima of the identified clusters. From the 778,199 points in the data series, 7364 points are above the chosen threshold, giving an estimate of the exceedance probability ζ of $\hat{\zeta} = 0.0095$. From the 7364 points above the threshold,

Table 12.4 Return level estimates and 95% confidence intervals for the POT model

	Oslo		Heimsjø		Honningsvåg	
$R = 1/p$	$\hat{z}_p$	CI	$\hat{z}_p$	CI	$\hat{z}_p$	CI
5	200.5	(189.3, 213.9)	329.5	(321.7, 338.8)	351.6	(338.1, 368.3)
10	211.1	(197.0, 228.5)	335.3	(325.9, 347.0)	360.5	(344.2, 380.9)
20	221.9	(204.7, 243.8)	340.7	(329.5, 354.8)	369.2	(350.1, 393.8))
100	247.6	(221.7, 282.2)	351.7	(336.5, 372.0)	388.8	(362.5, 424.6)
200	258.9	(228.8, 300.0)	355.8	(338.9, 378.9)	397.0	(367.4, 438.3)

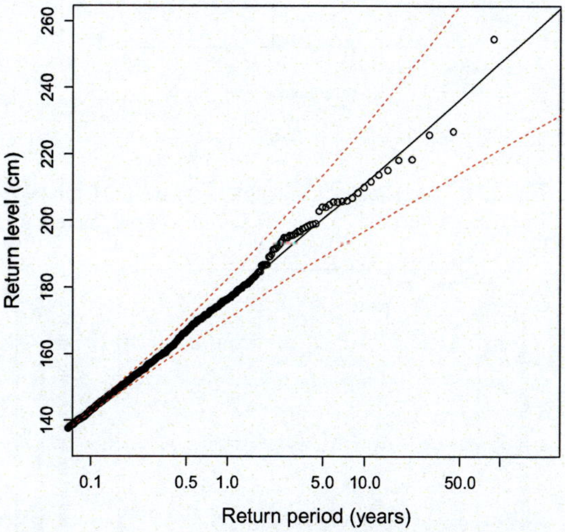

Fig. 12.7 Return levels (cm) for increasing return periods (years) for Oslo

1675 clusters were identified, giving an extremal index of $\hat{\theta} = 0.2275$. All this gives the return level estimates shown in Table 12.4.

The return level plot is shown in Fig. 12.7. All data points are contained within the confidence bounds of the fitted model, and most of them stay on or very close to the line. Points above 185 cm behave somewhat more erratically than those below, but not to a dramatic extent. This level corresponds approximately to the level at which the stability of parameters, as displayed in Fig. 12.5b, begins to wear off.

Heimsjø

Figure 12.8a shows the mean residual life plot for the Heimsjø data. Approximate linearity is found around the vertical line, which is placed at $u = 287$ cm. The stability plot in Fig. 12.8b shows the effect on the model parameters of threshold change around the 287 cm mark. Above approximately 280 cm and up to around 325–330 cm, the parameters are approximately constant.

An appropriate value of r is sought for the identification of independent clusters and the estimation of the extremal index. Figure 12.9a shows how the extremal index

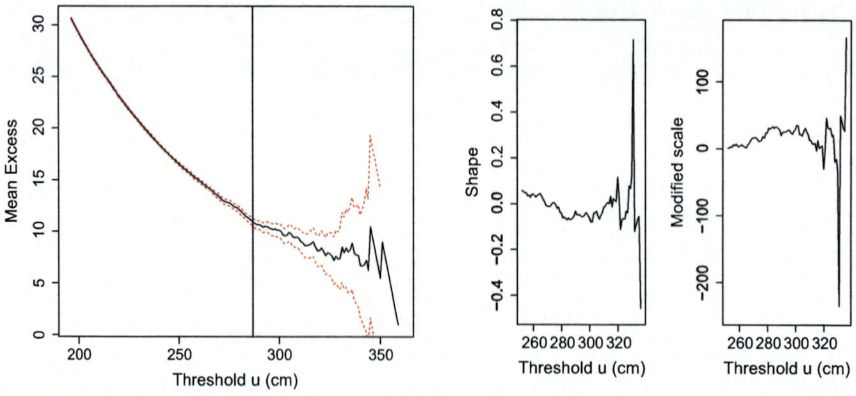

(a) (b)

Fig. 12.8 Threshold selection plots for the POT model for Heimsjø. (**a**) Mean of exceedances of increasing thresholds (cm). (**b**) Stability of parameters for increasing threshold (cm)

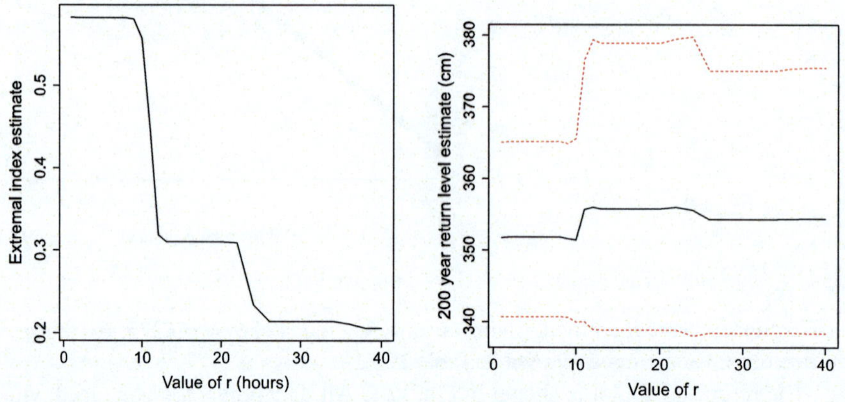

(a) (b)

Fig. 12.9 Change in extremal index and 200 year return level as calculated by the POT model for Heimsjø. (**a**) Extremal index for r between 1 and 40. (**b**) 200 year return level for r between 1 and 40

estimate behaves. It is close to piecewise constant, with significant drops at around $r = 10$ to 12 and at $r = 24$. Figure 12.9b shows that there is quite little difference between the estimate for $r = 1$ and larger r. The difference is somewhat larger than was found for Oslo, however, and r is chosen to include storms crossing from one tide to another. This leads to the choice $r = 13$.

With the 287 cm threshold, model parameters are estimated as $\hat{\sigma} = 13.709(0.653)$ and $\hat{\gamma} = -0.116(0.030)$. Probability and quantile plots show a fair agreement with the fitted model. However, for the highest sea levels, there is an issue with the model fit, which will be discussed below. From the 686867 points

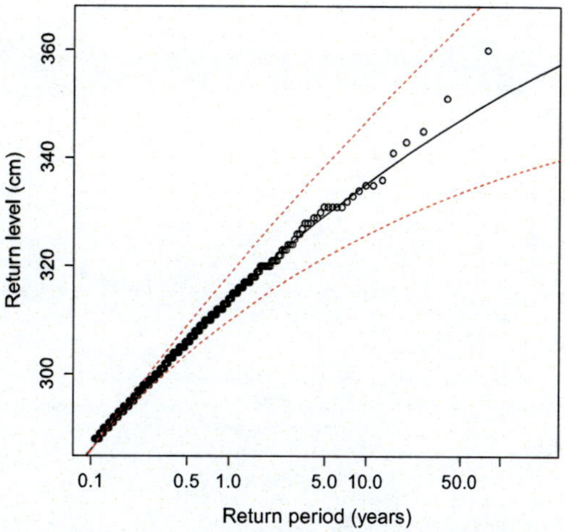

Fig. 12.10 Return levels (cm) for increasing return periods (years) for Heimsjø

in the Heimsjø data set, 2363 points are above 287 cm, giving an estimate of the exceedance probability ζ of $\hat{\zeta} = 0.0034$. From the points above the threshold, 733 clusters were identified, giving an extremal index of $\hat{\theta} = 0.310$.

Return level estimates are presented in Table 12.4. They go from slightly overshooting to slightly undershooting the published return levels, but all are certainly within confidence bounds.

As mentioned above, for the highest sea levels, there are some issues with the model fit. This is better seen by looking at the return level plot in Fig. 12.10, which shows irregularities after about 330 cm—corresponding to the level after which parameters in Fig. 12.8b are no longer near-constant. The five highest points are above the model line, and it seems possible that return levels for long return periods can be underestimated.

Honningsvåg

The mean residual plot in Fig. 12.11a shows tendencies of linearity within confidence bounds slightly before the 280 cm mark. A threshold of 278 cm is selected, indicated by the vertical black line. The stability plot in Fig. 12.11b is far from constant up to around 275–280 cm, but from there it is reasonable to call the parameters near-constant up to perhaps 310 cm. The threshold $u = 278$ is therefore barely within the acceptable area as far as the stability plot is concerned.

To find an appropriate level of r, Fig. 12.12a is inspected. Honningsvåg is tide-dominant like Heimsjø, and the same pattern is shown here as was observed there; the extremal index plot is approximately piecewise constant. Figure 12.12b shows a similar pattern as that for Heimsjø as well, but the differences in return levels are much larger. The values for $r \leq 10$ and $r > 24$ are about the same, however. The

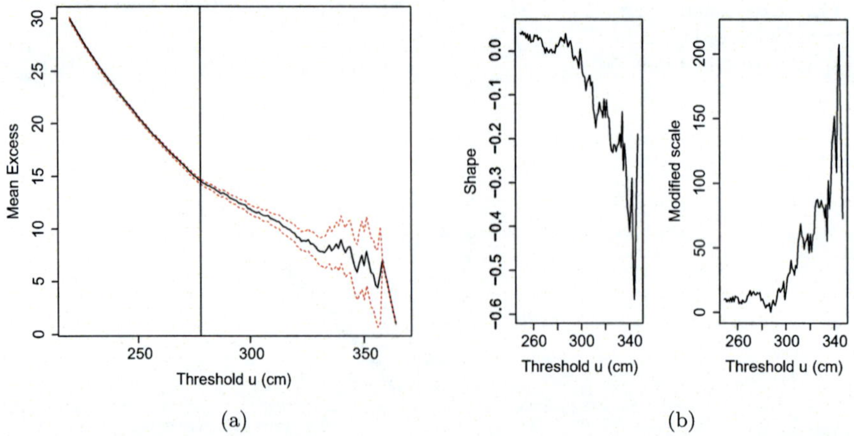

Fig. 12.11 Threshold selection plots for the POT model for Honningsvåg. (**a**) Mean of exceedances of increasing thresholds (cm). (**b**) Stability of parameters for increasing threshold (cm)

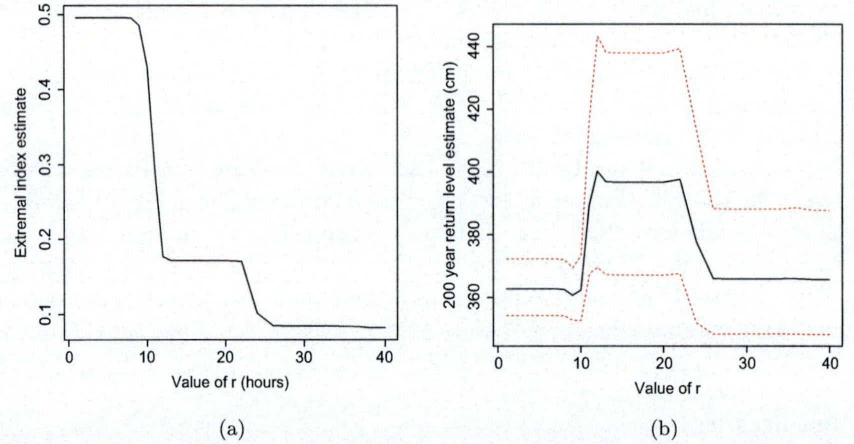

Fig. 12.12 Change in extremal index and 200 year return level as calculated by the POT model, for Honningsvåg. (**a**) Extremal index for r between 1 and 40. (**b**) 200 year return level for r between 1 and 40

difference is large enough to not be within the confidence intervals, and $r = 13$ is chosen since it was argued earlier for a $r \geq 12$.

A threshold of 278 cm gives estimated model parameters of $\hat{\sigma} = 14.692(0.530)$ and $\hat{\gamma} = -0.025(0.026)$. Probability and quantile plots show a fair agreement with the fitted model. However, as for Heimsjø, for the highest sea levels, there is an issue with the model fit. There are 330883 sea level measurements in the Honningsvåg data, 9264 of which are above 278 cm. This makes for an estimate of the exceedance

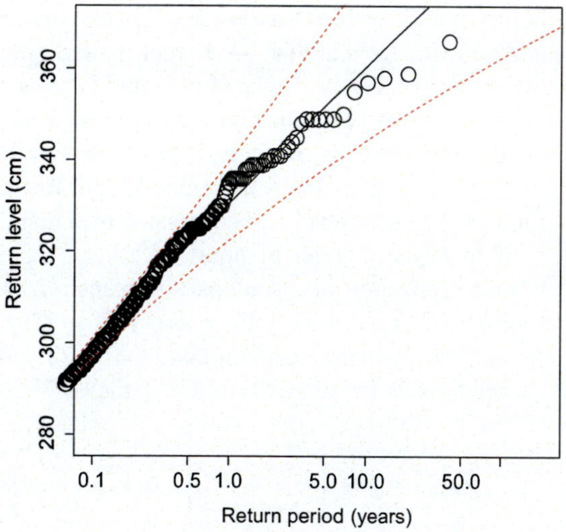

Fig. 12.13 Return levels (cm) for increasing return periods (years) for Honningsvåg

probability ζ of $\hat{\zeta} = 0.0280$. From the points above the threshold, 1590 clusters were identified, giving an extremal index of $\hat{\theta} = 0.172$.

Estimated return levels are presented in Table 12.4.

As indicated above, the quantile plot shows that a few of the most extreme data points are off the fitted model curve, with Fig. 12.13 giving a clearer impression. It looks likely that higher return levels may be overestimated, although all points are within confidence intervals.

12.5 Revised Joint Probabilities Method

The revised joint probabilities (RJP) method is an attempt by Tawn (1992) to improve the joint probabilities method employed by Pugh and Vassie (1980). Whereas they assume that hourly surge levels are independent, Tawn argues that this is clearly a false assumption. Instead, he uses that for a stationary sequence $Y_1, \ldots, Y_n, \ldots$,

$$\text{Prob}(\max\{Y_1, \ldots, Y_n\} < y) \approx [\text{Prob}(Y_1 < y)]^{n\theta} \tag{12.3}$$

for large y, and $0 < \theta \leq 1$. θ is the extremal index, and θ^{-1} is defined as the limit of $\theta^{-1}(y)$ as y tends toward the upper end point of the distribution of Y. $\theta^{-1}(y)$ is defined as the mean of the distribution of cluster sizes. The extremal index can be equal to 1 for both dependent and independent sequences, when the sequence behaves like an independent sequence at high levels. Unfortunately, as will be demonstrated in the section on the ACER method, for the data analyzed in this chapter, $\theta(y)$ does not seem to be a robust parameter relative to its dependence on y.

In fact, it will be seen that $\theta(y)$ may display significant dependence on y while there is clear evidence that $\theta(y) \rightarrow 1$ when y increases. However, in the present chapter, the extremal index is estimated as would be typically done in the RJP method.

The RJP method relies on a componentwise analysis of the sea level Z. It is divided into three components, $Z_t = M_t + X_t + Y_t$, where M_t is the mean sea level, X_t the tidal level and Y_t the surge level. For the purpose of this discussion, $M_t = 0$ for all t. The tidal level X_t is estimated by a numerical model.

If N is the number of hours in a year, the surge levels $Y_1, \ldots Y_N$ are taken to be a realization of a stationary sequence. It is then assumed that (Coles 2001) $\text{Prob}(\max\{Y_1, \ldots, Y_N\} \leq y) = F_s(y)^{N\theta_s} = G(y; \mu_s, \sigma_s, \gamma_s)$ for large values of y, say $y > u$. F_s is the marginal distribution for surges, while θ_s $(0 \leq \theta_s \leq 1)$ is the extremal index for surges. G refers to the GEV distribution model. It follows that,

$$F_s(y) = \exp\left\{-(N\theta_s)^{-1}\left[1 + \gamma_s \frac{y - \mu_s}{\sigma_s}\right]^{-1/\gamma_s}\right\} \quad \text{for } y > u, \tag{12.4}$$

on $\{y : 1 + \gamma_s(y - \mu_s)/\sigma_s > 0\}$, with $\sigma_s > 0$ and arbitrary γ_s and μ_s.

If T is the tidal cycle length in hours, it follows that (Tawn 1992),

$$\text{Prob}(\max\{Z_1, \ldots, Z_T\} \leq z) = \text{Prob}(\bigcap_{t=1}^{T} Y_t \leq z - X_t\} = \left\{\prod_{t=1}^{T} F_s(z - X_t)\right\}^{\theta}, \tag{12.5}$$

where θ is an hourly sea level extremal index $(\theta_s \leq \theta \leq 1)$. Combining Eqs. (12.4) and (12.5) gives the RJP distribution for annual maximum sea levels,

$$G(z) = \exp\left\{-\theta(T\theta_s)^{-1}\sum_{t=1}^{T}\left[1 + \gamma_s \frac{z - \mu_s - X_t}{\sigma_s}\right]^{-1/\gamma_s}\right\}, \tag{12.6}$$

where $z > u + \max\{X_1, \ldots, X_T\}$.

Equation (12.6) shows the case where the surge distribution $F_s(\cdot)$, and therefore its parameters μ_s, σ_s and γ_s, are independent of the concomitant tidal level. Unfortunately, the assumption that tide and surge are independent processes is poor in shallow water areas, where turbulent frictional processes on the sea bed cause the tide and surge components to interact. This causes effects such as surge values at high tides being damped and surges on the rising tide being amplified (Dixon and Tawn 1994). These effects vary from site to site however, so it is attempted to model the interaction on the residuals from the observed surges.

The tidal range from lowest observed tide (LAT) to highest observed tide (HAT) is split into n_b equi-probable bands, i.e., each band has an equal amount of observed measurements. For each of the tidal observations, there is a concurrent surge observation. This means that if there are n_b tidal bands and n_{obs} observations in total, there are n_{obs}/n_b observations of both the tide and surge in each band.

If the tide and surge were independent, an equal amount of points should also be expected to exceed a given level u. But if there is interaction, then the least number of points should be in the top band where surges are damped. Similarly, the largest number should be in the middle bands where surges are magnified. The amount of discrepancy between bands will therefore be a quantifiable measure of the tide-surge interaction.

The level u is now chosen to be a high empirical quantile of the surge distribution, z_q for $q = 0.9975$. For the independent case, there should then be $(1 - q) \cdot n_{obs}/n_b = v$ observations in each band. $n_b = 5$ is chosen as in Dixon and Tawn (1994), meaning $n_{obs} \cdot 0.0025/5 = v$ observations per tidal band. In actuality, since independence is a flawed assumption, N_i surges per band are observed, for $i = 1, \ldots, n_b$. To put the interaction into a quantifiable setting, a standard χ^2 test statistic is used, $\chi^2 = \sum_{i=1}^{5}(N_i - v)^2/v^2$. If $N_i \approx v$, then χ^2 will be small. Tide and surge are deemed to interact with 95 % confidence if the test statistic is above the associated 4 degrees of freedom table value of $\chi^2_{4,0.95} = 9.488$.

To account for the tide-surge interaction, the method used in Dixon and Tawn (1994) is adopted, where the surge series is location-scale normalized by $S_t^* = (Y_t - a(X_t))/b(X_t)$, where $a(X_t)$ and $b(X_t)$ are some tide-dependent functions. $\{S_t^*\}$ is then supposed to be stationary, and established methods can be used to estimate the associated model parameters μ_{s^*}, σ_{s^*} and γ_{s^*}. The parameter estimates for the original surge series are then given by $\mu_s(X) = \mu_{s^*}b(X) + a(X)$, $\sigma_s(X) = \sigma_{s^*}b(X)$, and $\gamma_s(X) = \gamma_{s^*}$.

Equation (12.6) is then modified to

$$G(z) = \exp\left\{-\theta(T\theta_s)^{-1}\sum_{t=1}^{T}\left[1 + \gamma_{s^*}\frac{z - \mu_{s^*}b(X_t) - a(X_t) - X_t}{(\sigma_{s^*}b(X_t))}\right]^{-1/\gamma_{s^*}}\right\}, \tag{12.7}$$

for $z > u + \max\{X_1, \ldots, X_T\}$. The estimation of $a(X)$ and $b(X)$ is discussed in Dixon and Tawn (1994).

12.5.1 Estimating Return Levels with the RJP Method

In the 1992 and 1994 reports (Tawn 1992; Dixon and Tawn 1994), the r-largest method is applied to estimate the parameters of the surge distribution. This method is a modification of the GEV annual maxima method, and uses the r largest values per year to estimate model parameters.

The threshold applied in the estimation of θ_s is now adopted, and the POT method is then used to estimate model parameters. This is a method which has the advantage of using much more data. The fact that for a POT model $\gamma_{POT} = \gamma_{GEV} = \gamma$ for the corresponding GEV model is used, with $\sigma_{POT} = \sigma_{GEV} + \gamma \cdot (u - \mu_{GEV})$ (Coles 2001).

In practice, the shape and scale are estimated with a maximum likelihood procedure by the function `fpot` in the *R* package `evd`, while the location parameter is calculated by the `fgev` function. This is also the case for the confidence intervals of model parameters. The model is fitted to the maxima of clusters, where the clusters are identified by the same rule that governs the estimation of θ.

The extremal indices θ and θ_s require two choices each; a threshold above which clusters are counted and an empirical clustering rule which says how many non-exceedances are allowed before the current cluster is terminated. For the first choice, the quantile that was found in the POT analysis is used. There a thorough analysis was done on where the threshold should be placed, and it makes sense using this same threshold to estimate θ. Applying the procedure of using the same quantiles for both extremal indices, as was done in Dixon and Tawn (1994), corresponding thresholds are obtained for the surge series and the estimation of θ_s as well.

For the second choice, it is necessary to decide what constitutes an independent storm. In Dixon and Tawn (1994), they found that $r = 30$ was a good choice, and that the ratio of extremal indices was not too dependent on this choice in any case. In the present analysis the results are more sensitive to this choice, since the POT method with clustering is used to estimate model parameters. Still this r value of 30 is adopted.

A practical challenge with the RJP method is that Eq. (12.7) is impossible to solve analytically for z. One may start out by defining $G(z_p) = 1 - p$ and get the relation,

$$- T\theta_s\theta^{-1}\ln(1-p) = \sum_{t=1}^{T}\left[1 + \gamma_{s*}\frac{z_p - \mu_{s*}b(X_t) - a(X_t) - X_t}{(\sigma_{s*}b(X_t))}\right]^{-1/\gamma_{s*}}, \tag{12.8}$$

but one cannot go further by algebraic manipulation. A numerical procedure is therefore developed where the right hand side of the equation is calculated for a range of relevant z values and then matched to the left side of the equation.

The tidal cycle length is 18.61 years, meaning that $T = 18.61 \cdot 8766 = 163135.3 \approx 163135$. Any span of consecutive 163135 observations should therefore approximately contain a full tidal cycle, and in the numerical procedure the last 163135 observations in the tidal series are taken. The numerical analysis code is written in R.

12.5.2 *Application to Water Level Measurements*

Oslo

A scrutiny of the sea level data for Oslo reveals that a large percentage of the observed sea level rises stem from surges. Locations with such characteristics are called surge-dominant, as opposed to tide-dominant where most of an observed sea level stems from the current tidal level. Areas where surge dominates usually do not have as much tide-surge interaction as tide-dominant areas, but the χ^2 test is still used to quantify the amount.

By using 5 tidal bands and cutting the data at the 99.75% quantile, 389 surges per band are expected. Instead 346, 380, 398, 431, and 386 data points per band are found, in order from the lowest tide level to the highest. This is not a very bad result, but there is still evidence of significant tide-surge interaction. The test statistic has a result of $\chi^2 = 9.778$, compared to the table value of $\chi^2_{4,0.95} = 9.488$.

Since the χ^2 test shows a significant level of interaction, it is modeled by using 5 tidal bands. Figure 12.14a shows the tide against surge data to the left, and tide against transformed surge data to the right, both showing only points above the 99.75% quantile. There seems to be very little difference between the left and the right plots except for the scales on the x axes.

Figure 12.14b shows how the ratio between the χ^2 test statistic and the corresponding $\chi^2_{n_b-1,0.95}$ value develops after 2–30 bands have been used to transform the surge data. $\chi^2/\chi^2_{n_b-1,0.95} = 1$ is indicated by the horizontal line, and ratios above this correspond to χ^2 tests showing significant interaction. It is seen that for $n_b = 2$ and $n_b = 15$ the interaction is insignificant, but for $n_b = 2$ the interaction is insignificant for the untransformed data as well, with a test statistic of 1.85 versus the corresponding table value of 3.84. Transforming the data with $n_b = 15$ seems needlessly complex for a model where there was hardly significant interaction in the original test. Furthermore, there is little actual difference in return levels. The same is true if they are compared to the return levels achieved using no tide-surge correction. For parsimony the model without correction is selected, based

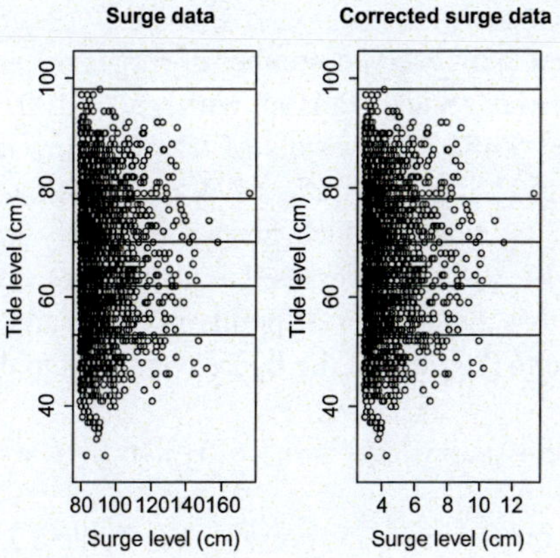

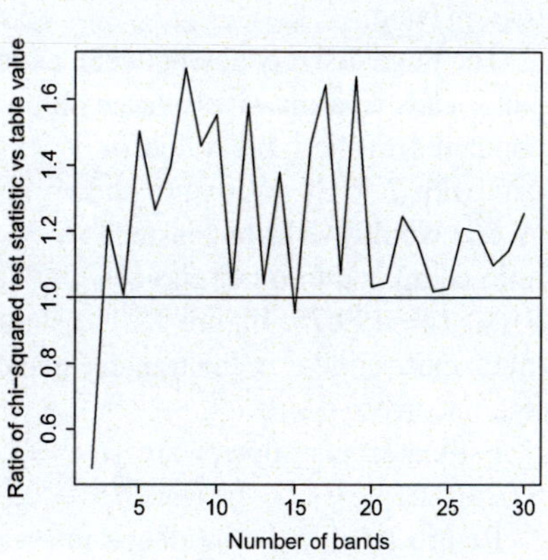

Fig. 12.14 Tide-surge interaction plots for Oslo. (**a**) Tide levels and concurrent surges above the 99.75% quantile. (**b**) Ratio between χ^2 test statistic and table value $\chi^2_{n_b-1,0.95}$

Table 12.5 Return level estimates and 95% confidence intervals for the RJP model

	Oslo		Heimsjø		Honningsvåg	
$R = 1/p$	$\hat{z}_p$	CI	$\hat{z}_p$	CI	$\hat{z}_p$	CI
5	194.5	(184.1, 207.6)	318.2	(314.6, 323.5)	332.4	(331.3, 335.6)
10	206.7	(192.5, 225.2)	325.4	(321.1, 336.1)	341.9	(339.1, 347.9)
20	217.9	(200.0, 242.5)	333.0	(327.0, 354.0)	350.9	(346.1, 360.4))
100	242.0	(215.0, 282.9)	352.5	(339.3, 422.4)	370.6	(360.2, 390.7)
200	251.8	(220.7, 300.7)	362.1	(344.1, 466.5)	378.7	(365.6, 404.4)

on Eq. (12.6) instead of Eq. (12.7). Estimated parameters are $\hat{\mu}_s = 95.655(2.313)$, $\hat{\sigma}_s = 17.576(1.053)$, $\hat{\gamma}_s = -0.049(0.039)$, $\hat{\theta} = 0.079$ and $\hat{\theta}_s = 0.068$.

Resulting return levels and confidence intervals for Oslo can be found in Table 12.5.

Heimsjø

Heimsjø is tide-dominant, and it is observed from the data that there is quite a small distance between the maximum of its tidal series and the maximum of observed sea levels. This also holds for the mean of annual maxima from tide calculations and observations. Tide-dominance usually implies larger tide-surge interaction than surge-dominance, and the χ^2 test for Heimsjø certainly lives up to this expectation. With 5 tidal bands and cutting at the 99.75% quantile, about 343 surge observations are expected per band. Instead, 790, 479, 288, 112 and 75 points are found. As expected for a tide-dominant location, the least number of points are in the highest tidal band, where the high tide dampens surges. By far the most points are in the bottom band.

The huge discrepancy between expected and observed number of surges in each band leads to a massively large value for the χ^2 test statistic, with $\chi^2 = 1007.3$ compared to the table value of $\chi^2_{4,0.95} = 9.488$. The results of transforming the data with 2 to 30 bands are shown in Fig. 12.15b. The ratio never drops below 1 as one would wish, but instead $n_b = 12$ is chosen, which gives a relatively low ratio compared to other choices. It gives $\chi^2 = 89.865$ versus the quantile value of $\chi_{11,0.95} = 19.675$. Figure 12.15a shows that the number of points in each band is much more similar in the transformed case to the right of the figure compared to the left, uncorrected side.

Estimated parameters are $\hat{\mu}_{s*} = 3.494(0.283)$, $\hat{\sigma}_{s*} = 1.671(0.160)$, $\hat{\gamma}_{s*} = 0.074(0.079)$, $\hat{\theta} = 0.106$ and $\hat{\theta}_{s*} = 0.063$.

Return level and confidence intervals for Heimsjø are presented in Table 12.5. The low return period levels are quite different from their published counterparts, but the upper confidence bound of the 20-year return level estimate contains the corresponding published value (Hansen and Roald 2000). This could perhaps be explained by the tide-surge interaction still present, but estimation of return levels

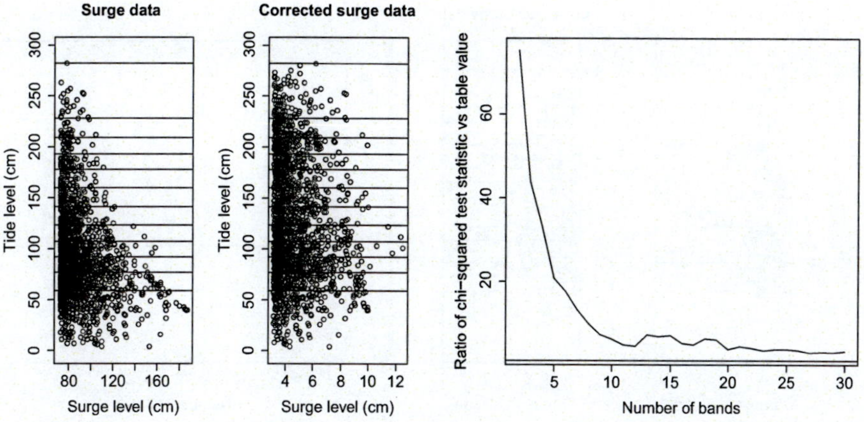

(a) (b)

Fig. 12.15 Tide-surge interaction plots for Heimsjø. (**a**) Tide levels and concurrent surges above the 99.75% quantile. (**b**) Ratio between χ^2 test statistic and table value $\chi^2_{n_b-1,0.95}$

without any tidal bands also show this apparent underestimation of low return period levels. The standard error of the estimated γ_s is larger than the estimate itself, and the estimate being positive means very large upper bounds for long return periods.

Honningsvåg

Honningsvåg is tide-dominant like Heimsjø, but the χ^2 test shows much less interaction than for Heimsjø. It is still significant on a 95% confidence level, with $\chi^2 = 23.652$. Figure 12.16b shows how the χ^2 ratio develops. After $n_b = 2$, it does not go below 1 until $n_b = 25$, but for $n_b = 16$ it is quite close and this is chosen. Here, $\chi^2 = 26.106$ against $\chi^2_{15,0.95} = 24.996$. Figure 12.16a shows the surge data before and after transformation with 16 tidal bands.

Estimated parameters are $\hat{\mu}_{s*} = 2.943(0.274)$, $\hat{\sigma}_{s*} = 1.617(0.124)$, $\hat{\gamma}_{s*} = -0.041(0.049)$, $\hat{\theta} = 0.085$ and $\hat{\theta}_{s*} = 0.043$.

Return levels and confidence intervals for Honningsvåg are shown in Table 12.5.

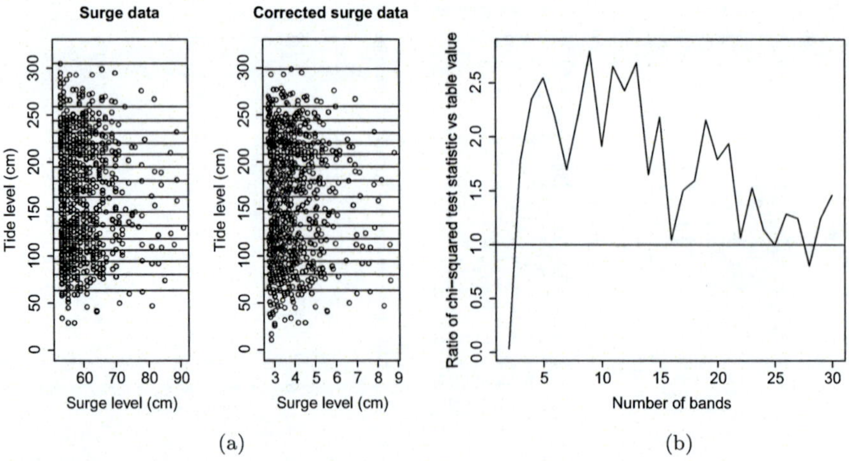

Fig. 12.16 Tide-surge interaction plots for Honningsvåg. (**a**) Tide levels and concurrent surges above the 99.75% quantile. (**b**) Ratio between χ^2 test statistic and table value $\chi^2_{n_b-1,0.95}$

12.6 The ACER Method

All calculations in this section were performed with the ACER package for Matlab (Karpa 2012).

12.6.1 Application to Water Level Measurements

Note that for estimation of the 5, 10, and 20-year return period levels, which for all stations are in sample estimates, the ACER method will provide unbiased estimates of the exact values. For estimation of the long return period levels, the first goal is to find a tail marker η_0 that represents a sufficiently high threshold. For the ACER method, this is found where the curves of the estimated ACER functions $\hat{\varepsilon}_k(\eta)$ start behaving regularly in the sense of Eq. (5.31). After having chosen such a point initially, one would typically go even further into the tail to verify robustness with respect to the estimated return levels. For the hourly sea level measurements studied in this chapter, it was found that a level for which the remaining data amounted to roughly 5–10% of the total data set gave stable numerical estimates.

Oslo

The curves in the ACER plot of Fig. 12.17a detail the effect dependence has on the ACER function estimates for Oslo. It can be seen that the independent case of

Fig. 12.17 ACER plots for Oslo. (**a**) Log plot of estimated ACER functions. (**b**) Fitted and extrapolated ACER_2 function

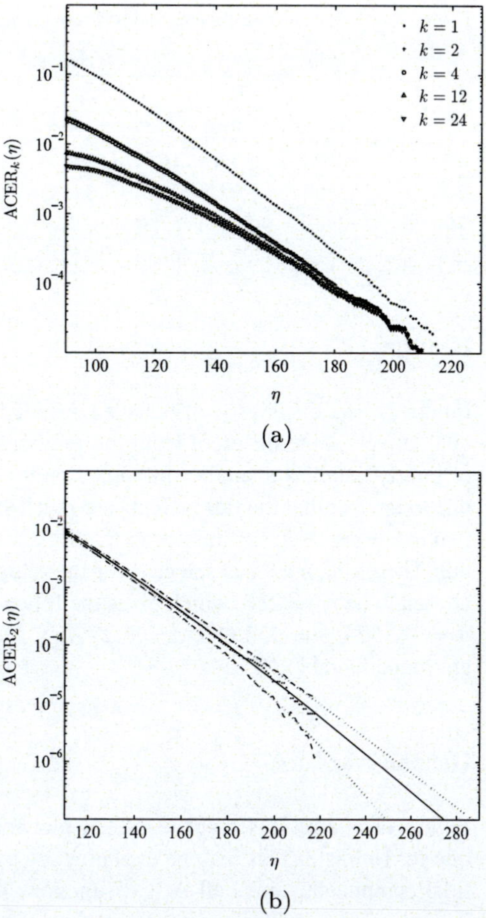

$k = 1$ has a curve which stays significantly above the rest, clearly demonstrating that hourly measurements are strongly dependent. The ACER plot also shows that there is a significant diurnal dependence effect, which is to be expected. But this effect is seen to vanish at the higher levels, showing that the ACER functions coalesce in the tail for $k \geq 2$. Thus, for estimation of long return period sea levels it is advantageous to choose $k = 2$, since this case allows for the use of more data than for $k > 2$, with the potential benefit of higher accuracy as a result.

With the selected estimated ACER function, i.e., $\hat{\varepsilon}_2(\eta)$, the optimized parameter estimates are $d = -4.048$, $b = 96.901$, $a = 0.037$, $c = 1.117$, which is based on a tail marker $\eta_0 = 100$. The return level estimates for Oslo are presented in Table 12.6.

Table 12.6 Return level estimates and 95% confidence intervals for the ACER model

	Oslo		Heimsjø		Honningsvåg	
$R = 1/p$	$\hat{z}_p$	CI	$\hat{z}_p$	CI	$\hat{z}_p$	CI
5	199.6	(191.0, 205.5)	330.5	(327.0, 333.6)	351.6	(346.4, 357.1)
10	210.1	(199.7, 216.9)	336.3	(332.4, 339.7)	357.7	(352.3, 363.8)
20	220.1	(207.7, 227.6)	341.6	(337.4, 345.3)	363.3	(357.8, 370.1))
100	242.3	(224.9, 251.8)	352.6	(347.7, 357.0)	375.3	(369.6, 383.8)
200	251.6	(232.0, 262.0)	357.0	(351.8, 361.7)	380.2	(374.4, 389.2)

Heimsjø

Figure 12.18a displays well-behaved ACER functions for Heimsjø, but the plot is qualitatively quite different from the previous one for Oslo. Now, the dependence of hourly data has a smaller influence on the ACER function values. This may be due to the fact that the tidal effects are much stronger for Heimsjø than for the other two locations. Also for this location, all ACER functions for $k \geq 2$ coalesce in the tail. Therefore, $k = 2$ is selected for the return level estimation. The tail marker is chosen to be $\eta_0 = 230$, which gives the following optimal values of the parameters: $d = -2.524$, $b = 154.689$, $a = 5.27 \cdot 10^{-6}$, $c = 2.755$. The return level estimates are found in Table 12.6.

Honningsvåg

Figure 12.19a reveals that the ACER plot for Honningsvåg is very similar to the one for Heimsjø. This may be explained by the fact that both stations have a strong tidal component. Since all ACER functions for $k \geq 2$ coalesce in the tail, $k = 2$ is again selected for the return level estimation. The tail marker is chosen to be $\eta_0 = 260$, which gives the following optimal values of the parameters: $d = -3.012$, $b = 231.209$, $a = 8.15 \cdot 10^{-4}$, $c = 1.907$. The return level estimates are given in Table 12.6.

12.7 Discussion of Results

12.7.1 Oslo

Oslo is surge-dominant and has the largest amount of data of the locations presented in this chapter. Both these factors mean that all the methods applied should perform quite well here. Surge-dominance means that the non-stochastic tidal component of the measured sea levels does not interfere to a large degree, and a large amount of data is naturally desirable. As for possible sources of error, the removal of the post-

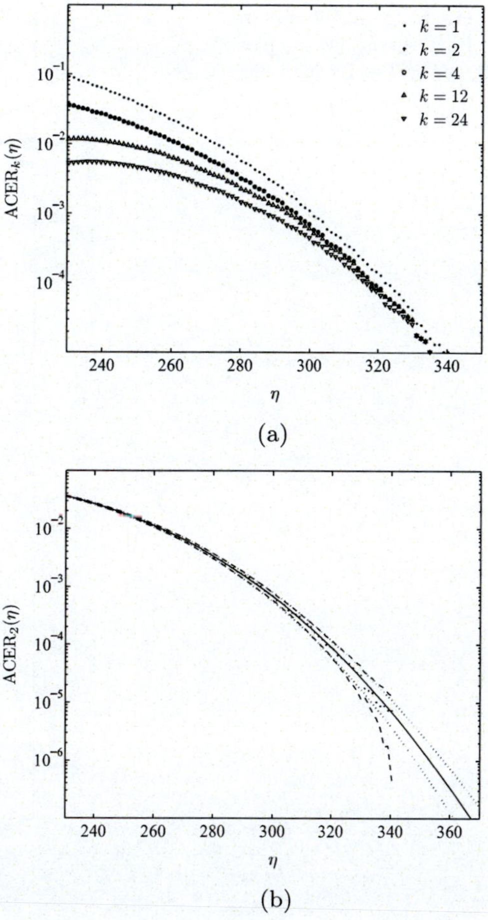

Fig. 12.18 ACER plots for Heimsjø. (**a**) Log plot of estimated ACER functions. (**b**) Fitted and extrapolated $ACER_2$ function

glacial rebound trend is certainly worth mentioning, and the same procedure as in the report by Hansen and Roald (2000) was followed.

Figure 12.20a shows that return levels are very similar for all methods for 5, 10 and 20-year periods, while larger differences arise for the 100 and 200 year periods. Note that the lines between the indicated points are only drawn for visual purposes.

Figure 12.20b shows the 200 year return level estimates, together with 95% confidence intervals. As seen, the Gumbel model is producing the largest estimate, at 266 cm, while the GEV model gives the smallest, at 249 cm. This difference is not large, and all methods contain the results from the other methods within their confidence interval. Also note that the ACER method has the shortest confidence interval, indicating perhaps a higher estimation accuracy.

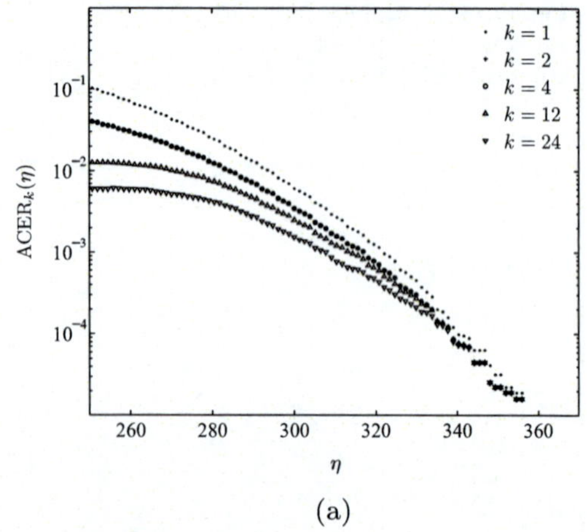

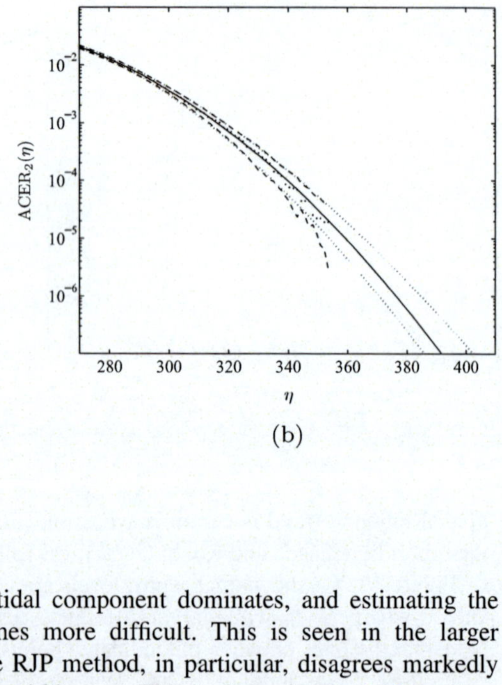

Fig. 12.19 ACER plots for Honningsvåg. (**a**) Log plot of estimated ACER functions. (**b**) Fitted and extrapolated ACER_2 function

12.7.2 Heimsjø

For Heimsjø, the non-stochastic tidal component dominates, and estimating the extreme value distribution becomes more difficult. This is seen in the larger differences between methods. The RJP method, in particular, disagrees markedly for the three first periods of 5, 10, and 20 years. This is shown in Fig. 12.21a. Since the ACER method provides unbiased estimates of the exact in sample values, it can be concluded that the RJP method estimates are inaccurate at these return periods. However, the RJP method performs much better for the 100 and 200 year levels, where it basically is in line with all models but the Gumbel one.

Figure 12.21b shows better the difference between the 200 year estimates, and highlights the extreme upper bound of the estimate obtained by the RJP method. The other confidence bounds look minuscule in comparison, but this is because of the scale of the bounds of the RJP method. The Gumbel model produces a significantly

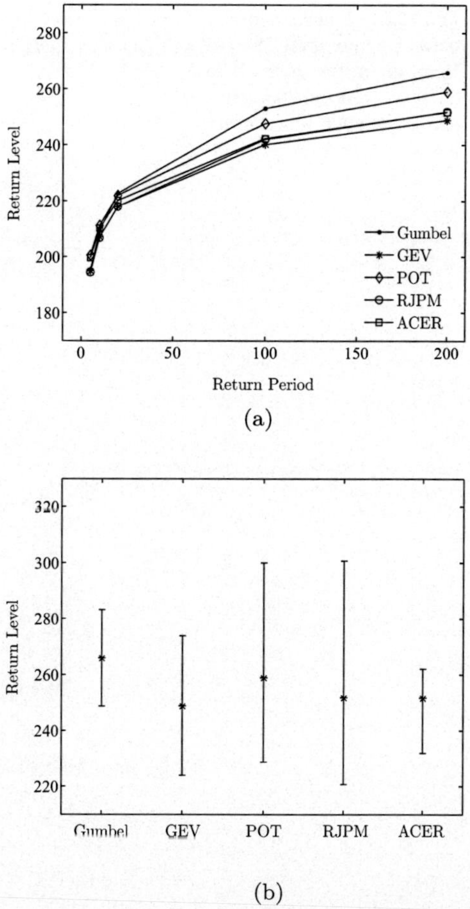

Fig. 12.20 Comparison between return levels in Oslo. (**a**) Return levels for 5, 10, 20, 100 and 200 year periods. (**b**) 200 year return levels and confidence bounds

higher return level compared to the other methods. However, the difference between the highest estimate (Gumbel) and the lowest (POT) for the longest return period is only 18 cm. It is seen that the ACER method again provides the smallest confidence interval.

12.7.3 Honningsvåg

Mostly the same pattern as in Heimsjø is shown for Honningsvåg, but with a greater scatter of the results. All methods except the RJP method agree at the 5 year return level, but diverge at the higher levels. The RJP method starts out estimating lower return levels, but ends up in the same region as the ACER estimate for the 200 year period. The Gumbel and POT methods give by far the largest estimates, while the GEV method produces by far the lowest.

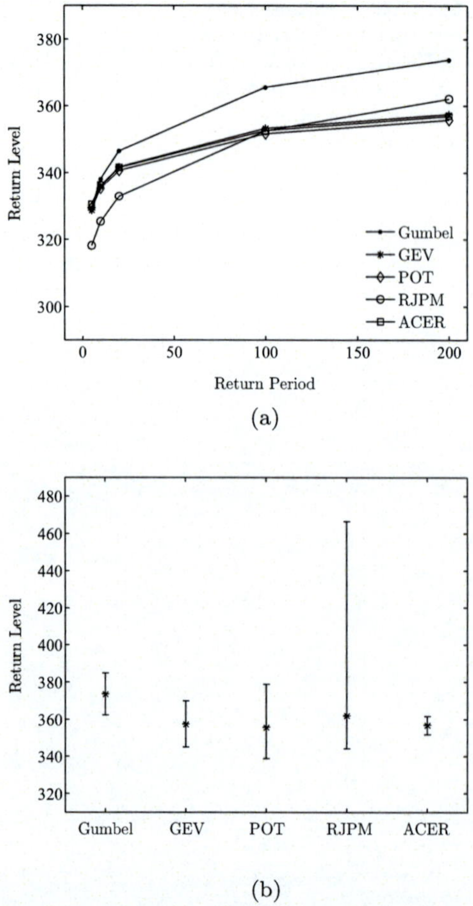

Fig. 12.21 Comparison between return levels in Heimsjø. (**a**) Return levels for 5, 10, 20, 100 and 200 year periods. (**b**) 200 year return levels and confidence bounds

Figure 12.22a compares the 5, 10, 20, 100 and 200 year estimates for all methods, while Fig. 12.22b shows the comparatively large difference between the 200 year estimates. The POT estimate for that period is 30 cm larger than the corresponding GEV estimate. Again the ACER method provides the smallest confidence interval.

12.7.4 Comments

The Annual Maxima method and the Peaks-over-Threshold method are both widely known and much applied methods. However, they both have some possible defects. As mentioned, the AM method throws away much of the data, and may end up fitting a poor model since all locations have less than 100 years of observed maxima to draw from. The POT method allows for the use of a great deal more data, but is subjected to two individual choices: the extreme threshold and the empirical

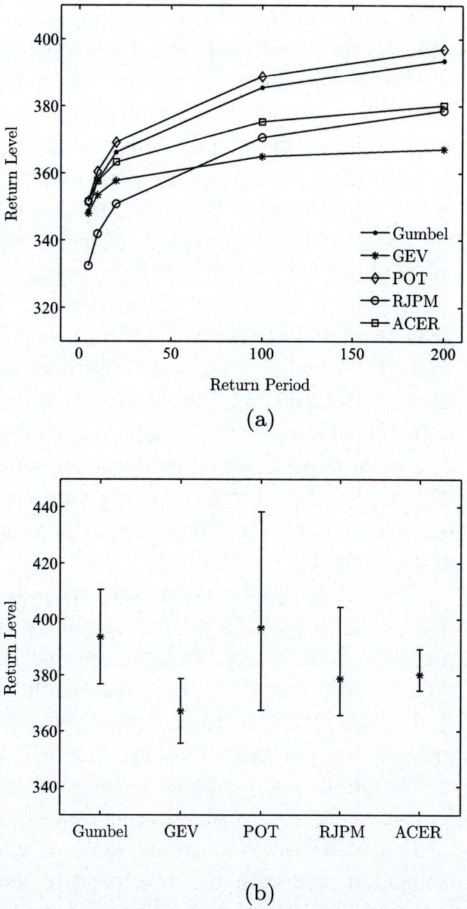

Fig. 12.22 Comparison between return levels in Honningsvåg. (**a**) Return levels for 5, 10, 20, 100 and 200 year periods. (**b**) 200 year return levels and confidence bounds

clustering rule. The former of the choices has available supportive literature (Coles 2001), but the selection of the clustering rule has little general theory, and some experience is a valuable asset in dealing with this problem.

The Revised Joint Probabilities (RJP) method is less widely applied than the first two methods, but has seen application to a number of British locations by the creators and proponents of the method, cf. Tawn (1992), Dixon and Tawn (1994), and Haigh et al. (2010). But because the literature is more sparse, and from fewer sources, it is more difficult to fully assess the reliability of this method.

As for the estimated return period levels by the RJP method, a notable feature is what appears to be underestimated short period return levels in the tide-dominant locations of Heimsjø and Honningsvåg. This is not an artifact from the tide-surge interaction modeling, since corresponding return levels were equally low when the estimation was performed without such modeling. The POT method is used to estimate surge parameters, which was not done by Dixon and Tawn (1994) in earlier work. This could perhaps affect results.

However, their methods are adopted in declustering the data using $r = 30$, which is quite different from the choice made in the POT analysis. It was done to emulate the method as used by Dixon and Tawn (1994) but could possibly have benefited from more careful consideration. Other choices that need to be made are the functions with which tide-surge interaction is corrected for and the quantiles that are deemed extreme. The correction functions were made to the specifications in Dixon and Tawn (1994), while the quantiles used were taken from the preceding POT section which provides a generally thorough explanation of the threshold selection.

In total, it is difficult to recommend a method that has the likely defect of underestimating short period return levels. Although it agrees more with the other methods for higher return levels, this is an area where there is a large margin of error anyway. The method is also somewhat difficult to implement, since Eq. (12.7) needs to be solved numerically. Tide measurements also need to be available. Using such data is, however, a strong point of the model, if applied correctly. More data are then used in the estimation. This attempt to incorporate sea level specific data and methodology into the return level estimation is arguably the most desirable quality of the method.

The ACER method is the most recently developed method, and it is therefore interesting to see that it produces results very similar to those by more conventional methods. In the return level plots in this section it is seen that the ACER method produces 200 year estimates in the middle of the range of the methods used here.

Two choices need to be made when estimating return levels with the ACER method; the tail marker η_0 and the ACER function ε_k used for the parameter estimation. It was found that as long as data around or above the 90-95% quantile were used, quite consistent results were obtained.

The ACER method seems to be a viable alternative for extreme sea level estimation, and with the developed methodology for parameter estimation and construction of confidence intervals it is quite easy to implement. One of the attractive features of the ACER method is its diagnostic power. By plotting the ACER functions of relevant order, it can be decided which order is necessary to capture the effect of dependence in the data on the extreme value statistics.

In summary, the ACER method seems to be an attractive method, with its quite easily implemented methodology and attractive statistical properties. When looking at 200 year return levels, it usually agrees with the other data-intensive methods, that is the RJP method and the POT method. While the empirical ACER functions provide a completely general nonparametric representation of the extreme value distribution given by the data, for the purpose of extrapolation to long return period levels a specific class of parametric distributions were introduced. This class of distributions specifically targets the cases where the Gumbel distribution is the appropriate asymptotic extreme value distribution. In the context of this chapter, this assumption is justified based on the underlying water level statistics, which invariably belongs to the domain of attraction of an asymptotic Gumbel distribution for the extremes.

References

Adler, R.J. 1981. *The Geometry of Random Fields*. New York: John Wiley & Sons, Ltd.

Adler, R.J., and J.E. Taylor. 2007. *Random Fields and Geometry*. New York: Springer.

Anderson, T.W. 1958. *An Introduction to Multivariate Statistical Analysis*. New York: John Wiley & Sons, Inc.

Argyris, J., and H.-P. Mlejnek. 1991. *Dynamics of Structures*. Amsterdam: Elsevier Science Publishers B.V.

Balakrishnan, N., and C.-D. Lai. 2009. *Continuous Bivariate Distributions*. New York: Springer, NY.

Batstone, C., M. Lawless, K. Horburgh, D. Blackman, and J.A. Tawn. 2009. Calculating extreme sea level probabilities around complex coastlines: A best practice approach. In *Proceedings Irish National Hydrology Conference*, 24–32. Hydrology Ireland.

Battjes, J.A. 1970. *Long Term Wave Height Distributions of Seven Stations around the British Isles* (Report A.44 ed.). Godalming: National Institute of Oceanography.

Baxevani, A., and I. Rychlik. 2006. Maxima for Gaussian seas. *Ocean Engineering* 33 (7): 895–911.

Beirlant, J., Y. Goegebeur, J. Segers, and J. Teugels. 2004. *Statistics of Extremes*. Chichester: John Wiley & Sons, Ltd.

Benfratello, S., M. Di Paola, and P.D. Spanos. 1998. Stochastic response of MDOF wind excited structures by means of Volterra series approach. *Journal of Wind Engineering and Industrial Aerodynamics* 74–76: 1135–1145.

Bollerslev, T. (1986). Generalized autoregressive conditional heteroskedasticity. *Journal of Econometrics* 31 (3): 307–327.

Bracewell, R.N. 1986. *The Fourier Transform and its Applications*. 2nd ed. New York: McGraw-Hill Inc.

Brockwell, P.J., and R.A. Davis. 2002. *Introduction to Time Series and Forecasting*. Berlin: Springer Science.

Bury, K.V. 1975. *Statistical Models in Applied Sciences*. New York: John Wiley & Sons, Inc.

Byström, H.N.E. 2005. Extreme value theory and extremely large electricity price changes. *International Review of Economics and Finance* 14 (1): 41–55.

Casella, G., and R. Berger. 2002. *Statistical Inference*. Boston: Cengage Learning.

Castillo, E., A.S. Hadi, N. Balakrishnan, and J.M. Sarabia. 2005. *Extreme Value and Related Models with Applications in Engineering and Science*. Hoboken: John Wiley & Sons, Ltd.

Cetin, A., and A. Naess. 2012. Toward a proper statistical description of defects. *International Journal of Fatigue* 38: 100–107.

A. Naess, *Applied Extreme Value Statistics*,
https://doi.org/10.1007/978-3-031-60769-1

Chan, K.F., and P. Gray. 2006. Using extreme value theory to measure value-at-risk for daily electricity spot prices. *International Journal of Forecasting* 22 (2): 283–300.
Chatfield, C. 1989. *The Analysis of Time Series–An Introduction*. London: Chapman and Hall.
Coles, S.G. 1994. A temporal study of extreme rainfall. In *Statistics for the Environment 2 - Water Related Issues*, ed. V. Barnett and K.F. Turkman, Chapter 4, 61–78. Chichester: John Wiley & Sons.
Coles, S.G. 2001. *An Introduction to Statistical Modeling of Extreme Values*. Springer Series in Statistics. London: Springer-Verlag.
Coles, S.G., and M.J. Dixon. 1999. Likelihood-based inference of extreme value models. *Extremes* 2: 5–23.
Coles, S.G., and J.A. Tawn. 1991. Modelling extreme multivariate events. *Journal of the Royal Statistical Society. Series B (Methodological)* 53 (2): 377–392.
Coles, S.G., and J.A. Tawn. 1994. Statistical methods for multivariate extremes: An application to structural design. *Journal of the Royal Statistical Society. Series C (Applied Statistics)* 43 (1): 1–48.
Contreras, J., R. Espinola, F.J. Nogales, and A.J. Conejo. 2003. ARIMA models to predict next-day electricity prices. *Power Systems, IEEE Transactions* 18 (3): 1014–1020.
Cook, N.J. 1982. Towards better estimation of extreme winds. *Journal of Wind Engineering and Industrial Aerodynamics* 9 (3): 295–323.
Cook, N.J. 1985. *The Designer's Guide to Wind Loading of Building Structures*. London: Butterworths.
Cramer, H. 1946. *Mathematical Methods of Statistics*. Princeton: Princeton University Press.
Cramer, H., and M.R. Leadbetter. 1967. *Stationary and Related Stochastic Processes*. New York: John Wiley & Sons.
Dahlen, K.E., P.B. Solibakke, S. Westgaard, and A. Naess. 2015. On the estimation of extreme values for risk assessment and management: The ACER method. *International Journal of Business* 20 (1): 33–51.
Davison, A.C., and D.V. Hinkley. 1997. *Bootstrap Methods and their Applications*. London: Cambridge University Press.
Davison, A.C., and R.L. Smith. 1990. Models for exceedances over high thresholds. *Journal of the Royal Statistical Society, B* 52 (3): 393–442.
de Haan, L. 1970. On regular variation and its applications to the weak convergence of sample extremes. Tract, Mathematical Centre, Amsterdam.
de Haan, L. 1994. Extreme value statistics. In *Extreme Value Theory and Applications*, ed. J. Galambos, J. A. Lechner, and E. Simiu. Dordrecht: Kluwer Academic Publishers.
de Haan, L., and J. de Ronde. 1998. Sea and wind: Multivariate extremes at work. *Extremes* 1: 7–45.
Dekkers, A.L.M., J.H.J. Einmahl, and L. de Haan. 1989. A moment estimator for the index of an extreme-value distribution. *The Annals of Statistics* 17 (4): 1833–1855.
Ditlevsen, O. 1971. *Extremes and First Passage Times*. Copenhagen: Dissertation (dr. techn.), Technical University of Denmark.
Ditlevsen, O. 2004. Extremes of random fields over arbitrary domains with application to concrete rupture stresses. *Probabilistic Engineering Mechanics* 19: 373–384.
Ditlevsen, O., and H.O. Madsen. 1996. *Structural Reliability Methods*. Chichester: John Wiley & Sons, Inc.
Dixon, M.J., and J.A. Tawn. 1994. Extreme sea-levels at the UK A-class sites: site-by-site analyses. Internal doc. no. 65, Proudman Oceanographic Laboratory.
Donley, M.G., and P.D. Spanos. 1990. *Dynamic Analysis of Non-Linear Structures by the Method of Statistical Quadratization*, vol. 57. *Lecture Notes in Engineering*. Berlin: Springer-Verlag.
Doob, J.L. 1953. *Stochastic Processes*. New York: John Wiley & Sons.
Draper, N.R., and H. Smith. 1998. *Applied Regression Analysis*. New York: Wiley-Interscience.
Eastoe, E.F., and J.A. Tawn. 2012. Modelling the distribution of the cluster maxima of exceedances of subasymptotic thresholds. *Biometrika* 99 (1): 43–55.

Efron, B., and R.J. Tibshirani. 1993. *An Introduction to the Bootstrap*. New York: Chapman and Hall.

Embrechts, P., C. Klüppelberg, and T. Mikosch. 1997. *Modelling Extremal Events*. New York: Springer.

Escribano, A., J.I. Pena, and P. Villaplana. 2002. Modeling electricity prices: International evidence. SSRN (Social Science Research Network).

Falk, M., J. Hüsler, and R.-D. Reiss. 2004. *Laws of Small Numbers: Extremes and Rare Events*. 2nd ed. Basel: Birkhäuser.

Ferro, C., A. T. and J. Segers. 2003. Inference for clusters of extreme values. *Journal of the Royal Statistical Society, B* 65 (2): 545–556.

Fisher, R.A., and L.H.C. Tippett. 1928. On the estimation of the frequency distributions of the largest or smallest member of a sample. *Proceedings of the Cambridge Philosophical Society* 24: 180–190.

Forristall, G.Z. 2000. Wave crest distributions: Observations and second-order theory. *Journal of Physical Oceanography* 30: 1931–1943.

Forristall, G.Z. 2006. Maximum wave heights over an area and the air gap problem. In *Proceedings 25th International Conference on Offshore Mechanics and Arctic Engineering*, OMAE–2006–92022. New York: ASME.

Forst, W., and D. Hoffmann. 2010. *Optimization - Theory and Practice*. New York: Springer.

Fréchet, M. 1927. Sur la loi de probabilité de l'écart maximum. *Annales de la Société Polonaise de Mathematique, Cracow* 6: 193–213.

Gaidai, O., G. Storhaug, and A. Naess. 2016. Extreme value statistics of large container ship roll. *Journal of Ship Research* 60 (2): 92–100.

Gaidai, O., G. Storhaug, and A. Naess. 2018. Statistics of extreme hydroelastic response for large ships. *Marine Structures* 61: 142–154.

Gaidai, O., A. Naess, O. Karpa, X. Xu, Y. Cheng, and R. Ye. 2019. Improving extreme wind speed prediction for North Sea offshore oil and gas fields. *Applied Ocean Research* 88: 63–70.

Garcia, R.C., J. Contreras, M.V. Akkeren, and J.B. Garcia. 2005. A GARCH forecasting model to predict day-ahead electricity prices. *Power Systems, IEEE Transactions* 20 (2): 867–874.

Gill, P., W. Murray, and M.H. Wright. 1981. *Practical Optimization*. London: Academic Press.

Gnedenko, B.V. 1943. Sur la distribution limite du terme maximum d'une série aléatoire. *Annals of Mathematics* 44: 423–453.

Grigoriu, M. 1984. Crossings of non-Gaussian translation processes. *Journal of Engineering Mechanics, ASCE* 110 (4): 610–620.

Grigoriu, M. 1995. *Applied Non-Gaussian Processes*. Englewood Cliffs: PTR Prentice Hall.

Gudendorf, G., and J. Segers. 2010. Chapter 6. Extreme-Value Copulas. In *Copula Theory and Its Applications: Proceedings of the Workshop Held in Warsaw, 25–26 September 2009*, ed. P. Jaworski, F. Durante, W. Härdle, and T. Rychlik. Berlin: Springer-Verlag.

Gumbel, E.J. 1958. *Statistics of Extremes*. New York: Columbia University Press.

Gumbel, E.J. 1960a. Bivariate exponential distributions. *Journal of the American Statistical Association* 55 (292): 698–707.

Gumbel, E.J. 1960b. Multivariate extremal distributions. *Bulletin de l'Institut International de Statistique* 37 (2): 471–475.

Gumbel, E.J. 1961. Bivariate logistic distributions. *Journal of the American Statistical Association* 56 (294): 335–349.

Gumbel, E.J., and C.K. Mustafi. 1967. Some analytical properties of bivariate extremal distributions. *Journal of the American Statistical Association* 62 (318): 569–588.

Haigh, I. D., R. Nicholls, and N. Wells. 2010. A comparison of the main methods for estimating probabilities of extreme still water levels. *Coastal Engineering* 57: 838–849.

Hansen, H., and L. Roald. 2000. Extreme water level analysis at sea at selected stations (in Norwegian). Doc. no. 11, Norwegian Water Resources and Energy Directorate.

Haring, R.E., and J.C. Heideman. 1978. Gulf of Mexico rare wave return periods. In *Proceedings of the Offshore Technology Conference*, Number OTC 3230. Houston.

Haver, S. 1980. *Analysis of Uncertainties related to the Stochastic Modelling of Ocean Waves*. Report UR-80-09, Division of Marine Structures, NTH (NTNU), Trondheim.

Haver, S. 2002. On the prediction of extreme wave crest heights. In *Proceedings of 7th International Workshop On Wave Hindcasting and Forecasting*. Banff: Meteorological Service of Canada, Environment Canada.

Haver, S., and K.A. Nyhus. 1986. A wave climate description for long term response calculations. In *Proceedings 9th International Conference on Offshore Mechanics and Arctic Engineering*. New York: ASME.

Haver, S., and S.R. Winterstein. 2008. Environmental contour lines: A method for estimating long term extremes by a short term analysis. *Transactions - Society of Naval Architects and Marine Engineers* SMTC-067-2008.

Heffernan, J.E., and J.A. Tawn. 2004. A conditional approach for multivariate extremes (with discussion). *Journal of the Royal Statistical Society, Series B* 66 (2): 497–546.

Henrici, P. 1977. *Applied and Computational Complex Analysis*. vol. II. New York: John Wiley & Sons, Inc.

Hosking, J.R.M., and J.R. Wallis. 1987. Parameter and quantile estimation for the generalized Pareto distribution. *Technometrics* 29: 339–349.

Hosking, J.R.M., and J.R. Wallis. 1997. *Regional Frequency Analysis: An Approach Based on L-Moments*. London: Cambridge University Press.

Hosking, J.R.M., J.R. Wallis, and E.F. Wood. 1985. Estimation of the generalized extreme-value distribution by the method of probability-weighted moments. *Technometrics* 27: 251–261.

Hougaard, P. 1986. A class of multivariate failure time distributions. *Biometrika* 73 (3): 671–678.

Hsing, T. 1987. On the characterization of certain point processes. *Stochastic Processes and Applications* 26: 297–316.

Hsing, T. 1991. Estimating the parameters of rare events. *Stochastic Processes and Applications* 37: 117–139.

Jahns, H.O., and J.D. Wheeler. 1972. Long term wave probabilities based on hindcasting of severe storms. In *Proceedings of the Offshore Technology Conference*, Number OTC 1590. Houston.

Johnson, N.L., and S. Kotz. 1970. *Distributions in Statistics: Univariate Continuous Distributions - 2*. New York: John Wiley & Sons.

Kac, M., and A.J.F. Siegert. 1947. On the theory of noise in radio receivers with square law detectors. *Journal of Applied Physics* 18: 383–397.

Karpa, O. 2012. *ACER User Guide and Program*. NTNU: Freely available at the internet address: https://folk.ntnu.no/arvidn/ACER.

Karpa, O. 2014. *ACER 2D Matlab Programs*. NTNU: Freely available at the internet address: https://folk.ntnu.no/arvidn/ACER2D.

Karpa, O., and A. Naess. 2013. Extreme value statistics of wind speed data by the ACER method. *Journal of Wind Engineering and Industrial Aerodynamics* 112: 1–10.

Karunakaran, D., S. Haver, M. Baerheim, and N. Spidsoe. 2001. Dynamic behaviour of the Kvitebjørn jacket in the North Sea. In *Proceedings of the 20th International Conference on Offshore Mechanics and Arctic Engineering*, OMAE2001/OFT1184. Rio de Janeiro: ASME.

Krogstad, H.E. (1985). Height and period distributions of extreme waves. *Applied Ocean Research* 7 (3): 158–165.

Krogstad, H.E., J. Liu, H. Socquet-Juglard, K.B. Dysthe, and K. Trulsen. 2004. Spatial extreme value analysis of nonlinear simulations of random surface waves. In *Proceedings 23rd International Conference on Offshore Mechanics and Arctic Engineering*, OMAE–2004–51336. New York: ASME.

Langley, R.S. 1984. The statistics of second order wave forces. *Applied Ocean Research* 6 (4): 182–186.

Larrabee, R.D., and C.A. Cornell. 1981. Combination of various load processes. *Journal of the Structural Division, ASCE* 107: 223–239.

Leadbetter, M.R. 1983. Extremes and local dependence in stationary sequences. *Zeitschrift für Wahrscheinlichkeitstheorie und verwandte Gebiete* 65: 291–306.

Leadbetter, M.R. 1995. On high-level exceedance modeling and tail-inference. *Journal of Statistical Planning and Inference* 45: 247–260.

Leadbetter, M.R., G. Lindgren, and H. Rootzén. 1983. *Extremes and Related Properties of Random Sequences and Processes*. New York: Springer-Verlag.

Ledford, A.W., and J.A. Tawn. 1996. Statistics for near independence in multivariate extreme values. *Biometrika* 83: 169–187.

Madsen, H.O., S. Krenk, and N.C. Lind. 1986. *Methods of Structural Safety*. New Jersey: Prentice-Hall Inc.

Maes, M., and K. Breitung. 1997. Direct approximation of the extreme value distribution of non-homogeneous Gaussian random fields. *Journal of Offshore Mechanics and Arctic Engineering* 119: 252–256.

Makkonen, L. 2006. Plotting positions in extreme value analysis. *Journal of Applied Meteorology and Climatology* 45 (2): 334–340.

Makkonen, L. 2008. Problems in the extreme value analysis. *Structural Safety* 30: 405–419.

MATLAB. 2009. *version 7.9.0.529 (R2009b)*. Natick: The MathWorks Inc.

McKay, M.D., W.J. Conover, and R.J. Beckman. 1979. A comparison of three methods for selecting values of input variables in the analysis of output from a computer code. *Technometrics* 21: 239–245.

McNeil, A.J., and R. Frey. 2000. Estimation of tail-related risk measures for heteroscedastic financial time series: An extreme value approach. *Journal of Empirical Finance* 7 (3–4): 271–300.

Melchers, R.E. 1999. *Structural Reliability Analysis and Prediction*. 2nd ed. West Sussex: John Wiley & Sons. ISBN 0471987719.

Moan, T., Z. Gao, and E. Ayala-Uraga. 2005. Uncertainty of wave-induced response of marine structures due to long term variation of extratropical wave conditions. *Journal of Marine Structures* 18 (4): 359–382.

Montgomery, D.C., E.A. Peck, and G.G. Vining. 2002. *Introduction to Linear Regression Analysis*. Amsterdam: Elsevier Science Publishers B. V.

Murakami, Y. 2002. *Metal Fatigue: Effects of Small Defects and Nonmetallic Inclusions*. 2nd ed. London: Academic Press.

Naess, A. 1983. Prediction of extremes of Morison type loading - an example of a general method. *Ocean Engineering* 10 (5): 313–324.

Naess, A. 1984. On the long-term statistics of extremes. *Applied Ocean Research* 6 (4): 227–228.

Naess, A. 1985a. The joint crossing frequency of stochastic processes and its application to wave theory. *Applied Ocean Research* 7 (1): 35–50.

Naess, A. 1985b. Statistical analysis of second-order response of marine structures. *Journal of Ship Research* 29 (4): 270–284.

Naess, A. 1986. The statistical distribution of second-order slowly-varying forces and motions. *Applied Ocean Research* 8 (2): 110–118.

Naess, A. 1987. The response statistics of non-linear second-order transformations to Gaussian loads. *Journal of Sound and Vibration* 115 (1): 103–129.

Naess, A. 1990a. Approximate first-passage and extremes of narrow-band Gaussian and non-Gaussian random vibrations. *Journal of Sound and Vibration* 138 (3): 365–380.

Naess, A. 1990b. Statistical analysis of nonlinear, second-order forces and motions of offshore structures in short-crested random seas. *Probabilistic Engineering Mechanics* 5 (4): 192–203.

Naess, A. 1998a. Estimation of long return period design values for wind speeds. *Journal of Engineering Mechanics, ASCE* 124 (3): 252–259.

Naess, A. 1998b. Statistical extrapolation of extreme value data based on the peaks over threshold method. *Journal of Offshore Mechanics and Arctic Engineering, ASME* 120: 91–96.

Naess, A. 1999. Extreme response of nonlinear structures with low damping subjected to stochastic loading. *Journal of Offshore Mechanics and Arctic Engineering, ASME* 121: 255–260.

Naess, A. 2000a. Characteristic functions related to quadratic transformations of Gaussian processes. Technical Report R-1-00, Department of Structural Engineering, Norwegian University of Science and Technology, Trondheim.

Naess, A. 2000b. Crossing rate statistics of quadratic transformations of Gaussian processes. Technical Report R-2-00, Department of Structural Engineering, Norwegian University of Science and Technology, Trondheim.

Naess, A. 2010. Estimation of extreme values of time series with heavy tails. Preprint Statistics No. 14/2010, Department of Mathematical Sciences, Norwegian University of Science and Technology, Trondheim.

Naess, A. 2011. A note on the bivariate ACER method. Preprint Statistics No. 01/2011, Department of Mathematical Sciences, Norwegian University of Science and Technology, Trondheim.

Naess, A., and O. Batsevych. 2010. Space-time extreme value statistics of a Gaussian random field. *Probabilistic Engineering Mechanics* 25 (5): 372–379.

Naess, A., and O. Batsevych. 2012. Space-time extreme value statistics of non-Gaussian random fields. *Probabilistic Engineering Mechanics* 28: 169–175.

Naess, A., and P.H. Clausen. 1999. Statistical extrapolation and the peaks-over-threshold method. In *Proceedings 18th International Conference on Offshore Mechanics and Arctic Engineering*, OMAE–99–6422. New York: ASME.

Naess, A., and P.H. Clausen. 2001. Combination of peaks-over-threshold and bootstrapping methods for extreme value prediction. *Structural Safety* 23: 315–330.

Naess, A., and O. Gaidai. 2008. Monte Carlo methods for estimating the extreme response of dynamical systems. *Journal of Engineering Mechanics, ASCE* 134 (8): 628–636.

Naess, A., and O. Gaidai. 2009. Estimation of extreme values from sampled time series. *Structural Safety* 31: 325–334.

Naess, A., and J.M. Johnsen. 1992. An efficient numerical method for calculating the statistical distribution of combined first-order and wave-drift response. *Journal of Offshore Mechanics and Arctic Engineering* 114 (3): 195–204.

Naess, A., and J.M. Johnsen. 1993. Response statistics of nonlinear, compliant offshore structures by the path integral solution method. *Probabilistic Engineering Mechanics* 8 (2): 91–106.

Naess, A., and H.C. Karlsen. 2004. Numerical calculation of the level crossing rate of second order stochastic Volterra systems. *Probabilistic Engineering Mechanics* 19 (2): 155–160.

Naess, A., and O. Karpa. 2015a. Statistics of bivariate extreme wind speeds by the ACER method. *Journal of Wind Engineering and Industrial Aerodynamics* 139: 82–88.

Naess, A., and O. Karpa. 2015b. Statistics of extreme wind speeds and wave heights by the bivariate ACER method. *Journal of Offshore Mechanics and Arctic Engineering, ASME* 137: 021602–1–7.

Naess, A., and U. Machado. 2000. Response statistics of large compliant offshore structures. In *Proceedings 8th ASCE Specialty Conference on Probabilistic Mechanics and Structural Reliability*. New York: ASCE.

Naess, A., and T. Moan. 2013. *Stochastic Dynamics of Marine Structures*. New York: Cambridge University Press.

Naess, A., and J.O. Royset. 2000. Extensions of Turkstra's rule and their application to combination of dependent load effects. *Structural Safety* 22: 129–143.

Naess, A., H.C. Karlsen, and P.S. Teigen. 2006. Numerical methods for calculating the crossing rate of high and extreme response levels of compliant offshore structures subjected to random waves. *Applied Ocean Research* 28 (1): 1–8.

Naess, A., O. Gaidai, and S. Haver. 2007. Efficient estimation of extreme response of drag dominated offshore structures by Monte Carlo simulation. *Ocean Engineering* 34 (16): 2188–2197.

Naess, A., O. Gaidai, and P.S. Teigen. 2007. Extreme response prediction for nonlinear floating offshore structures by Monte Carlo simulation. *Applied Ocean Research* 29 (4): 221–230.

Naess, A., O. Gaidai, and O. Karpa. 2013. Estimation of extreme values by the average conditional exceedance rate method. *Journal of Probability and Statistics*. https://doi.org/10.1155/2013/797014.

Neal, E. 1974. Second order hydrodynamic forces due to stochastic excitation. In *Proceedings 10th ONR Symposium*, Cambridge.

Nelsen, R.B. 2006. *An Introduction to Copulas*. Springer Series in Statistics. New York: Springer, NY.

Newland, D.E. 1991. *An Introduction to Random Vibrations and Spectral Analysis*. 2nd ed. London: Longman.

Norwegian Meteorological Institute. 2012. Climate Data Web Services. http://eklima.met.no/wsKlima/start/start_en.html. Accessed 29 Dec 2011.

Numerical Algorithms Group. 2010. *NAG Toolbox for Matlab*. Oxford: NAG Ltd.

Palutikof, J.P., B.B. Brabson, D.H. Lister, and S.T. Adcock. 1999. A review of methods to calculate extreme wind speeds. *Meteorological Applications* 6: 119–132.

Perrin, O., H. Rootzen, and R. Taesler. 2006. A discussion of statistical methods used to estimate extreme wind speeds. *Theoretical and Applied Climatology* 85 (3–4): 203–215.

Pickands, J. 1981. Multivariate extreme value distributions. *Bulletin of the International Statistical Institute* 49: 859–878.

Piterbarg, V.I. 1996. *Asymptotic Methods in the Theory of Gaussian Processes and Fields*. Providence: American Mathematical Society.

Pugh, D.J., and J.M. Vassie. 1980. Applications of the joint probability method for extreme sea-level computations. In *Proceedings of the Institution of Civil Engineers, Part 2(69)*, 959–975.

Reiss, R.-D., and M. Thomas. 2007. *Statistical Analysis of Extreme Values*. 3rd ed. Basel: Birkhäuser.

Rice, S.O. 1954. Mathematical analysis of random noise. In *Selected Papers on Noise and Stochastic Processes*, ed. N. Wax, 133–294. New York: Dover Publications, Inc.

Robert, C.Y. 2009. Inference for the limiting cluster size distribution of extreme values. *Annals of Statistics* 37 (1): 271–310.

Robinson, M.E., and J.A. Tawn. 2000. Extremal analysis of processes sampled at different frequencies. *Journal of the Royal Statistical Society, B* 62 (1): 117–136.

Sagrilo, L.V.S., Z. Gao, A. Naess, and E.C.P. Lima. 2011. A straightforward approach for using single time domain simulations to assess characteristic extreme responses. *Ocean Engineering* 38: 1464–1471.

Sarpkaya, T., and M. Isaacson. 1981. *Mechanics of Wave Forces on Offshore Structures*. New York: Van Nostrand Reinholdt.

Schall, G., M.H. Faber, and R. Rackwitz. 1991. The ergodicity assumption for sea states in the reliability estimation of offshore structures. *Jornal of Offshore Mechanics and Arctic Engineering, ASME* 113 (3): 241–246.

Schetzen, M. 1980. *The Volterra and Wiener Theories of Nonlinear Systems*. New York: John Wiley & Sons, Inc.

Segers, J. 2005. Approximate distributions of clusters of extremes. *Statistics and Probability Letters* 74: 330–336.

Shinozuka, M. 1974. Digital simulation of random processes in engineering mechanics with the aid of FFT techniques. In *Stochastic Problems in Mechanics*, 277–286. Waterloo: University of Waterloo Press.

Shinozuka, M., and G. Deodatis. 1991. Simulation of stochastic processes by spectral representation. *Applied Mechanics Review* 44 (4): 191–203.

Shinozuka, M., and C.-M. Jan. 1972. Digital simulation of random processes and its application. *Journal of Sound and Vibration* 25 (1): 111–128.

Sinsabvarodom, C., A. Naess, B.J. Leira, and W. Chai. 2022. Extreme value estimation of Beaufort sea ice dynamics driven by global wind effects. *China Ocean Engineering*. https://doi.org/10.1007/s13344--022--0046--3.

Skjong, M., A. Naess, and O.B. Naess. 2013. Statistics of extreme sea levels for locations along the Norwegian coast. *Journal of Coastal Research* 29 (5): 1029–1048.

Sklar, A. 1959. Fonctions de répartition à *n* dimensions et leurs marges. *Publications de l'Institut de statistique de l'Université de Paris* 8: 229–231.

Smith, R.L. 1992. The extremal index for a Markov chain. *Journal of Applied Probability* 29: 37–45.

Smith, R.L., J.A. Tawn, and S.G. Coles. 1997. Markov chain models for threshold exceedances. *Biometrika* 84 (2): 249–268.

Socquet-Juglard, H., K. Dysthe, K. Trulsen, H.E. Krogstad, and J. Liu. 2005. Probability distributions of surface gravity waves during spectral changes. *Journal of Fluid Mechanics* 542 (Nov 2005): 195–216.

Sun, J. 1993. Tail probabilities of the maxima of Gaussian random fields. *Annals of Probability* 21 (1): 34–71.

Swider, D.J., and C. Weber. 2007. Extended ARMA models for estimating price developments on day-ahead electricity markets. *Electric Power Systems Research* 77 (5): 583–593.

Tawn, J.A. 1988. Bivariate extreme value theory: Models and estimation. *Biometrika* 75 (3): 397–415.

Tawn, J.A. 1990. Discussion of paper by A. C. Davison and R. L. Smith. *Journal of the Royal Statistical Society, B* 52 (3): 393–442.

Tawn, J.A. 1992. Estimating probabilities of extreme sea-levels. *Applied Statistics* 41 (1): 77–93.

Tawn, J.A., and J.M. Vassie. 1989. Extreme sea levels: The joint probabilities method revisited and revised. In *Proceedings of the Institution of Civil Engineers, Part 2(87)*, 429–442.

Teigen, P.S., and A. Naess. 1999. Stochastic response analysis of deepwater structures in short-crested random waves. *Journal of Offshore Mechanics and Arctic Engineering, ASME* 121: 181–186.

Teigen, P.S., and A. Naess. 2003. Extreme response of floating structures in combined wind and waves. *Journal of Offshore Mechanics and Arctic Engineering, ASME* 125: 87–93.

Tiago de Oliveira, J. 1982. *Bivariate Extremes: Models and Statistical Decision*. North Carolina: Center for Stochastic Processes, University of North Carolina, Chapel Hill. Technical report No. 14.

Tiago de Oliveira, J. 1984. Bivariate models for extremes; Statistical decision. In *Statistical Extremes and Applications*, 131–153. Dordrecht: Springer.

Toffoli, A., E. Bitner-Gregersen, M. Onorato, and A. Babanin. 2008. Wave crest and trough distributions in a broad-banded directional wave field. *Ocean Engineering* 35 (17–18): 1784–1792.

Toro, G.R. 1984. *Probabilistic Analysis of Combined Dynamic Responses*. Report no. 65, Stanford University, Palo Alto.

Tromans, P.S., and L. Vanderschuren. 1995. Response based design conditions in the North Sea: Application of a new method. In *Proceedings of the Offshore Technology Conference*, Number OTC 7683. Houston.

Turkstra, C.J. 1970. *Theory of Structural Safety*. Ontario: SM Study No. 2, Solid Mechanics Division, University of Waterloo.

Vanmarcke, E.H. 1975. On the distribution of the first-passage time for normal stationary random processes. *Journal of Applied Mechanics* 42: 215–220.

Vanmarcke, E.H. 1983. *Random Fields: Analysis and Synthesis*. Cambridge: The MIT Press.

Vinje, T. 1983. On the statistical distribution of second-order forces and motions. *International Shipbuilding Progress* 30: 58–68.

Waal, D.J., and P.H.A.J.M. van Gelder. 2005. Modelling of extreme wave heights and periods through copulas. *Extremes* 8: 345–356.

WAFO-group. 2000. *WAFO - A Matlab Toolbox for Analysis of Random Waves and Loads - A Tutorial*. Lund: Math. Stat., Center for Math. Sci., Lund University.

WAMIT. 2008. *WAMIT Inc.* www.wamit.com.

Watson, G.S. 1954. Extreme values in samples from m-dependent stationary stochastic processes. *The Annals of Mathematical Statistics* 25 (4): 798–800.

Wen, Y.-K. 1990. *Structural Load Modeling and Combination for Performance and Safety Evaluation*. Amsterdam: Elsevier Science Publishers B.V.

Wen, Y.K., and H.T. Pearce. 1981. Recent developments in probabilistic load combinations. In *Proceedings on Probabilistic Methods in Structural Engineering*, St. Louis: ASCE.

Weron, R. 2006. *Modeling and Forecasting Electricity Loads and Prices: A Statistical Approach*. New Yok: John Wiley & Sons, Inc.

Winterstein, S.R. 1985. Non-normal responses and fatigue damage. *Journal of Engineering Mechanics, ASCE* 111 (10): 1291–1295.

Winterstein, S.R. 1988. Nonlinear vibration models for extremes and fatigue. *Journal of Engineering Mechanics, ASCE* 114: 1772–1790.

Winterstein, S.R., and C.A. MacKenzie. 2013. Extremes of nonlinear vibration: Comparing models based on moments, L-moments, and maximum entropy. *Journal of Offshore Mechanics and Arctic Engineering* 135 (2): 021602.

Winterstein, S., T. Ude, C.A. Cornell, P. Bjerager, and S. Haver. 1993. Environmental parameters for extreme response: Inverse FORM with omission factors. In *Proceedings 6th International Conference on Structural Safety and Reliability (ICOSSAR'93)*. Innsbruck: Balkema.

Wong, E., and B. Hajek. 1985. *Stochastic Processes in Engineering Systems*. New York: Springer, NY.

Xu, X., F. Wang, O. Gaidai, A. Naess, Y. Xing, and J. Wang. 2022. Bivariate statistics of floating offshore wind turbine dynamic response under operational conditions. *Ocean Engineering* 257: paper 111657.

Yu, S., W. Wu, B. Xie, S. Wang, and A. Naess. 2020. Extreme value prediction of current profiles in the South China Sea based on EOFs and the ACER method. *Applied Ocean Research* 105: paper 102408.

Yue, S. 2000. The Gumbel mixed model applied to storm frequency analysis. *Water Resources Management* 14: 377–389.

Yue, S. 2001a. A bivariate extreme value distribution applied to flood frequency analysis. *Nordic Hydrology* 32 (1): 49–64.

Yue, S. 2001b. The Gumbel logistic model for representing a multivariate storm event. *Advances in Water Resources* 24 (2): 179–185.

Yue, S., and C.Y. Wang. 2004. A comparison of two bivariate extreme value distributions. *Stochastic Environmental Research and Risk Assessment* 18: 61–66.

Yue, S., T. Ouarda, B. Bobée, P. Legendre, and P. Bruneau. 1999. The Gumbel mixed model for flood frequency analysis. *Journal of Hydrology* 226 (1–2): 88–100.

Yun, S. 1998. The extremal index of a higher-order stationary Markov chain. *Annals of Applied Probability* 8: 408–437.

Yun, S. 2000. The distribution of cluster functionals of extreme events in a d'th-order Markov chain. *Journal of Applied Probability* 37: 29–44.

Zachary, S., G. Feld, G. Ward, and J. Wolfram. 1998. Multivariate extrapolation in the offshore environment. *Applied Ocean Research* 20 (5): 273–295.

Zhang, X.-Y., Y.-G. Zhao, and Z.-H. Lu. 2019. Unified Hermite polynomial model and its application in estimating non-Gaussian processes. *Journal of Engineering Mechanics* 145 (3): 04019001.

Index

A. Naess, *Applied Extreme Value Statistics*,
https://doi.org/10.1007/978-3-031-60769-1